A HANDBOOK TO ENGLISH RO

Also by Jean Raimond

ROBERT SOUTHEY: L'homme et son temps, l'oeuvre, le rôle
LE PRÉROMANTISME ANGLAIS (*with Pierre Arnaud*)
VISAGES DU ROMANTISME ANGLAIS
LE ROMAN ANGLAIS AU XIXᵉ SIÈCLE (*with P. Coustillas and J.-P. Petit*)

Also by J. R. Watson

EVERYMAN'S BOOK OF VICTORIAN VERSE (*editor*)
ENGLISH POETRY OF THE ROMANTIC PERIOD, 1789–1830
WORDSWORTH'S VITAL SOUL
THE POETRY OF GERARD MANLEY HOPKINS
PRE-ROMANTICISM IN ENGLISH POETRY OF THE EIGHTEENTH
CENTURY (*editor*)

A Handbook to English Romanticism

Edited by

Jean Raimond
Professor of English, University of Reims

and

J. R. Watson
Professor of English, University of Durham

St. Martin's Press

First edition 1992
Reprinted 1994

Published in Great Britain by
THE MACMILLAN PRESS LTD
Houndmills, Basingstoke, Hampshire RG21 2XS
and London
Companies and representatives
throughout the world

A catalogue record for this book is available
from the British Library

ISBN 0–333–46951–8 hardcover
ISBN 0–333–60706–6 paperback

Printed in Hong Kong

First published in the United States of America 1992 by
Scholarly and Reference Division,
ST. MARTIN'S PRESS, INC.,
175 Fifth Avenue,
New York, N.Y. 10010

ISBN 0–312–07914–1

Library of Congress Cataloging-in-Publication Data
A Handbook to English romanticism / edited by Jean Raimond and J. R.
Watson.
p.   cm.
Includes index.
ISBN 0–312–07914–1
1. English literature—19th century—History and criticism-
-Handbooks, manuals, etc.   2. English literature—18th century-
-History and criticism—Handbooks, manuals, etc.   3. Romanticism-
-Great Britain—Handbooks, manuals, etc.   I. Raimond, Jean.
II. Watson, J. R. (John Richard), 1934–
PR457.H26   1992
820.9'145—dc20   91–41426
                                                            CIP

Contents

Preface   xiii

Notes on the Contributors   xiv

The Abolition of the Slave Trade
VINCENT NEWEY   1

Akenside, Mark (1721–70)
WILLIAM RUDDICK   4

The American Revolution
TOM FURNISS   5

The Anti-Jacobin
J. R. WATSON   10

Architecture
PIERRE ARNAUD   11

Association
J. R. WATSON   14

Austen, Jane (1775–1817)
W. A. CRAIK   16

Beattie, James (1735–1803)
PIERRE MORÈRE   18

Blake, William (1757–1827)
FRANÇOIS PIQUET   19

Bonaparte, Napoleon (1769–1821)
J. R. WATSON; BERNARD BEATTY   31

Burke, Edmund (1729–97)
TOM FURNISS   37

Burns, Robert (1759–96)
DONALD A. LOW   42

Byron, George Gordon, sixth Lord (1788–1824)
BERNARD BEATTY   45

Campbell, Thomas (1777–1844)
J. R. WATSON   56

Cartoons
J. R. WATSON   57

Chatterton, Thomas (1752–70)
GEORGES LAMOINE   58

Clare, John (1793–1864)
MARK STOREY   61

Cobbett, William (1763–1835)
J. R. WATSON   66

Coleridge, Samuel Taylor (1772–1834)
DAVID JASPER   69

Collins, William (1721–59)
MICHÈLE PLAISANT   78

Constable, John (1776–1837)
J. R. WATSON   80

Cowper, William (1731–1800)
VINCENT NEWEY   83

Crabbe, George (1754–1832)
MARK STOREY   92

De Quincey, Thomas (1785–1859)
J. R. WATSON   94

Drama
J. R. WATSON   95

Dreams
CHRISTIAN LA CASSAGNÈRE   97

The Fantastic
DAVID JASPER   103

The French Revolution
JEAN RAIMOND   105

German Literature
KEVIN HILLIARD   113

German Philosophy and Criticism
KEVIN HILLIARD   115

Godwin, William (1756–1836)
DAVID JASPER   119

Goldsmith, Oliver (1728–74)
WILLIAM RUDDICK   121

The Gothic Novel
PIERRE ARNAUD   123

Gray, Thomas (1716–71)
WILLIAM RUDDICK   128

Greece and the Philhellenes
VINCENT NEWEY   130

Hazlitt, William (1778–1830)
WILLIAM RUDDICK   134

Hogg, James (1770–1835)
JACQUES BLONDEL   137

Hunt, James Henry Leigh (1784–1859)
MARY WEDD   138

Imagination
DENISE DEGROIS   141

The Industrial Revolution
J. R. WATSON   143

The Jacobin Novel
J. R. WATSON   147

Keats, John (1795–1821)
JEAN-CLAUDE SALLÉ   149

Lamb, Charles (1775–1834)
MARY WEDD   162

Landor, Walter Savage (1775–1864)
PIERRE VITOUX   166

Landscape
SUZY HALIMI   167

Macpherson, James (1736–96)
DONALD A. LOW   171

The Marvellous and Occult
J. R. WATSON   172

Maturin, Charles Robert (1780–1824)
CLAUDE FIEROBE   175

Mediaevalism
GEORGES LAMOINE   177

Metre and Form
J. R. WATSON   180

Moore, Thomas (1779–1852)
THÉRÈSE TESSIER   182

Nature
MARCEL ISNARD   185

Negative Capability
JEAN-CLAUDE SALLÉ   187

Nelson, Horatio (1758–1805)
J. R. WATSON   189

Orientalism
J. R. WATSON   192

Paine, Thomas (1737–1809)
TOM FURNISS   196

Painting
J. R. WATSON   199

Palmer, Samuel (1805–81)
J. R. WATSON   202

Pantheism
MARCEL ISNARD   204

Peacock, Thomas Love (1785–1866)
W. A. CRAIK   206

Periodicals
DEREK ROPER   209

Peterloo
BERNARD BEATTY   211

The Picturesque
SUZY HALIMI   213

Political Thought
DEREK ROPER   215

Religious Thought: Wesley, Swedenborg
DAVID JASPER   219

Robespierre, Maximilien (1758–94)
JACQUES BLONDEL   221

Rogers, Samuel (1763–1855)
J. R. WATSON   223

Rousseau, Jean-Jacques (1712–78)
JACQUES BLONDEL   225

Scott, Walter (1771–1832)
HUBERT TEYSSANDIER   228

The Self
CHRISTIAN LA CASSAGNÈRE   237

Sensibility
MICHÈLE PLAISANT   242

Shelley, Mary (1797–1851)
J. R. WATSON   246

Shelley, Percy Bysshe (1792–1822)
JEAN PERRIN   249

Smart, Christopher (1722–71)
WILLIAM RUDDICK   260

Southey, Robert (1774–1843)
JEAN RAIMOND   262

The Sublime
SUZY HALIMI   269

Thomson, James (1700–48)
MICHÈLE PLAISANT   272

Turner, Joseph William Mallord (1775–1851)
J. R. WATSON   275

Warton, Joseph (1722–1800)
VINCENT NEWEY   279

Warton, Thomas, the younger (1728–90)
VINCENT NEWEY   280

Waterloo
BERNARD BEATTY   281

Wellington, Arthur Wellesley, first Duke of (1769–1852)
J. R. WATSON   283

Wollstonecraft, Mary (1759–97)
TOM FURNISS   284

Wordsworth, Dorothy (1771–1855)
J. R. WATSON   288

Wordsworth, William (1770–1850)
MARCEL ISNARD   290

Young, Edward (1683–1765)
VINCENT NEWEY   306

Index   309

Preface

This handbook is designed as a guide to the Romantic movement in English literature, and it is intended for the student who wishes to obtain information, quickly and easily, about the principal literature of the period. There are so many books on English Romanticism, starting from so many different critical positions, that a student may be forgiven for registering some initial bewilderment; and it is often difficult, amid the excitement of different critical expositions, to discover the factual background to the discussion. This handbook is intended to make it easier to find out when something happened, or who someone was, or when a novel or a poem was published.

In addition to such factual material, this handbook provides brief essays on certain concepts which were of importance to the English Romantic writers, such as dreams, or the self, or orientalism: these should be helpful to students who are trying to get their bearings in the period. There are also brief essays on the historical background, for those who require a quick guide to the French Revolution, or to the campaign in Britain for the abolition of the slave trade, or to the career and influence of Napoleon Bonaparte. Through essays on political writers, such as Paine and Burke, and discussions of changes under the heading of the Industrial Revolution, it becomes easier to see the great writers of the Romantic period in the context of their time.

This handbook is the work of a number of British and French scholars, and it would not have been possible without the co-operation and enthusiasm of contributors from both countries. In an age when educational co-operation in Europe is advancing faster than the Channel Tunnel, it is a pleasure to present this handbook as evidence of working together. The British and French editors, working from Durham and Reims respectively, are grateful to all the contributors for their hard work and their patience with the lengthy processes of editing and printing. They are also grateful to the British Council for financial support to enable a preliminary planning-visit to take place, and for the encouragement of Dr Kenneth Churchill at the Paris office.

Notes on the Contributors

Pierre Arnaud, Lecturer in English at the Université Lumière Lyon II, is President of the Société d'Etudes Anglo-Américaines des XVIIᵉ et XVIIIᵉ Siècles and editor of *Tropismes*. His special field of study is the eighteenth century, English Romanticism and modern literary criticism. He has published *Ann Radcliffe et le fantastique, Essai de psychobiographie* and *Le préromantisme anglais* (with Jean Raimond) and has written articles on Gothic fiction, Goldsmith, Mrs Radcliffe, Godwin, James Hogg and Iris Murdoch.

Bernard Beatty is Senior Lecturer in English at the University of Liverpool. He is joint editor of *Literature of the Romantic Period, 1750–1850* and *Byron and the Limits of Fiction*; author of *Byron's 'Don Juan'* and *Byron's 'Don Juan' and Other Poems*; and academic editor of the *Byron Journal*.

Jacques Blondel was, until his retirement in 1980, Professor of English in the University of Clermont-Ferrand II, where he initiated a Centre du Romantisme Anglais. He is the author of *Emily Brontë, expérience spirituelle et création poétique*, of *Le paradis reconquis*, and of *Emerveillement et profanation chez William Blake* (1968). His translations include works by Blake, Emily Brontë, R. L. Stevenson, and C. S. Lewis.

W. A. Craik retired recently from the English Department of the University of Aberdeen. She is the author of *Jane Austen: The Six Novels, The Brontë Novels* and *Elizabeth Gaskell and the English Provincial Novel*.

Denise Degrois studied in Paris at the Sorbonne and the Ecole Normale Supérieure, and is now a professor of English at the Sorbonne Nouvelle (Université de Paris III). Her publications include a thesis on Coleridge's aesthetics, regular contributions to the volumes of the Centre du Romantisme Anglais, Clermont-Ferrand, and articles on Coleridge and Romanticism in periodicals (*Etudes anglaises, Wordsworth Circle, Romantisme*, etc.), and in miscellanies bearing on the history of ideas (for instance, *Genèse de la conscience moderne*).

Claude Fierobe, Professor of English and Head of the English Department at the University of Reims, is the author of *C. R. Maturin, l'homme et l'oeuvre*. He has written articles on a number of Irish and English subjects, on the 'Gothic' novel and on the fantastic in literature. He is also co-translator of Yeats's uncollected prose. He is President of the Société Française d'Etudes Irlandaises.

Tom Furniss teaches English at the University of Strathclyde. He has a particular interest in the work of Burke and Paine, and in the political writings of the Romantic period.

Suzy Halimi, born in Algeria, studied in Paris at the Ecole Normale Supérieure and the Sorbonne. She then taught for two years in a secondary school. She is now a professor in the English Department at the Sorbonne Nouvelle and is President of her university. A specialist in eighteenth-century Britain, its civilisation and literature (particularly the novels of the period), she has published five books and some thirty articles on various aspects of the social scene, the history of ideas and the artistic developments of the time.

Kevin Hilliard is a lecturer in German at the University of Durham. He teaches a course on German philosophy from Kant to Nietzsche, and has published work on German literature from the eighteenth to the twentieth century.

Marcel Isnard, Maître de Conférences at the Sorbonne Nouvelle, teaches English literature and is the author of articles on various writers, including Maine de Biran, Sir James Mackintosh and Wordsworth.

David Jasper is Director of the Centre for the Study of Literature and Theology, University of Glasgow. He is the author of *Coleridge as Poet and Religious Thinker*, *The New Testament and the Literary Imagination* and *The Study of Literature and Religion*, and founded the Durham Conferences on Literature and Religion. He has edited the proceedings of the first three conferences (under the titles *Images of Belief in Literature*, *The Interpretation of Belief* and *The Critical Spirit and the Will to Believe*); has co-edited (with R. C. D. Jasper) *Language and the Worship of the Church*; and is editor of the journal *Literature and Theology*.

Christian La Cassagnère was Assistant Professor at the Sorbonne and Paris–Vincennes, 1965–71, and taught English literature as a guest professor at the California State University, Los Angeles. He is now Professor of English at the University B. Pascal, Clermont-Ferrand, and President of the Centre du Romantisme Anglais. Author of a book on Shelley, *La mystique de 'Prometheus Unbound'*, of a psychocritical study of Coleridge's poetry, and of various articles, he has also edited several collections of essays on romanticism, including *Visages de l'angoisse* and *Le double dans le romantisme anglo-américain*.

Georges Lamoine has taught English literature at the University of Toulouse–Le Mirail since 1970. He has edited texts of Defoe and Fielding and has translated Swift's *Gulliver's Travels* and *A Tale of a Tub* into French. Combining his interests in British criminal justice and in the Bristol area in the eighteenth century, he has edited *Bristol Gaol Delivery Fiats 1741–99*, and is now preparing *Charges to the Grand Jury 1688–1800*.

Donald A. Low is Senior Lecturer in English at the University of Stirling, and has edited *Robert Burns: The Critical Heritage* and *Critical Essays on Robert Burns*. He has also published *That Sunny Dome*, a study of Regency England; and *Thieves' Kitchen*, on the unusual topic of the Regency underworld.

Pierre Morère is Professor of English Literature at the Université Stendhal, Grenoble, and founded the Centre d'Etudes Ecossaises, supported by the Centre National de la Recherche Scientifique. His main areas of interest are literature and the history of ideas in eighteenth-century Britain. He has published studies of Scottish philosophers, novelists and fiction in this period, and is currently engaged on research into the Scottish School of Common Sense.

Vincent Newey is Professor of English at Leicester University. He is the author of *Cowper's Poetry* and has edited symposia on Bunyan, Byron, and on literature and nationalism. His other publications include articles on Puritan writers and on Romantic poetry, and he is currently working on a study of late-Victorian fiction. He is one of the editors of the *Byron Journal*, and of the *Bulletin* of the British Association for Romantic Studies.

Jean Perrin is Professor of English and Head of the Department of English and American Studies at the Université Stendhal, Grenoble. His main interests are Romantic and Shakespearean studies, with particular attention to imagery, symbolism and myth. He has published a book on Shelley (*Les structures de l'imaginaire Shelleyen*) and a number of articles in French and English in *Essays and Studies, Etudes anglaises, Corps écrit*, and other reviews.

François Piquet is Professor of English Literature and vice-president in charge of research at the University of Lyon III. He is the author of 'Blake et le sacré', based on his doctoral thesis; and of some twenty articles on Blake. He has also published studies on Burke, Fuseli, Wordsworth and Coleridge. He is currently working on an *Anthologie de la poésie anglaise* for la Pléiade, Gallimard.

Michèle Plaisant is Senior Professor of Eighteenth-Century English Literature and Civilisation at Lille University, and has been Vice-Chairman of the Société d'Etudes Anglo-Américaines since 1974. Formerly director of the Lille University centre for research into eighteenth-century England (1974–81), she is the author of *La sensibilité dans la poésie anglaise au début du XVIIIe siècle: évolution et transformations*, and has published widely on eighteenth-century England, poetry and women's writing.

Jean Raimond is Professor of English Literature at the University of Reims. He was Chairman of the Société des Anglicistes de l'Enseignement Supérieur from 1986 to 1990 and has been President of his university since 1987. He is the author of *Robert Southey, Poésie et philosophie dans l'oeuvre de Charles Morgan* and *Visages du romantisme anglais*, and co-author of *Le préromantisme anglais* (with Pierre Arnaud) and *Le roman anglais au XIXe siècle*. He is currently working on a French edition of Kipling's works for la Pléiade, Gallimard.

Derek Roper is Senior Lecturer in English at the University of Sheffield. His publications include an edition of *Lyrical Ballads, 1805*; and *Reviewing before the 'Edinburgh', 1788–1802*.

William Ruddick taught English in the University of Manchester until his retirement in 1990. He is an authority on prose writers of the early nineteenth century; edited John Gibson Lockhart's *Peter's Letters to his Kinsfolk* (1977); and is editor of the *Charles Lamb Bulletin*.

Jean-Claude Sallé teaches English at the University of Burgundy. He has published articles on Sterne, Hazlitt and Keats.

Mark Storey is Professor of English Literature at the University of Birmingham. His publications as author and editor include *Clare: The Critical Heritage*, *The Poetry of John Clare: A Critical Introduction*, *Poetry and Humour from Cowper to Clough*, *The Letters of John Clare*, *Byron and the Eye of Appetite*, Gissing's *Private Papers of Henry Ryecroft*, and *Poetry and Ireland since 1800*.

Thérèse Tessier is Professor of English Literature at the University of Paris XII, and President of the French Byron Society. She is the author of *La poésie lyrique de Thomas Moore*, part of which was translated as *The Band of Erin: A Study of Thomas Moore's 'Irish Melodies'*. She has written numerous articles on Byron, on Thomas Moore, and on other Romantic poets and Irish writers.

Hubert Teyssandier is Professor of English Literature at the Institut du Monde Anglophone, Sorbonne Nouvelle. His publications include *Les formes de la création romanesque à l'époque de Jane Austen et de Walter Scott* and articles on Henry James, Patrick White and Nabokov. He is the editor of *Polysèmes*, a review that specialises in the relations between literature and the visual arts.

Pierre Vitoux is Professor of English Literature at the Université Paul Valéry, Montpellier. His publications include his doctoral dissertation, *L'oeuvre de Walter Savage Landor; L'histoire des idées en Grande Bretagne*; and articles on Keats and Shelley, twentieth-century fiction (Huxley, D. H. Lawrence, Joyce), and problems of narratology.

J. R. Watson is Professor of English at the University of Durham. His publications include books on Wordsworth and Hopkins, and *English Poetry of the Romantic Period, 1789–1830*.

Mary Wedd was Principal Lecturer in English at Goldsmiths' College, University of London. She was for twelve years editor of the *Charles Lamb Bulletin*, has been closely associated with the annual international Wordsworth Summer Conference at Grasmere since its inception, and is a regular lecturer at the Wordsworth Winter School there.

The Abolition of the Slave Trade

The extent of the slave trade in the eighteenth century can be measured by the fact that in the hundred years to 1786 over 2 million negroes were imported into America and the British West Indian colonies alone. The number taken annually from the African continent by the ships of various European countries about the year 1790 has been estimated at 74,000, with the British responsible for more than half the traffic and the French for more than a quarter. Religious and humanitarian opposition, however, had become increasingly widespread as the century progressed and as the public grew more fully aware of the colonial system of slavery itself and the nature of the commerce in flesh that supported it. Pope, Thomson and Dr Johnson were among the several well-known writers to denounce the slave trade, but the most influential attack was probably Cowper's emotive indictment of 'human nature's broadest, foulest blot' at the beginning of book II of *The Task* (1785), which followed upon his equally hard-hitting lines on 'cargoes of despair' and those who 'Trade in the blood of innocence, and plead / Expedience as warrant for the deed' in 'Charity' (1782).

In the passage from *The Task* Cowper refers to the fact that any slave who set foot in the British Isles became immediately free. This significant ruling had been made in 1772 in the celebrated case of the negro Somerset, but the first parliamentary motion against the slave trade was introduced in 1776 by David Hartley (son of the author of *Observations on Man*), who asserted that it was 'contrary to the laws of God and the rights of man'. Though unsuccessful, Hartley's effort prepared the way for subsequent united political action, as did the example of the Quakers, who, on both sides of the Atlantic, had regularly issued appeals and taken practical steps against the system. The publication of Thomas Clarkson's *Essay on the Slavery and Commerce of the Human Species* in 1786 gave major impetus to the movement, and brought Clarkson into association with other leading abolitionists, above all William Wilberforce, a member of Parliament for Yorkshire, whose name has rightly become synonymous with the evangelical-Christian concern and indefatigable dedication which underpinned and drove forward the anti-slavery cause in Britain. Using evidence gathered by a committee formed in 1787 under the presidency of Granville Sharp, Wilberforce put forward a bill for abolition in the House

1

of Commons in 1791. After some reversals, but by an inexorable process of overall advance, the necessary legislation was eventually passed in 1806 during the ministry of Lord Grenville and Charles James Fox and became law in 1807.

Denmark had preceded Great Britain in 1792 as the first European power to abolish the slave trade. The others were to follow in due course; and in the United States an act prohibiting the importation of African slaves came into force in 1808. In France abolition had been widely pressed for in the 1780s, but discontent in the colony of Saint-Domingue, and representations from the planters there, led an alarmed Paris Assembly to exclude the colonies from the constitution framed by the Declaration of the Rights of Man. This brought about the bitter revolt of the negroes and free mulattos of Saint-Domingue in 1791, which produced the great leader Toussaint l'Ouverture and ensured the eventual establishment of the independent republic of Haiti. Though the decree of 1790 was repealed in the wake of the rebellion, French abolition itself emerged by fits and starts. At the Congress of Vienna, after Napoleon's abdication in 1814, it was agreed that the slave trade should be ended as soon as possible, but the aim of the French government seems to have been to delay this until a fresh stock of slaves could be introduced into Haiti if the colony were recovered. Bonaparte enacted abolition during the Hundred Days of his restoration in 1815. This was ratified at the second Peace of Paris in the same year, and confirmed by French legislation in 1818.

French abolitionists in particular had always been as concerned to wipe out slavery directly as to stop the supply of slaves. It was again in Britain, however, that the greater goal was effectively pursued. Here Wilberforce worked alongside Thomas Fowell Buxton, who first put the question to Parliament in 1823. Exactly ten yeas later slavery was abolished in the British colonies. The rest of Europe, and some of the American states, gradually took up the challenge.

Romantic literature often reflects or promotes the ideals of the anti-slavery movement. Cowper's outstanding advocacy reaches beyond the aggressive rendering of general truths, noted above, to the subtler technique of dramatising the feelings of the victim himself so as to convince the reader of his suffering and humanity. This can be seen in the portrait of 'the sable warrior' in *Charity*, or in the monologue 'The Negro's Complaint' (1788), which was widely reprinted in newspapers and magazines (and has an ironical counterpart in the strange, jaunty 'Sweet Meat Has Sour Sauce', where an imaginary slave-trader sings a resentful farewell to his profession of whips, padlocks and chains). A similar but more complex psychological approach appears in Blake's 'The Little Black Boy', which, whatever else, destabilises the accepted hierarchy and relations of 'black' and 'white' by making the black boy lovingly desire to shield the English boy from God's heat while yet still fearfully wanting to be loved. Blake, prophet against empire, touches frequently elsewhere on relevant aspects of man's inhumanity to man; and in the 1790s Robert Southey published a whole series of 'Poems concerning the Slave Trade', including six sonnets (1794), 'To the Genius of Africa' (1795), and 'The Sailor' (1798), a dramatic monologue based on actual recorded observations of the horrors of physical oppression.

In these poets radical Christian sensibility intersects with the Romantic theme of the 'noble savage' which had developed in the eighteenth century as Anson's *Voyages* and other travel books had encouraged an interest in the remote and unfamiliar. There is little evidence in the Romantic period of any precise understanding of distant cultures, and it is noticeable that Cowper's and Southey's anti-slavery poems accuse Western society from the standpoint of Western morality and values rather than insight into the otherness of its victims. Yet the primitive did provide generalised images of goodness and innocence which called the direction of contemporary civilisation in question. Cowper's reflections on Omai 'the gentle savage' in book I of *The Task*, for example, celebrate the simple, paradisal pleasures of the natural state and oppose these to the evils of acquisitive commercialism, although the same passage interestingly pities the islander for the loss of the 'sweets' of civilised life which he had briefly experienced during his stay in England. It is in Rousseau's *Discours sur les sciences et les arts* (1750) and *Discours sur l'origine de l'inégalité* (1755) that we find the clearest protest against the corruptness of society and most ambitious expression of the cult of the noble, happy savage.

In *The Prelude* (1805 text, X.201–26) the political debate stirred up by Wilberforce's first Bill forms part of the background to Wordsworth's personal drama after his return from France in 1792, though the current failure of the 'just intent' and 'righteous hope' of good men disturbs him less when he thinks of how the future triumph of the principles of the French Revolution will bring about the fall of *every* 'rotten branch of human shame'. The sonnet beginning 'Clarkson! it was an obstinate hill to climb', written by Wordsworth in 1807, commemorates both the individual achievement of the man, 'true yoke-fellow of Time', and a far-reaching victory which will belong to 'all nations'. Byron's praise, in *Don Juan*, XIV.82–5, is reserved for Wilberforce, the 'Washington of Africa', who struck down an 'immense Colossus'. The encomium is tinged with the vigorous scepticism characteristic of late Byron, and is given edge by the poet's dislike of Wilberforce's Tory connections: 'You have freed the *blacks* – now pray shut up the whites'. But it is sincere. Wilberforce's place in history was secure.

VINCENT NEWEY

Akenside, Mark (1721–70)

Akenside was the son of a butcher. He was born and educated at Newcastle upon Tyne. He studied theology at Edinburgh but soon changed to medicine, in which subject he achieved distinction at Edinburgh, Leiden and Cambridge. He practised successfully and, in 1761, became a physician to Queen Charlotte.

Akenside began writing poetry while still a schoolboy at Newcastle, publishing a short poem in Spenserian stanzas, 'The Virtuoso', in the *Gentleman's Magazine* during 1737: it antedates the Spenserian imitations of Shenstone and Thomson which made the mode fashionable shortly afterwards. In 1738 Akenside began a long poem in blank verse entitled *The Pleasures of Imagination* (a title drawn from Addison's *Spectator*, essays 411–21), which was published by Dodsley in 1744 on Pope's recommendation. A political poem, the *Epistle to Curio* (attacking William Pulteney's abandonment of Whig principles), appeared in the same year, and the *Odes on Several Subjects* in 1745, when Akenside was still only twenty-four.

In 1746 Akenside wrote the blank verse 'Hymn to the Naiads' (published in *Dodsley's Miscellany* in 1758) but after this his literary activities were few and unimportant. He concentrated on medicine almost entirely, though from 1757 onwards he devoted some time to revising *The Pleasures of Imagination*. The posthumously published final text of 1772 includes the beginning of a fourth book in which the Northumbrian scenery of Akenside's youth is recalled. It shows an awareness of how specific landscapes can affect the mind, moving beyond the more general relationship between pleasure and the energised imagination studied in the earlier parts of the poem; and provides a link with the localised studies of the imagination's working in connection with particular times and places which characterise the poetry of Wordsworth.

Akenside's *Pleasures of Imagination* is a didactic poem. It investigates the nature of the intellectual powers which make up the imagination, the ways in which imaginative perception and invention create pleasure, the connections between imagination and the passions, and the way in which the mind works when the imagination is engaged. These investigations build upon the recognition of the imagination as a creative principle by Shaftesbury (and, on occasion, Addison), already touched on in verse by David Mallet's *The Excursion* (1728). Akenside's treatment prefigures central interests of Wordsworth, Coleridge and the second generation of Romantic poets. His carefully wrought neo-classicism anticipates Landor, but the chief interest of his *Odes on Several Subjects* lies in their contribution to the revival of the ode form in the 1740s by Gray, Collins and Joseph Warton: together they liberated the ode to deal with specific times, places and emotions through processes of personal association, in which private reflections or the flow of feelings are linked together by observable energies of the imagination without moral didacticism or a necessary subservience of lyricism to formal structures.

WILLIAM RUDDICK

The American Revolution

Mention of the American Revolution invariably summons images of a group of townsmen dressed as Mohawks tipping tea into Boston Harbour in late 1773 – an event which somehow led to the Declaration of Independence on 4 July 1776. There is in fact some truth in this sketch, although these events only become intelligible within a more complex historical and political narrative. On its American front, the Seven Years' War (1756–63) between Britain and France consisted of a struggle for dominance of the North American continent. Britain's victory, confirmed by the Treaty of Paris in 1763, left it with the most powerful and extensive empire in the world (encompassing India to the east and North America to the west). At the same time, since the expenses of the war had almost doubled Britain's national debt, and since the government decided it needed to garrison British troops to protect the colonies, George Grenville, George III's first minister, levied new taxes on the colonists. Those who were making the best profits from the colonial relation with Britain – the merchants and planters – were hit hardest by the Sugar Act of 1764 and a Stamp Tax, introduced in 1765, on newspapers, pamphlets, advertisements and legal documents. A congress of hitherto completely independent colonies met at Albany and issued a declaration of 'no taxation without representation'. This was backed up with a boycott of British goods and the formation of politically active groups called 'Sons of Liberty'. Rockingham's short-lived ministry repealed the Stamp Act in 1766, but Parliament simultaneously declared its sovereign right to tax and legislate for the colonists. In 1767, with Rockingham out of office, Charles Townsend, Chancellor of the Exchequer and responsible for colonial affairs, introduced what have become known as the 'Townsend Acts' – imposing duties on the colonists' imports of paper, lead, paint, glass and tea – in order to solve problems within the British economy. Crowds gathered under 'Liberty Trees' and rioted against customs officers, making the duties impossible to collect. British troops who were moved from the frontiers and billeted on towns were inevitably seen as a threat to liberty, and the 'Boston Massacre' of March 1770 – in which a crowd throwing snowballs at troops guarding a custom house were fired on, with the loss of five lives – provoked reactions throughout the colonies. By coincidence, on the same day as the massacre, Lord North's new administration repealed all the Townsend duties – except that on tea.

In 1773, however, as Edward Countryman notes, 'Parliament once again turned to the American colonies to resolve problems that had arisen elsewhere' (*The American Revolution*, 1987). In order to rescue the East India Company – which 'was failing either to return a profit to its shareholders or to consolidate Britain's hold on India' – Parliament introduced the Tea Act, which allowed the East

India Company 'to market its tea directly to America . . . and bypass the network of auctions, wholesalers, and colonial merchants through which its tea previously had been sold'. No longer having to be shipped via England, the company's tea was thus subject only to the Townsend duty of 3d a pound. Thus, Parliament thought, Americans would buy the cheaper tea – and therefore, at last, pay a tax it had levied – and simultaneously solve its problems in India and America. But Parliament's systematic attempt to limit the prosperity of the colonies and to exploit them solely for its own benefit backfired. For what this myopic vision seems not to have reckoned with was that the American merchants, 'who might be crushed by the East India Company's newfound strength', might react and might find support from a majority of the colonists. Once again, with its 'Tea Party' of December 1773, Boston was at the centre of the action: 'After a meeting . . . of some 2000 Bostonians to discuss the best course of action, a group of men disguised as Mohawk Indians boarded the tea-carrying ships in the harbour and threw the cargoes overboard' (Clive Emsley, *An Atlantic–Democratic Revolution?*, 1971).

Britain responded with the 'Coercive Acts' – which the colonists christened the 'Intolerable Acts' – of 1774, which closed Boston's port and revoked the Charter of Massachusetts. This served merely to prompt the colonists towards greater unity through a common support for Boston; delegates from twelve of the colonies met as the First Continental Congress in Philadelphia in September, producing a Declaration of Rights to be sent to George III. That the Congress should still consider addressing the King of the British Empire is indicative that, as yet, a large number of the delegates still regarded themselves as British subjects and felt that their grievances were not with the King but with a parliament which had denied them the rights due to all such subjects. At the same time, this manifesto insisting on the colonists' rights was much more aggressive than the conciliatory approach of moderates, who were narrowly defeated in a close vote. The Continental Association which emerged out of the first Congress is indicative that 'The first steps towards destroying British power and towards creating a revolutionary government had been taken' (Countryman).

But the situation altered significantly between this Congress and that of the following year since, in the meantime, skirmishes between British troops and colonists at Lexington and Concord signalled the beginning of war. But, even so, the Second Congress (with Franklin and Jefferson now in its ranks) remained divided between radicals who looked towards independence and moderates who sought reconciliation with George III. Thus, 'while, on the one hand, Congress elected George Washington Commander-in-Chief of the Continental Army, on the other it endorsed Dickinson's "Olive Branch Petition"' (Emsley). Such a division characterises a difference of interests and loyalties throughout the colonies – many of whose people still thought of themselves as British, and some of whom profited from relations with the imperial power. But, instead of responding to the olive branch, George III declared the colonists to be 'rebels' and acted accordingly – blockading American ports and trying to incite black slaves to rebel and Indian tribes to go on the warpath.

Still the colonists held back from revolution and independence. But Tom Paine's *Common Sense*, published in January 1776 and achieving unprecedented sales, seems to have offered them the courage, the inspiration, and the vision: 'Tom Paine offered more than just a vivid summary of what Americans were against. He gave them something to be for: a republic. . . . *Common Sense* . . . changed the terms of American debate. No longer would the questions be how reconciliation might be won and how the British Constitution might be applied to American reality. Henceforth the issues were the coming of independence and the kind of republic it would bring' (Countryman). Whether or not a single pamphlet can do so much, everyone knows that on 4 July 1776 Congress issued the Declaration of Independence – a document which attacks George III for the first time and no longer appeals to the rights of British subjects but to the 'natural' rights of man.

The Declaration of Independence did not, of course, end the war, but it changed the terms 'of what was being fought for. Between 1776 and the American victory in 1783, the colonists whom the British had despised for their poor military prowess in the Seven Years' War developed into an army capable of defeating that of the world's greatest power. The question which then remained was what kind of country should emerge now that the colonists had the unique task of creating a political system from scratch. The main problem was that, having won their independence, individual colonies were reluctant to give political power to a centralised system. Each colony therefore became a separate state through creating its own constitution; at the same time, however, the loose grouping of the states under the Articles of Confederation of 1777 proved unable to deal with the commercial problems which faced the country as a whole. The third of a series of conferences held to address this problem assembled in 1787 and – through an undemocratic and illegal process – so amended the Articles of Confederation as virtually to eliminate them altogether. From the lengthy debates which resulted emerged the Constitution of the United States and the Bill of Rights drafted by the 'Founding Fathers'.

The American Revolution was, by definition, also a British – or even European – event. Its documents and actions have been seen as significant influences on the French Revolution – especially through the participation of French soldiers in the American war and the publication in France of the Declaration of Independence. Indeed, Tom Paine's vision of a new political order in France and Britain was grounded in active participation in the American Revolution. In Britain, many were sympathetic to the colonists' cause before the Revolution. Edmund Burke, spokesman for the Whig opposition, urged conciliation with the colonies for the same reasons as he urged reform in Britain – to protect commercial interests and avoid revolution. Burke warned the government that if it continued its unjust policies towards the colonists in the name of an abstract sovereignty it would 'teach them by these means to call that sovereignty itself into question' ('Speech on American Taxation', 1774). This too is why Burke initially supported British radicals who 'sought the reform of Parliament and the English constitution in the name of the same natural and historical rights as those to which the Americans appealed. They, too, demanded representation in Parliament for those taxed by Parliament' (Isaac Kramnick, Introduction to Paine's *Common Sense*, 1976). But it is possible to see the actual Revolution as a turning-point: it proved not only that the British government could tyrannise its own subjects, but that it could be successfully resisted and the British constitution itself called into question. Americans had preferred the rights of men to the rights of British subjects and thus offered an example, and a developed rhetoric of liberty, to British radicals who wanted to go beyond mere reform. After the Revolution America became – in the eyes of British radicals and poets – the land of freedom: during Pitt's 'reign of Terror' after the French Revolution Coleridge thought of escape to 'Content and Bliss on

Transatlantic shore' ('On the Prospect of Establishing Pantisocracy in America', *c*.1794), while Blake would celebrate that 'Empire is no more!' (*America*, 1793, pl. 6).

TOM FURNISS

The Anti-Jacobin

The Anti-Jacobin was a weekly publication which ran from 20 November 1797 to 9 July 1798. Its principal founders and chief contributors were the Tory politician George Canning (1770–1827) and his friend George Ellis (1753–1815), a lively conversationalist to whom Scott dedicated the introduction to canto V of *Marmion*. The editor was William Gifford, subsequently editor of the *Quarterly Review*. It appeared on Mondays, having been printed on Sunday on the presses of the *Sun* newspaper.

As its title implies, it was an attempt to counter the continuation of pro-revolutionary sentiment: after the fall of Robespierre and the end of the Terror, sympathy for France under the Directory was considerable, led by the radicals of the Godwin circle but also shared by the Whigs. *The Anti-Jacobin* described the Whig as

> A steady Patriot of the World alone,
> The Friend of every Country – but his own.

During a period of French success during the war, it appealed to a crude sense of British patriotism and robustly denounced all French sympathisers. Its scorn for what it saw as absurd libertarian notions is seen in the attack on British supporters of the Revolution, such as Paine, Godwin, Helen Maria Williams and Thomas Holcroft, as 'creeping creatures, venemous [sic] and low'; they are joined in the following extract by Coleridge, Southey, Charles Lloyd and Charles Lamb, who are sarcastically urged to praise Reveillère-Lepeaux, a reformer who tried to introduce a new revolutionary religion based on the love of humanity;

> C——DGE and S—TH-Y, L— d, and L—BE and Co.
> Tune all your mystic harps to praise LEPEAUX.

Praise him each Jacobin, or Fool, or Knave,
And your cropp'd heads in sign of worship wave!
All creeping creatures, venemous and low,
PAINE, W–LL—MS, G–DW–N, H–LC—FT; praise LEPEAUX.

The Anti-Jacobin was scurrilous, reactionary and sometimes abusive, but it was well-timed. It 'caught the tide of public opinion at the turn' (Wendy Hinde, *George Canning*, 1973), and prepared the British public for the anti-French sentiment which emerged when the French invaded Switzerland in 1798.

J. R. WATSON

Architecture

The eighteenth century is known as the age of reason and classicism, but it was also the age of sentiment and sensibility in art and literature. Alongside a taste for classical art there was a renewal of interest in the Middle Ages, and Gothic architecture in particular, as the seeds of Romanticism began to develop.

The Gothic style, characterised by the pointed arch, emerged in England towards the end of the twelfth century. Early Gothic emphasises the vertical, as does the later, Perpendicular style – a kind of national mediaeval style created in England in the fourteenth century. Early Perpendicular work can be seen in the nave of Canterbury and in the choir of Gloucester Cathedral, where a late-Gothic decorative grille encases the original Romanesque structure. Perpendicular work was still being carried out at Oxford in the seventeenth century, but by then the style was generally regarded as barbaric.

In the 1730s, after the first twenty years of Baroque experiment and innovation, a vogue for what became known as Palladianism began. This style, particularly associated with country houses, was based on the villas built by Palladio in the Veneto during the sixteenth century. In England, leading examples of the style are Lord Burlington's house at Chiswick (1729, a modified version of the Villa Rotonda, Vicenza); Mereworth Castle, Kent; Houghton

Hall, Norfolk; and Prior Park, Bath, by John Wood. In the second half of the eighteenth century the tradition continued but became less homogeneous. Sir William Chambers (1726–96) built in a more eclectic style which incorporated contemporary French influences, while Robert Adam (1728–92), the most famous British architect in this tradition, designed exteriors in a modified Palladian style. He has also been called 'the father of the Classical Revival', on account of his interest in classical Roman buildings. But terms such as 'Classical Revival', 'classicism' and 'neo-classicism' can be misleading, because they invite us 'to conceive the style as having been opposed to Romanticism – a conception foreign to the eighteenth century' (Hugh Honour, *Neo-Classicism*, 1968).

Another trend of the early eighteenth century was for what has been called 'architectural romanticism', which originally took the form of sham ruins. At first these were classical, since they were chiefly inspired by Italian landscape. But soon fashionable taste came to prefer Gothic, if only because that was the usual style of authentic ruins in Britain. The first sham Gothic ruins were erected before 1745. Horace Walpole mentioned some built by Sanderson Miller at Hagley, Worcestershire, and wrote Richard Bentley an enthusiastic letter about them in 1758. Sham ruins soon became so fashionable that Miller and others were asked to design all sorts of ruined castles for the aristocracy and the richer gentry. The surviving examples are known as 'follies'.

Thus during the eighteenth century the Middle Ages gradually took the place of classical antiquity as an ideal in architecture as well as in literature. A taste for the Gothic, and not just Gothic ruins, was closely bound up with Romanticism; and at first its chief advocates were not architects, but poets and antiquarians. County histories and volumes of 'local antiquities' helped fuel a craze to which poets such as Thomas Gray (1716–71), Joseph Warton (1722–1800) and Thomas Warton (1728–90) also in large measure contributed. In 1754 Thomas Warton, in his *Observations on The Faerie Queene*, was the first to write about the origin and development of Gothic architecture. His treatise has been called 'the real signal for the Revival of the Gothic' (Kenneth Clark, *The Gothic Revival*, 1928), but in 1762 Horace Walpole published his *Anecdotes of Painting in England*, which was an even stronger signal for the Gothic Revival. In his chapter on medieval architects he wrote, 'The pointed arch . . . was certainly intended as an improvement on the circular. . . . It is difficult for the noblest Grecian temple to convey half so many impressions to the mind as a cathedral does of the best Gothic taste.' At that time some writers attempting to determine the essence of the Gothic found in it an inspiration towards the sublime and infinite (Edmund Burke, 1757), while others associated it with the picturesque (William Gilpin, 1768).

In 1749 Walpole bought the property called Strawberry Hill, near the Thames at Twickenham. He began to transform it into a kind of Gothic castle with the help of William Robinson and Richard Bentley. A rounded tower was added here, a chapel thrown out there, stained glass placed in the windows; old armour and weapons were distributed in suitable spots; an altar was turned into a mantlepiece. Reconstruction and the collection of suitable material lasted twenty years, but still Walpole was not satisfied: he disparaged the result as 'a toy castle', and instead turned his efforts to novel-writing. *The Castle of Otranto* (1764) was the second outcome of a dream suggested by the author's thoughts about mediaeval structures. Walpole thus became the father of Gothic fiction. By the time Strawberry Hill was finished, 'the taste for Gothic had spread from a few eccentrics to the mass of fashionable country gentlemen and had turned into a real craze. Walpole, as the unconscious instrument of Romanticism, was largely responsible for this diffusion' (Clark).

William Beckford of Fonthill, Wiltshire, was the author of *Vathek* (1786), whose hero is an Oriental version of Walpole's Manfred. When he inherited Fonthill it was a classical mansion, rebuilt after 1754 by his father in the Palladian style. But he soon grew tired of it, all the more so as all the world was admiring Strawberry Hill and talking 'grim Gothic'. As early as 1790 he thought of embellishing his estate with some form of romantic ruin, a convent, where he could retire. In 1796 he asked James Wyatt (1746–1813) to design a ruin in the Gothic style to crown the highest point of the estate (Wyatt had for many years devoted himself to the Gothic; and the cathedrals of Salisbury, Hereford, Lincoln and Lichfield sadly testify to his 'improvements and restorations'). The first plan was discarded and a larger one was developed during 1796–7. In its earlier conception it comprised merely the Great Tower and the south and west wings, containing the Fountain Court. During the next few years a great wing and an octagon tower were added to the plan; Beckford perpetually changed his ideas, putting up and pulling down towers, proposing additions and alterations as the work progressed. By 1807 all was ready. The building consisted of an octagon tower more than 276 feet high with a hall 120 feet high; from this stretched, north and south, two wings about 400 feet long and 25 feet broad. The wings to the east and west were of more ordinary proportions. 'Fonthill can hardly be considered as more than stage scenery. But as scenery it was superb. All that the eighteenth century demanded from Gothic was present in Fonthill, and present more lavishly than in real medieval buildings' (Clark). But Fonthill Abbey was built with incredible speed and Wyatt's ignorance of Gothic methods of construction made it highly in-secure. The true instability of Fonthill only became clear after Beckford had sold it. One night in 1825 the tower collapsed, destroying part of the 'Abbey'. Little trace of the huge structure now remains.

Thanks to Walpole and Beckford, mediaeval antiquities became extremely popular. In 1811 Sir Walter Scott himself asked William Atkinson to turn his farmhouse into a modern Gothic building in the Scottish baronial style: Abbotsford was Scott's romantic resi-dence from 1812 until his death in 1832.

PIERRE ARNAUD

Association

The phrase 'association of ideas' appears in the fourth edition of Locke's *Essay concerning Human Understanding* (1700), although the theory that the memory works through some kind of similarity or contiguity goes back to Aristotle. In the eighteenth century the theory was used to explain the development of the intellectual, moral and spiritual senses: beginning with sensations, the mind proceeds to simple ideas associated with those sensations, and then to complex ideas, which are compounded from simple ideas under the influence of association. The writer who most strongly propounded this theory was David Hartley (1705–57), whose *Observations on Man, his Frame, his Duty, and his Expectations* (1749) was an attack on the idea that the moral state is instinctive or innate. Hartley attempted to show that ideas come from simple sensations, and that human beings rise from sensations of pain and pleasure towards the contemplation and understanding of higher things, until the mind is capable of sympathy, theopathy and the moral sense. Hartley thought that the higher qualities would supersede the lower ones, until the mind's ultimate happiness would be found in the contemplation of the love of God. For some time his optimism greatly appealed to Coleridge, who named his son Hartley, and who acknowledged Hartley's work in a note to the following passage from 'Religious Musings', describing the human soul as

> From Hope and firmer Faith to perfect Love
> Attracted and absorbed: and centred there
> God only to behold, and know, and feel,
> Till by exclusive consciousness of God
> All self-annihilated it shall make
> God its Identity: God all in all!

Coleridge, and the other Romantic writers, were also attracted by Hartley's apparently consistent explanation of how an individual grows and develops. As an epistemological theory, it seemed to be complete, and universal: as Coleridge remarked in chapter 5 of *Biographia Literaria*, 'the law of association being that to the mind, which gravitation is to matter'. Some Romantic writers were also enchanted by his style: De Quincey described Hartley's work as 'a monument of absolute beauty'. Samuel Rogers used Hartleian ideas extensively in *The Pleasures of Memory*, and William Godwin quoted Hartley in his *Enquiry concerning Political Justice*, book IV, chapter 7: 'Of the Mechanism of the Human Mind'. It has been argued by Arthur Beatty (*William Wordsworth, his Doctrine and Art in their Historical Relations*, 1927) that Wordsworth was also deeply influenced by Hartley.

One of the reasons for Hartley's early influence may have been his attempt to provide a physical explanation for the operations of the mind. He was a physician, and put forward an account of the neurophysiology of the brain which involved the action of 'vibrations' or 'vibratiuncles' on the 'medullary substance'. Hartley was therefore described by Coleridge in 'Religious Musings' as

> he of mortal kind
> Wisest, he first who marked the ideal tribes

> Down the fine fibres from the sentient brain
> Roll subtly-surging. . . .

In the longer term this explanation, which was probably intended by Hartley to give respectability to his theories, helped to undermine them, and as early as 1775 Joseph Priestley omitted them in his abridged edition of the *Observations on Man*. In the opening paragraph of chapter 6 of *Biographia Literaria*, for example, Coleridge showed how far he had moved from his original Hartleian position by speaking scornfully of 'Hartley's hypothetical vibrations in his hypothetical oscillating ether of the nerves'. Coleridge deals with the theory of association in chapters 5–7 of *Biographia Literaria*. His principal objections to Hartley's theory are twofold. In the first place, if the formation of the soul is determined by its sense impressions, and the chance association of accidental ideas, then there is no free will, and no necessary development from lower to higher, as Hartley had suggested; and in the second place, it becomes impossible to think about the first cause, or God, because the sense impressions and associated ideas do not encourage the contemplation of that which lies beyond them: 'The existence of an infinite spirit, of an intelligent and holy will, must, on this system, be mere articulated motions of the air' (ch. 7). The argument leads Coleridge forward to his subsequent discussion of the imagination in chapter 13, in which it appears that Coleridge is celebrating, in contrast to Hartley, the God-like creativity of both the primary and secondary imaginations.

J. R. WATSON

Austen, Jane (1775–1817)

Jane Austen stands at the ebb-tide between the eighteenth-century and the Romantic novel; modern criticism emphasises this view of her work more than it does the earlier idea of her as the upholder, in form and attitude, of traditional moral and social values, the supreme technician of high comedy in the novel of manners. Her short career of only six complete novels – *Sense and Sensibility* in

1811 (three years before Walter Scott's *Waverley*) to the posthumously published *Northanger Abbey* and *Persuasion* in 1818 – is misleading. She wrote steadily from her teenage years onwards. While it is clear that her last three novels (*Mansfield Park*, 1814; *Emma*, 1816; *Persuasion*, 1818), were written in the one or two years before publication, her earlier novels are less certainly datable: *Northanger Abbey* (originally called *Susan*) dates from 1798–9 and was sold to a Bristol publisher in 1803; *Pride and Prejudice* (as *First Impressions*) and *Sense and Sensibility* (as *Elinor and Marianne*) existed in versions as epistolary novels before 1797. No manuscripts of any of these survive.

A number of works remained unpublished until the twentieth century: two incomplete manuscript drafts of novels, *The Watsons* (*c*.1805) and *Sanditon* (1817); one short, possibly aborted, novel in letters, *Lady Susan* (*c*.1805); and various collections of juvenilia, mainly from 1790–3, which ridicule and burlesque to the point of farce the conventions of the contemporary sentimental novel and drama. They show that Jane Austen could say, with her character Elizabeth Bennet, 'follies and inconsistencies do divert me, I own, and I laugh at them whenever I can'.

This spirit is manifested in all her novels, from the most 'light and bright and sparkling', *Pride and Prejudice*, to the most serious, *Mansfield Park*, and to the most open to the influence of Romanticism, *Persuasion*, with its references to Byron, and its enthusiasm for picturesque landscape.

Living apart from the mainstream of literary life, Jane Austen, youngest daughter of a Hampshire clergyman, spent her life, unmarried, as one of a large family of naval men and clerics and their wives and children. She lived wholly in the south of England (at Steventon, Southampton, Chawton and Winchester – where she died – in Hampshire; at Bath, in Somerset; and at Lyme Regis in Dorset), and travelled largely within the same region (to London, Kent, and so on). Her novels inhabit the same general territory. She belonged to the country gentry who comprise most of her characters: 'the middling classes of society' whose delineation Sir Walter Scott praised in his review of *Emma* for the *Quarterly Review* (Oct 1815). Her plots are the conventional ones of the novel of manners, of the heroine whose experience in society charts a moral and social education, leading through mistakes and confusions to social and moral reordering emblematised in happy marriage. Writing from first-hand observation, she yet claimed not to copy from life, being 'too proud of her gentlemen to admit they were only Mr A. or Colonel B.' (quoted in J. E. Austen-Leigh, *Memoir of Jane Austen*, 1870).

She published anonymously, and wrote in the first instance for her appreciative family and friends, a fact which accounts for her precise, narrow-gauge irony and a tone the opposite of what Scott called his 'Big Bow-wow strain'. His praise for her 'exquisite touch which renders ordinary common-place things and characters interesting from the truth of the description and the sentiment' (Journal, 14 Mar 1826) is as revealing as her own metaphor of the 'little bit (two inches wide) of ivory on which I work with so fine a Brush, as produces little effect after much labour' (letter to J. Edward Austen, 16 Dec 1816).

Though not a well-known author in her own time, even though her last publisher was the influential John Murray, publisher to Byron, Moore and Scott, she soon became the object of enthusiastic appreciation, as her readership increased. The line of enthusiasts stretches from Archbishop Whately (1821) to A. C. Bradley (1911) and beyond. Reservations concern her narrow range, trivial content and lack of passion, as in Charlotte Brontë's remark that Miss Austen, being 'without "sentiment"', without poetry, maybe *is* sensible, real (more *real* than *true*) but she cannot be called great' (letter to G. H. Lewes, 18 June 1848), and as in D. H. Lawrence's distaste for her 'sharp knowing in apartness' (*A Propos of Lady Chatterley's Lover*, 1930). F. R. Leavis (*The Great Tradition*, 1948) places her as the first of the line that leads to George Eliot, Henry James and Joseph Conrad. Her high reputation continues to be reiterated: J. F. Burrows' recent study *Computation into Criticism* (1987) reinforces her reputation as a stylist and master of speech idiom, by the use of computer and statistical analysis.

W. A. CRAIK

Beattie, James (1735–1803)

James Beattie was a Scottish poet and philosopher. Professor of Moral Philosophy at Marischal College, Aberdeen, he was one of the leading figures of the School of Common Sense, supporting a traditional current of realism in opposition to the sceptical tenets of David Hume. His major philosophical treatise, *An Essay on Truth* (1770), won him a royal pension granted by George III. Beattie held traditional religious views which tended to reconcile the more moderate branches of the Church of England and of the Presbyterian Church of Scotland. A staunch defender of faith, he was highly suspicious of such eighteenth-century notions as the essential trustworthiness of human reason and the inevitability of progress. At the same time, he shared his contemporaries' heightened sensibility and fascination with nature. Beattie's miscellaneous poems include odes, epitaphs, elegies and fables written in the classical mode. He advocated the principle of the imitation of nature and the use of poetic diction. His most famous poem, *The Minstrel* (1771–3), composed in Spenserian stanzas, brought him renown and helped pave the way for Wordsworth's *Prelude* and Clare's 'The Village Minstrel'. The poem traces the 'progress of genius', the growth of Edwin's mind from childhood to maturity. The wild scenes of Scotland fill the young hero's mind with wonder, and the first book of the poem is an image of Eden on earth. However, the conversation with a hermit living in a secluded cave warns the boy of the unexpected dangers of the world. Nature, glamorous as it may seem, is but a Paradise Lost in which man must be guided by reason and instruction. The consummate poet, a man of genius, must be able to combine the pleasures of the imagination with the requirements of mild reason. Thus anticipating Coleridge's well-known distinction between fancy and imagination, Beattie, though a late-Augustan poet, also ranks among the early Romantics.

PIERRE MORÈRE

Blake, William (1757–1827)

Blake was born in 1757 to a hosier and apprenticed at fifteen to the engraver James Basire, for whom he made drawings at Westminster Abbey, thereby acquiring a taste for the Gothic. In 1780 he witnessed the Gordon Riots and the burning of Newgate Prison. He married Catherine Boucher in 1782, and the following year his *Poetical Sketches* were printed at the expense of John Flaxman and the Revd. A. S. Mathew but not published. He opened a print shop in partnership the following year. In 1789 he and his wife attended the first London meeting of the Swedenborgian New Church, and he engraved *Thel* and *The Songs of Innocence*. The *French Revolution* was set in type but never published, and by March 1791 the poet had moved across the Thames to 13 Hercules Buildings, Lambeth. In 1793, when Louis XVI was executed and war declared between Britain and France, Blake engraved *America* and *The Visions of the Daughters of Albion*. By this time he was clearly distressed at the course of the French Revolution and wrote in his notebook, 'I say I shan't live five years, and if I live one it will be a wonder.' 1794 saw the publication of the combined volume *The Songs of Innocence and of Experience*, of *Europe* and of *The First Book of Urizen*. Along with twelve colour prints, including *Newton* and *Nebuchadenezzar*, the last three illuminated prophecies of the Lambeth period were produced in 1795. They are *The Song of Los*, *The Book of Los* and *The Book of Ahania*, whose last plate, featuring a heap of guillotined heads, suggests an increasingly pessimistic view of the revolutionary process. From 1797 onwards, using atlas-sized paper provided for the project of illustrating Young's *Night Thoughts*, Blake composed *Vala*, his first long poem, which he revised over a ten-year period until it grew into *The Four Zoas*.

From 1800 to 1803, the Blakes lived at Felpham, Sussex, under the patronage of William Hayley, minor poet and *bel esprit*. Blake worked on *The Four Zoas* by night and Hayley's assignments by day, producing engravings and miniatures with increasing personal dissatisfaction. In April 1803 he was accused by Scofield, a drunken soldier whom he had evicted from his garden, of uttering seditious threats – a serious accusation in view of the war hysteria that prevailed in Britain at the time. Blake's trial for sedition in January 1804 at Chichester Quarter Sessions ended in acquittal to the cheers of the spectators. Meanwhile the poet had taken rooms in South Molton Street, London, had completed *Milton* and had begun *Jerusalem*.

The next twenty years of Blake's life were marked by increasing poverty, obscurity and loneliness. A notebook entry reads, 'Tuesday Jan. 20, 1807, between Two and Seven in the Evening – Despair'. His first and only exhibition, which was held in 1809 in his brother's shop and for which he wrote *A Descriptive Catalogue*, proved a failure. *The Examiner* called him 'an unfortunate lunatic'. He was reduced in 1815 to engraving figures of dishes for a catalogue of Wedgwood's pottery and later on to selling his collection of prints. Flaxman and Thomas Butts, however, remained friends of Blake, with Butts commissioning a series of watercolours of Job in 1818. It was at this time that a young painter, John Linnell, and a group of his friends, calling themselves 'the Ancients', became Blake's admirers and supporters. The first copy of *Jerusalem* was engraved in 1820 and Linnell commissioned the Dante illustrations in 1824. In the same year Blake became acquainted with Samuel Palmer. On April 1827 he wrote to George Cumberland, 'I have been very near the Gates of Death & have returned very weak & an Old Man feeble & tottering but not in Spirit & Life, not in the Real Man, the Imagination which Liveth for Ever.' He died on 12 August of that year, singing about what he saw in heaven.

Modern criticism of Blake begins with Northrop Frye and David V. Erdman. Frye's *Fearful Symmetry* (1947), which has passed into general literary discourse as no other study of Blake has done, presents Blake as a central rather than a peripheral figure in English literature and defends *Jerusalem* as his ultimate attempt at stating by means of an explicit myth the simultaneity of cosmic, historical and psychological events. A remarkably biblical poet (even by the standards of English literature) Blake develops a central myth whose boundaries are creation, fall, redemption and apocalypse. Man is a Luvah, or form of life, subject to two impulses: one the prophetic impulse, leading him forward to vision; the other the natural impulse, which drags him back to unconsciousness and finally to death. There is, however (and somewhat paradoxically), an Aristotelian side to Frye's approach and he seems to forget that he is himself, at times, the source of the highly patterned reality he perceives. Erdman's *Blake: Prophet against Empire*, subtitled 'A Poet's Interpretation of the History of his Own Time' (1954), attends not to archetype but exclusively to the politico-historical element in Blake: not only the revolutionary and Napoleonic wars, but also domestic repression in Britain, persecutions in Ireland and the French Terror are shown to be part of the fabric of Blake's poems. Erdman – whose *Illuminated Blake* (1974), a plate-by-plate commentary of the corpus, is invaluable – is particularly instructive on *America* and *Europe*. But to read Blake's work as covert political allegory is finally reductive, as it excludes the archetypal element completely. At the opposite extreme (though both consider a Blakean plate as a cryptic document to be deciphered), Kathleen Raine's approach is fundamentally apolitical. Her *Blake and Tradition* (1968), despite its massive learning, suffers from an over-literal application of supposed occult sources. Morton Paley's two studies *Energy and the Imagination* (1970) and *The Continuing City: Blake's Jerusalem* (1983), while taking into account the enormous development of Blake scholarship, provide a background against which the unique figure of Blake stands clearly outlined.

America

As indicated on the title page, the poem was composed at Lambeth (where Blake had settled by March 1791) and printed in 1793. Its mood originally partook of the joyful certainty typical of the years 1790–1 that, with the advent of revolution in America and France, a Golden Age was dawning. But by 1793, at a time when the counter-revolution was building up and the British government had (on 21 May) passed an Act 'against diverse wicked and seditious writings', it took some courage to give one's address on the front-page of so controversial a book. *America* was the first of Blake's illuminated books designated as a 'prophecy'. He no longer attempted in it, as he had done in *The French Revolution* (1791), to dramatise history. Instead, and this is in keeping with his conception of a prophet's duty, he recorded the formula of all revolutions, making use of the American material without regard for chronology.

For specifically historical references Blake drew information from Joel Barlow's *Vision of Columbus* (1787) and its epic pageant of American history viewed prophetically by Columbus from a Mount of Vision. Blake extended the stage to include both the British and American coasts, swelled the machinery to a large cast of angels and demons, and replaced Columbus with the King of England, who, looking westward from Albion, trembles at the vision of the Insurgents.

An eighteen-page folio booklet in 'illuminated printing' (*Prospectus*, 10 Oct 1793), *America* deals with a cycle of history that begins with the birth and rise of an Orc or human serpent of Independence during the American War and Revolution, and concludes with the end of the war and that Spirit's momentary repression. The Preludium (which was probably written last) envisions in the form of myth an outbreak of repressed energy: Orc breaks his chains and has violent intercourse with a nameless female principle. An identity is thus established between psychological and historical expressions of rage and lust. Orc is the pent-up energy in human nature and human history that is at first held down but finally erupts or explodes. The prophecy proper begins with the defiance of Washington but continues as a wordy contest and then a battle between Orc, the Spirit of Freedom, supported by the guardian angels of the thirteen colonies, and their enemy Albion's Angel, the tutelary Spirit of Britain. Then war begins in heaven, casting its shadows on the earth beneath. Urizen, the aged tyrant God of storm and snow, freezes the entire action. The prophecy ends with the promise that Orc's fire will ultimately melt Urizen's frost.

In *America* Orc is a wholly positive figure. In the context of a so-called 'Orc cycle' Northrop Frye has nevertheless described the Orcean figures in Blake's early poetry as partly negative, obeying a pattern obvious in many revolutions by which 'the iron hand' which 'crushed the Tyrant's head . . . became a Tyrant in his stead' (*Fearful Symmetry*). This pregnant imagery, however, belongs to 'The Grey Monk', a text which was written after 1800. Such a reading as Frye's clearly undermines the radicalism of Blake's first prophecy, his faith in the possibility of a new social order: the cycle of history prophetically examined in *America* is not that of rebellion–repression–tyranny, but is one of enslavement–liberation–re-enslavement (see Erdman in *Blake's Visionary Forms Dramatic*, 1970). The first indication of another kind of cycle, in which the revolutionary figure is corrupted by success, appears in *The Book of Ahania*, whose protagonist is Fuzon, Urizen's rebellious son.

The Four Zoas

Blake's first attempt at a long epic exists in the form of a much revised manuscript on which he worked from 1795 to 1808. The text is of central importance to an understanding of the growth of the poet's mind. It affords the reader a rare glimpse into the workshop of a creative genius. In the course of these years Blake's life and thought underwent enormous changes, which are reflected in his manuscript. Regarded by Frye as 'the greatest abortive masterpiece in English literature' (*Fearful Symmetry*), *The Four Zoas* is the only mature prophecy that Blake abandoned. The poem was originally entitled *Vala, or the Death and Judgement of the Ancient Man, A Dream of Nine Nights* (1797). When the poet decided to make extensive revisions of this first version written in elegant copper-plate script he used discarded proof sheets from the grand commercial project (1796–7) of engraving Edward Young's *Night Thoughts*. But his method of revision was to add and expand rather than delete and compress: during the years that saw the transformation of *Vala* into *The Four Zoas*, the manuscript collected enough changes to bewilder even Blake himself (see G. E. Bentley, Jr, '*Vala*', or '*The Four Zoas*, 1963). Some of the sketches in the manuscript were erased by Linnell, who was shocked by their erotic inspiration.

After he had completed *The First Book of Urizen*, Blake made a heroic attempt at co-ordinating the separate stories told by earlier books into one grand story of mankind from the origins to the end of time. The action takes place simultaneously within the consciousness of the human race and the mind of each individual during his lifetime. Like *Night Thoughts* the poem is in nine 'Nights', and it describes the nightmare of a cosmic man, Albion, now fallen into division, who once embodied the divine and the human, male and female, subject and object, mind and nature. Although the fall is a highly overdetermined event, its main cause appears at first to be man's seduction by Vala, the cruel, designing, sensual, selfish female. Hence the original title of the work. Vala's very name and the veil she wears suggest the veil of material appearance that obscures spiritual reality. But in the process of revising *Vala* Blake came to attribute the fall to a war among Albion's four primary attributes, his Zoas (the 'living creatures' who in Ezekiel 1:15, 19–23 surround God's throne), each of them representing a major faculty in man: Urizen embodies conscious thought, Luvah passion and desire, Urthona imagination, Tharmas compassion. The fallen situation is further complicated by the fact that each of the Zoas has become separated from his Emanation or female counterpart.

Serious criticism of the poem has been made possible by H. M. Margoliouth's 1956 edition of *Vala* ('*Vala*': *Blake's Numbered Text*): before him no one had bothered to sort out *Vala* from *The Four*

Zoas. Bentley has conclusively shown that all the directly Christian references in *The Four Zoas* were late additions. Throughout Albion's sleep, Jesus's loving concern manifests itself in seven historical phases ('The Eyes of God') not fully identified till night VIII. They record a development of man's images of God, from a primitive conception of God as a baby-eating idol to God as a self-sacrificing Saviour. In the course of the apocalyptic ninth night, each of the Zoas is reconciled to his Emanation and Albion is able to resume his life in Eternity.

Jerusalem

The date 1804 on the title page of Blake's last and longest illuminated prophecy (it numbers 100 plates) cannot be the date of the poem's completion and is certainly not that of its engraving. In mid-1807 George Cumberland wrote, 'Blake has eng[raved] 60 Plates of a new Prophecy' (G. E. Bentley, Jr, *Blake Records*, 1969). As *Milton* has only fifty plates, he very probably refers to *Jerusalem*, most of which seems to have been written between 1804 and 1808, with the engraving beginning some time in 1805 and lasting some fifteen years. It was a time of loneliness and political disillusionment for Blake. The 'Public Address' drafted in the Notebook after the failure of his only exhibition alludes to a poem 'relating the manner in which I have routed a nest of Villains'. Such villains, who play their parts in the text of *Jerusalem*, include Scofield, the soldier whom Blake expelled from his garden and who accused him of seditious threats, for which Blake was tried at Chichester in 1804; and the Hunt brothers, whose *Examiner* savaged Blake in 1808–9 and who appear in the Prophecy as the three-headed Giant Hand, leader of the Sons of Albion.

As early as *The Four Zoas* one senses in Blake a drive towards an inclusive myth. In the opinion of Morton D. Paley (*The Continuing City*, 1983), Blake's entire creative life up to *Jerusalem* was a prelude to that great work. Dedicated to the Public, to the Jews, to the Deists and to the Christians, *Jerusalem* is encyclopaedic, at once Blake's *Paradise Lost* and his *Divine Comedy*. It presents both the mythic history and the visionary future of mankind, all devolving on the present moment. Albion subsumes all Britons, together with their island and history. His Emanation, Jerusalem, derives from the biblical image of the New Jerusalem popular with non-conformists and evangelicals. As in the Bible, Blake's Jerusalem is both a city and a woman. As a city, she is surrounded by enemies; as a woman, she either is led a captive to Babylon or is a bride in Eternity. Trapped by mistaken laws of righteousness, Albion scorns Jerusalem's love and allows himself to be seduced by Vala, Jerusalem's shadow and the seductive goddess of nature. He falls into a torpid sleep of despair which in contemporary life takes the form of war, empire and exploitation. Inextricable from this sexual myth is the visionary myth of Los. Working within Ulro, the hell of the material world or of the material body, he and his sons labour at their furnaces, giving permanent form to all things on earth. He ardently wishes to save Albion, who is unaware of his existence. Los must also meanwhile struggle against himself, against his own Spectre, 'all the forces within the artist which urge him against creativity'. Los, keeper of the Divine Vision in times of trouble, is the true hero of the poem. After a last offensive of Antichrist, he brings the awakening of Albion and his reunion with Jerusalem.

Apart from Southey's gibe about 'a perfectly mad poem called *Jerusalem*, Oxford Street is in Jerusalem' (letter to Henry Crabb Robinson, 21 July 1811), the poem received no critical notice during Blake's lifetime. As late as 1946, it could still be regarded as 'a fantastic jumble' (A. Kazin, *The Portable Blake*). Northrop Frye's great contribution to the study of the poem was his demonstration of its mythic structure. Arguing that, like Coleridge), Blake regarded the Bible as a repository of archetypal situations repeated through history, he showed how Blake related these symbols to one another and to a literary tradition: 'In reading *Jerusalem*, there are only two questions to consider: how Blake interpreted the Bible and how he placed that interpretation in an English context' (*Fearful Symmetry*). *Jerusalem* is no longer treated as a quarry for Blakean 'philosophy' but is investigated as a poetic structure whose generic elements are now coming into focus.

The Marriage of Heaven and Hell

The title page gives neither author nor place nor date, but the book was advertised for sale in 1793. It was probably begun in 1789 and completed in the course of the following year, except for the final 'Song of Liberty', which contains allusions to events of 1792–3. In one copy on plate 3 Blake has put the date 1790 after the first words of the sentence 'As a new heaven is begun and it is now thirty-three years since its advent, the Eternal Hell revives.' This ostensibly refers to the New Jerusalem announced by Emmanuel Swedenborg, a Swedish engineer-turned-visionary, for the year 1757. It happened to be the year of Blake's birth. The enthusiasm and gusto characterising *The Marriage* may also be ascribed to the fact that Blake at this Christological age felt he had mastered the difficult process of 'illuminated printing'.

A prophetic manifesto, *The Marriage* is also a satire, perhaps 'the epilogue to the golden age of English satire' (Frye, *Fearful Symmetry*), in its parodies both of the Bible and of the religious teachings of Swedenborg (which Blake had come to like less and less) and in its assaults on popular images of heaven and hell. Blake's text is written in a mixed style. It can best be defined as a condensed version of what Frye calls an 'anatomy', characterised by variety in subject matter and an intense concern with intellectual error. It consists of a poem as prologue or Argument (possibly the first piece of free verse in English), exposing the way in which false religion has invaded the paths of truth; six chapters in prose; and a Song as epilogue. Each chapter includes a series of dogmatic statements followed by a fanciful dithyramb illustrating the foregoing statements and based on accounts of conversations and adventures with angels and devils. Such a deliberate medley of exposition and anecdote requires readers who are not too literal-minded or liable to take every assertion of the text as a straightforward opinion of the author. Blake's zestful desire to shock clearly suggests that he intended to undermine simplistic, uninspired systems in which the passive 'good' is valued over active 'evil'. His praise of uninhibited vigour of thought and action in the 'Proverbs of Hell' can be summed up into one diabolical formula: sexual excess leads to antinomian perception. The risen body becomes the expanding imagination: 'Against the supernaturalist Blake asserts the reality of the body as being all of the soul that the senses can perceive; against the naturalist he asserts the greater reality of the imaginative over the given body' (H. Bloom, *The Visionary Company*, 1962). Section II (plate 3) proclaims that 'Without Contraries is no Progression'. Although it is tempting to read the aphorism along Hegelian lines and apply to it the dialectic of *Aufhebung*, which simultaneously annuls each state and raises it to a higher one, Blake's Contraries neither progress, disappear nor alternate because they polarise human life (see Martin K. Nurmi, *Blake's 'Marriage'*, 1972: 'Whoever tries to reconcile them seeks to destroy existence. Religion is an endeavour to reconcile the two').

It does not take long to recognise that the *Marriage* has nothing to do with the simple inversion of moral good and evil which is known as Satanism and which forms an important aspect of Romanticism (see A. C. Swinburne, *William Blake*, 1868). Hell occupies centre stage in the *Marriage* only because politically and spiritually Blake was in an angel-dominated culture. After eighteen centuries during which Christianity had come to identify itself with the interests of the ruling class, it was time to recall the disruptive action of Jesus, the true antinomian. Swedenborg, Blake's immediate target, is relegated to the position of the Angel sitting in the tomb: 'His writings are the linen clothes folded up.'

Milton

This poem, in two books, was composed from 1800 to 1804 and engraved c.1809. The title page, when first etched, announced twelve books. In its final form, *Milton* has fifty plates. It was partly written at Felpham and continued in London. In his 'Public Address' (1810), Blake refers to 'a Poem concerning my three years' Herculean Labours at Felpham, which I will soon Publish' (*Complete Writings*, p. 592).

The narrative falls into three parts.

1 In the Bard's Song, Milton, though in heaven, expresses his unhappiness. The song allegorises Blake's difficulties with his patron Hayley, presented as Satan. Milton realises that Satan's errors are his own.

2 Milton resolves to return to earth and Eternal Death in order to renew his spirit. He drops like a shooting star into Blake's left foot. This part of the narrative records a single moment of transcendent awareness, a moment in and yet out of time, such as Eliot described in *Four Quartets*: Milton is represented as existing on four separate levels. His 'Redeemed Portion' shapes Urizen out of clay. His 'Mortal Part' is frozen on the rock of the fallen world. His 'Real Humanity' walks above in majesty. His 'elect' Spectre separates from Blake. This points to the increas-

ingly soteriological nature of the text's central myth: we are not so much regenerated as rescued from ourselves.

3 Ololon, Milton's Emanation (who is called 'sixfold' because of his three wives and three daughters), is reconciled with Milton in Blake's own cottage at Felpham. The poem ends with the casting-off of Satan–Hayley, as Blake, inspired by Milton, is taken into Golgonooza, Los's city of arts, the outpost of heaven in Ulro.

The subject of *Milton* is the awakening of the imagination – Milton's, Blake's, and the reader's – to its full human–divine potential. It is also an analysis of the errors of the intellect which thwart or limit such an awakening and to which John Milton showed himself subject (David Fuller, *Blake's Heroic Argument*, 1988). For the Romantics, what Milton had said and done was felt to be immediately relevant to any individual concerned with his own salvation, with that of the English nation as a whole and that of other nations, since, according to Blake, they were originally one within Albion. Milton's ability to depict eternal states, to oppose tyranny, to purify as best he could within his historical and personal limitations the Christianity of his time accounted for his prestige as a true poet. Yet he did not escape error. His self-righteous religion had interfered with his vision. *Paradise Lost* had shown a false image of God as a law-giving tyrant who insisted on blood sacrifice as the price to be paid for sin. Also, some aspects of Milton's characterisation of Eve and Dalila seemed anti-feminist even by eighteenth-century standards. His failure to grasp the spiritual nature of love was a missing link in his poetic argument. He must therefore return to earth to annihilate his moralistic self-hood and reconcile himself with that part of his imagination he had rejected and which assumes the form of Ololon. *Milton* is concerned with prophetic succession in so far as Blake draws together what Milton had put asunder. According to Harold Bloom, the author of *Milton* is to be regarded as 'the most profound and original theorist of revisionism and an inevitable aid in the development of a new theory of poetic influence' (*The Anxiety of Influence*, 1973). He engaged in fierce, brazen, but creative misinterpretation of his mentor, the purpose of which was to assert his own integrity.

Songs of Innocence and of Experience

In 1784–5 Blake was a partner in a print shop and it was felt that an illustrated volume of verse of high quality would be a useful addition to the stock. The very first Songs of Innocence were composed at this period. Three of them – 'Nurse's Song', 'The Little Boy Lost' and 'Holy Thursday' – appeared first in the boisterous satire *An Island in the Moon* (1784). The illuminated volume *Songs of Innocence* was first published in 1789, and, though its contents were incorporated in *Songs of Innocence and of Experience* in 1794, as late as 1808 Blake was still producing the original volume, as customers ordered it. In part, the form of the songs derives from Isaac Watts, whose *Divine and Moral Songs in Easy Language* (1715), a chapbook intended for children, was still popular in Blake's time. Watts's didactic and moralistic intentions, however, are subtly subverted by Blake's adaptations.

The subtitle of the combined volume, 'Shewing the Two Contrary States of the Human Soul', clearly suggests that the relationship of innocence and experience is one not of direct, static contrast, but of shifting tensions. Although it is a norm by which experience is evaluated, the state of innocence does not preclude dangers, corruption, change: wolves lurk at nightfall and summer dies into autumn; innocence is no pastoral idyll but a condition of mind and spirit. While printing and selling *Songs of Innocence*, Blake was drafting *Songs of Experience* in his Notebook. He constantly reconsidered the interplay between the two states of the soul and the interrelationships between his songs conveying those states: he thus transferred 'The Little Girl Lost', 'The Little Girl Found', 'The Schoolboy' and 'The Voice of the Ancient Bard' from *Innocence* to *Experience* and added 'To Tirzah' to *Experience* after 1805, when he was in the throes of composing his great prophecies. The Bard who 'Present, Past and Future sees' and who introduces us to experience is aware that the fallen condition need not be permanent. Yet, as 'Earth's Answer' suggests, the outlook proposed by *Experience* is so bleak as to exclude the possibility of a way out. The most sacred links seem corrupted in *Experience* by the poisonous Tree of Mystery that grows in the human brain. But contrasting poems such as the two versions of 'Holy Thursday' point to the importance of realising that each poem's presentation of the world is coloured by the state that the individual speaker is in. This distinction adumbrates a further distinction Blake will operate, between the states themselves and individuals in those states.

Songs of Innocence and of Experience is the only one of Blake's books that attracted the admiration of his fellow writers during his lifetime. Wordsworth had 'Laughing Song', 'Holy Thursday' and 'The Tyger' copied out in a commonplace book (see Bentley, *Blake Records*). Coleridge, who had borrowed from C. A. Tulk a copy of the *Songs*, commented upon it in a letter of 12 February 1818, where he declared that 'The Divine Image' and 'Night' pleased him in the highest degree, but that he was somewhat perplexed by the 'Nurse's Song' of *Experience* and by 'The Little Girl Lost' (*Blake Records*). In another letter he hazarded a guess about the author: 'He is a man of Genius – and I apprehend a Swedenborgian – certainly a mystic, emphatically' (*Collected Letters*, ed. E. L. Griggs, 1956–71, IV, 883).

FRANÇOIS PIQUET

Bonaparte, Napoleon (1769–1821)

Life

Napoleon Bonaparte was born at Ajaccio, Corsica, in 1769. He was educated in France at the military school at Brienne, and in 1784 went to the *école militaire* at Paris, from which he was commissioned as a lieutenant in 1785. He had a chequered career in the early years of the Revolution, and was for a time associated with Robespierre and the Jacobins: he first distinguished himself militarily when he commanded the artillery at the siege of Toulon (then held by the British) in December 1793. His great opportunity came with the insurrection against the National Convention in Paris, on 4 October 1795, when he (nominally second-in-command) fired on the crowd and dispersed them, thus saving the Convention. He was made commander of the army of Italy: in the ensuing campaign against the Austrians he became famous at the crossing of the bridge at Lodi. His conquest of northern Italy was crowned by the treaty of Campo Formio (Oct 1797), which ceded to France Belgium and the left bank of the Rhine (and also extinguished the Venetian Republic, thus occasioning Words-worth's sonnet).

He then turned his attention to Egypt and Syria, hoping for more spectacular victories; but he was frustrated by the resistance of Admiral Sidney Smith at Acre ('that man made me miss my destiny') and by Nelson's naval victory at Aboukir. He returned to France, then governed by the Directory, at an opportune moment seizing power in the *coup d'état* of 18 Brumaire (9 Nov 1799). Three consuls were established, and Napoleon, as First Consul, set in motion a series of administrative reforms, made peace with the Roman Catholic Church, and centralised the powers of govern-ment.

His first military campaign as First Consul saw him crossing the Alps by the Great St Bernard Pass (an event commemorated in a romantic painting by Jacques-Louis David), and winning a great victory over the Austrians at the battle of Marengo (14 June 1800). He was made Consul for life in 1802, and Emperor in 1804; meanwhile, after the brief Peace of Amiens (1802–3) he was pros-ecuting the war against Great Britain. His plans for an invasion, with an army at Boulogne, caused great alarm along the south coast: they failed because the French navy could not gain control of the English Channel. Abandoning the camp at Boulogne, Napo-leon decided to return to fighting the continental powers: he marched across Europe and won the battle of Ulm (19 Oct 1805); two days later, Nelson destroyed the French fleet at Trafalgar.

Six weeks after Ulm, Napoleon defeated the Russians and Aus-trians at Austerlitz (2 Dec 1805); another great victory at Jena (Oct 1806) disposed of the Prussians, with further victories at Eylau (Feb 1807) and Friedland (June 1807). At the Peace of Tilsit, later that year, Napoleon came to a partition agreement with the Russian Tsar Alexander I, and found himself master of Western Europe.

The stability of the Napoleonic Empire was short-lived. Re-bellion broke out in Spain (1808), and the British (with Wellington as second-in-command) landed in Portugal. Encouraged by this, the Austrians re-entered the war, only to be defeated again at Wagram (July 1809). Napoleon, concerned at the potential threat of the rival power of Russia, decided to engage in an all-out campaign against the Tsar, and assembled a vast army for the campaign of 1812. When he invaded Russia, he made rapid progress: the Russians retreated, fought a delaying action at Borodino, and then burned Moscow. Napoleon, caught without adequate supplies, was forced to turn back in the middle of the Russian winter, and his retreating army was harassed by Russian cavalry. Only fragments of the great army survived. Meanwhile Wellington was advancing through Portugal and Spain, and by 1814 the allies were in France. Napoleon abdicated (11 Apr 1814), and the Bourbon monarchy (in the person of Louis XVIII) was restored. Napoleon was forcibly retired to Elba.

Frustrated by his inactivity, and by the state of Europe, he escaped from Elba and landed in the south of France (1 Mar 1815). Armies sent to capture him fell under the spell of his charismatic leadership, and by 20 March he was in Paris, with the king an exile in Ghent. Napoleon assembled an army and marched towards Belgium, intending to meet separately the British under Wellington and the Prussians under Blücher: the battle of Waterloo (18 June 1815) was won by Wellington's masterly and heroic defensive tactics, and by the timely arrival of the Prussians; it forced Napoleon into a second abdication, and on 15 July he surrendered to a British naval officer, Commander Maitland of the *Bellerophon*. He was sent to St Helena, in the south Atlantic, where he died in March 1821.

J. R. WATSON

Napoleon Bonaparte and English Romanticism

Napoleon was a complex figure in his personality and achievements. The most perspicacious of his contemporaries (for example Byron, and even Pitt) saw this; but for the most part he was interpreted according to a series of stereotypes which varied according to the commentator. Thus he was an ignoble scourge of the world (Scott), a monstrous barbarian (Wordsworth), a beneficent genius (Godwin), a despicable tyrant (Coleridge), a warmonger (Shelley), a charlatan and coward (*The Times*), the scourge of reactionary monarchs and friend to liberty (Hobhouse, Hazlitt, Lord Holland). He was lauded or despised as 'the child and champion of Jacobinism' (Pitt's famous phrase in a speech to the House of Commons, 11 Feb 1800) or celebrated or condemned as its destroyer. His fall from power was the end of his significance for most people; or, for Byron, a haunting confirmation of the emblematic character of his life. Napoleon's imprisonment on St Helena was, for some, an appropriate humiliation; for others it fell short of the execution he should have received; for yet others it was an ignoble act by the British government.

Napoleon's character and the success or failure of his exploits were the subject of constant comments in letters, diaries, newspapers and poems of the period. He was presented at some length in major poems by Southey, Scott and especially Byron. Both Hazlitt and Walter Scott wrote biographies of him.

Early reactions in Britain to Napoleon are bound up with existing attitudes to the French Revolution. Apart from Scott, the early Romantics (Blake, Wordsworth, Coleridge, Southey) sympathised with the Revolution and were appalled by the use of British armies against the revolutionary government. However, French success in these early wars, Napoleon's aggressive policy towards Switzerland and the German states, and the power vested in Napoleon by the Sieyès constitution (1798) made it difficult to maintain enthusiasm for Napoleon as the defender of liberty. Coleridge, for instance, reviewed the constitution in the *Morning Post* (26, 31 Dec 1798) and declared that liberty was at an end because all power now derived from Bonaparte. Southey, similarly, could admire the revolutionary general but not the dictatorial consul: 'Why', he wrote, 'had not the man perished before the walls of Acre in his greatness and his glory?' (undated letter, London, 1801). During the peace negotiations that led to the treaty of Amiens (1802) English public opinion was more sympathetic to Napoleon, interpreting his victories as the precursors of peace. Many took advantage of the treaty to flock to Paris and gaze at Napoleon in person. The resumption of war in 1803 was largely seen as caused by Napoleon's ambition; and, unlike the earlier wars, was interpreted as national rather than ideological conflict. In these circumstances, Wordsworth, Coleridge and Southey became bitterly opposed to Napoleon's France and, by sympathising with the British government's maintenance of hostilities, became, step by step, converted to Tory principles and policies. The few who stood out against this almost universal condemnation of Napoleon were now grouped together as 'the Napoleonists'. The principal figures here were Samuel Whitbread, Elizabeth Inchbald, Lord Holland, Charles Fox, Thomas Holcroft, William Godwin, William Hazlitt, Leigh Hunt, John Thelwall, Capel Lofft, John Cam Hobhouse and Lord Byron. These intellectual figures seemed to enjoy the position of superiority to ordinary public opinion that their undisguised admiration for Napoleon, in their view, demonstrated. However, even their admiration for Napoleon was strained by his coronation as Emperor (1804) and, particularly, by the undoubtedly popular nature of the Spanish uprising against Napoleon's seizure of Spain (1808). Mrs Inchbald defended Napoleon's intervention but, for the most part, the events united Tory, Whig and radical opinion, albeit for different reasons, against Napoleon's aggression. Both Byron and Wordsworth were opposed to the Convention of Cintra (1808). Byron, who owned a bust of Napoleon whilst he was at Harrow School, was himself in Spain during the Peninsular War and lamented the deaths of Spanish patriots:

And must they fall? the young, the proud, the brave,
To swell one bloated Chief's unwholesome reign?
(*Childe Harold*, I.liii)

The 'bloated Chief' was also, nevertheless, 'the Scourger of the world' who tumbled 'feebler despots from their sway' (I.lii). It was this capacity of Napoleon to frighten established governments and settled principles that ensured the loyalty of most of 'the Napoleonists', such as Hazlitt, despite occasional doubts. But there was also (though the Code Napoléon attracted little comment in Britain) some admiration for him as the embodiment of enlightenment rationality (Godwin). There was captivation also by his enigmatic personality, glamour, and the military genius shown especially at Austerlitz (1805), Jena (1806) and Friedland (1807). His apparent invincibility delighted his admirers but underlay the almost hysterical nature of British abuse, as in Coleridge's letters to *The Courier* in 1809, or in Southey's 'Ode Written during the Negotiations with Buonaparte, 1814', a lofty piece of political invective inspired by the Laureate's abhorrence of imperial France.

The 'Napoleonists' faced a different test after the failure of Napoleon's Russian campaign (1812). Leigh Hunt became physically ill when he heard of the abdication. Hazlitt, after Waterloo, unwashed and unshaved, took to alcohol. Byron was appalled by the events and by Napoleon's unheroic survival of defeat. His *Ode to Napoleon Bonaparte* reached its tenth edition in the year of publication (1814). It was written in apparent horror that Napoleon could fall at all, and fall in such a way. Napoleon was an 'evil spirit', an emblem of ambition's fall, a brooding prisoner trapped for ever in his own dreadful memories. Such a specifically Byronic analysis reveals Byron's continuing identification with Napoleon.

The similarity between the two figures is a commonplace: Byron's own 'exile' from Britain in 1816 increased the identification and he travelled in a coach copied from that of Napoleon. In the last two cantos of *Childe Harold* (III.xvii–xlv; IV.lxxxix–xcii) he gives two brilliant and fully considered portraits of Napoleon that show how deeply he had brooded on Napoleon's career and personality. He extends the critique of Bonaparte made in section xvii of Scott's *The Field of Waterloo*. He sees him as an over-reacher who is made and destroyed by his own tempestuous restlessness. Napoleon had 'a deaf heart which never seem'd to be / A listener to itself' (IV.xci). Byron was prepared to accept, too, the charge that Napoleon was a conqueror indifferent to the vast shedding of blood upon which his conquests depended. He kept close track of the reports of Napoleon's behaviour and circumstances on St Helena. After Napoleon's death, Byron produced a lengthy portrait of him in ll. 43–259 of *The Age of Bronze* (1822), which again presents his life as a moral lesson much as Charles XII is used by Johnson in his *The Vanity of Human Wishes*. Byron now calls him 'a noble creature', a 'Prometheus', ascribes Wellington's victory at Waterloo to 'blunder' and 'treachery', and finally says,

> A single step into the right had made
> This man the Washington of worlds betrayed:
> A single step into the wrong has given
> His name a doubt to all the winds of heaven . . . .
>
> (ll. 233–6)

Byron's is the most complex of Romantic attempts to understand and represent Napoleon's life. In general, his British admirers are forced into some degree of complexity by the inherent tensions between Napoleon's roles as monarch, revolutionary, liberator and conqueror. Those who consistently disliked him did so for simpler reasons. Shelley, friend of revolution, always sees Napoleon as its betrayer. Earth's judgement in 'Lines Written on Hearing the News of the Death of Napoleon' (1821) is also Shelley's opinion:

> Napoleon's fierce spirit rolled,
> In terror and blood and gold,
> A torrent of ruin to death from his birth.
>
> (ll. 33–6)

Walter Scott consistently hated Napoleon from the opposite view-point to Shelley's. He saw Napoleon as a self-made man who displaced evolved forms of government and practice by irrational conquest and over-rational methodisation. Scott approved, however, of the impetus given to the future unification of both Italy and Germany by Napoleon's actions. Scott always needed to set events back in time, as he does in his poems and novels, in order to understand their causes and consequences. His massive *The Life of Napoleon Bonaparte* (9 vols, 1827) is an impressive compilation of material and events but lacks this larger perspective. The phenomenal energy and tawdriness of Napoleon's personality and achievements are more vividly rendered in Byron's verse.

BERNARD BEATTY

Burke, Edmund (1729–97)

Edmund Burke is one of the most important and complex political figures of late-eighteenth-century Britain, and his influence on English Romanticism is still being discovered. Burke's seminal place in the history of political thought in Britain has never been in question, but, as C. B. Macpherson shows, the exact nature of his political position and influence is still a matter of dispute. For, if – as the most eloquent proponent of the Whig principles of 1688–9 and of a free-market economics which closely resembles Adam Smith's – Burke has been claimed as a founding father of British liberalism, his attack on the radical philosophy of the French Revolution and defence of the British constitution has also allowed him to be seen as a central figure in the development of British conservatism. As Macpherson puts it, 'the central Burke problem which is still of considerable interest in our own time is the question of the coherence of his two seemingly opposite positions: the defender of a hierarchical establishment, and the market liberal' (C. B. Macpherson, *Burke*, 1980). Macpherson tries to resolve this 'inconsistency' by arguing that the traditional order Burke defends had been a capitalist one since 1688 and that Burke

employs natural-order rhetoric in order to disguise capitalism's
true nature. To see him as a free-market capitalist making use of
traditional forms in this way is thus to differ significantly from the
commonly accepted view of Burke as the defender of an aristocratic past (see, for example, Raymond Williams, *Culture and Society 1780–1950*, 1961).

This dispute reflects what seemed to Burke's contemporaries a
central inconsistency in his life and work – epitomised by the fact
that he could defend the rights of the American colonists suffering
under the tyranny of the British government and then 'betray' his
reforming principles in *Reflections on the Revolution in France* (1790).
That he opposed the Revolution at all – and especially with such
violence – thus came as a surprise to Whigs and radicals alike.
Various theories have been offered as to why Burke should have
reacted in the way he did. Mary Wollstonecraft thought he was
really an envious revolutionary; some, including Marx, accused
him of accepting a government bribe; some said he had lost his
reason; Isaac Kramnick ascribes his reaction to the Revolution as a
sign of Oedipal anxieties (*The Rage of Edmund Burke*, 1977); Macpherson argues that Burke saw the Revolution not as promoting
capitalism but as destroying a doctrine of natural subordination
essential for its success. Conor Cruise O'Brien explains Burke's
dread of and fascination with the Revolution by pointing out that
as an Irishman Burke would at once have realised that revolution
was a very real threat, but also have seen a parallel between the
French treatment of the Catholic Church and British domination in
Ireland (see his introduction to *Reflections on the French Revolution*,
1968).

Born in Dublin to a Catholic mother and a father who seems to
have had himself confirmed in the Established Church for professional reasons, Burke necessarily had an ambiguous relation to
a British establishment in which he became prominent but never
pre-eminent. Although himself baptised and educated in the Anglican Church of Ireland, Burke fought institutionalised discrimination against Catholics throughout his life. To be a Catholic was
to be part of a disadvantaged minority in England, while to be Irish
was to witness Britain's imperialist and often brutal repression of
Catholicism in one's homeland.

Burke went to London in 1750 to read for the Bar, gave up the
law for literature and politics, and published his first works – a
satirical *Vindication of Natural Society* in 1756 and *A Philosophical*

Enquiry into the Origin of Our Ideas of the Sublime and Beautiful in 1757. The latter is now seen as one of the most significant and influential treatises on aesthetics in the eighteenth century, theorising what would become the pivotal aesthetic assumptions of the Gothic novel and influencing, among others, Johnson, Wordsworth, Kant, Turner, Ruskin and Thomas Hardy. In 1765 Burke became private secretary to the Marquess of Rockingham, who was then forming a Whig administration as First Lord of the Treasury. In 1766 he was 'elected' to Parliament (via a borough seat) and supported Rockingham's short-lived government. In 1770 he published *Thoughts on the Cause of the Present Discontents*, a pamphlet which attacked the excessive influence of the court in Parliament and in order to counter this developed a notion of political party which has been seen as foreshadowing the present-day party political system. Burke was properly elected as a Member of Parliament for Bristol in 1774 and there joined Fox's opposition to Lord North's administration – especially in his criticism of the latter's policy towards the American colonies. On losing his Bristol seat in 1780, Burke once again accepted a borough seat, this time in Rockingham's control, and held it until he retired from the House of Commons in 1794.

But Burke – a 'self-made' Irishman with neither wealth nor title – was never to achieve the political position his considerable talents deserved. In Rockingham's brief administration of 1782 Burke was given not a place in the Cabinet but the position of paymaster-general of the armed forces. Rockingham's death in 1782 eventually allowed Pitt to form a government which initiated almost fifty years of Tory rule and put Burke into opposition for the rest of his life. Burke's major preoccupation over the next years was his lengthy but eventually unsuccessful impeachment of Warren Hastings for the corrupt way in which the East India Company administered India. In the Regency Crisis of 1788, which revolved around the principles and party political interests at stake in the appointment of a regent for the temporarily insane George III, Burke once again found an occasion for great rhetorical performances in the House of Commons. But when Fox assembled a possible government, in case Pitt's should fall over the crisis, Burke was again passed over. As Conor Cruise O'Brien puts it, 'By the time of the French Revolution, then, Burke – whose party in any case remained in opposition – was a rather isolated figure, unpopular, frustrated, hard-pressed by exhausting labours, and to some extent already estranged from his old parliamentary friends and colleagues.'

Burke, then, was sixty years old when the French Revolution drove him to produce the book which brought him lasting fame, eclipsed all his previous writings, and earned him his reputation as a conservative defender of traditional order. The full title of Burke's reaction – it tells us that it also contains reflections *On the Proceedings of Certain Societies in London Relative to that Event* – indicates that Burke's 'rage' was not so much against the Revolution itself as against the suggestion that Britain should follow suit. For not only had Pitt and Fox spoken with admiration of the Revolution in Parliament, but a group calling itself the Revolution Society – founded in 1788 to commemorate the 'Glorious Revolution' of 1688 – had sent a letter of congratulation to the National Assembly and was suggesting that the settlement of 1688 gave the English people a right to 'cashier' their monarch much as the French had done theirs. Much of *Reflections* was taken up with a refutation of this claim through an exposition of the Whiggish principles of that settlement, while its principal target for abuse was Dr Richard Price – a dissenter and member of the Revolution Society whose *Discourse on the Love of our Country* (1789) greeted the Revolution as a 'blaze' which would spread to all the countries of Europe and liberate its people from monarchical despotism.

The 'memorable centrepiece' (J. T. Boulton, *The Language of Politics in the Age of Wilkes and Burke*, 1963) of *Reflections* is Burke's impressionistic description of the events of 6 October 1789 when the King and Queen of France were taken by a revolutionary crowd from Versailles to Paris. Burke presents the palace as 'swimming in blood' and has an 'almost naked' Marie-Antoinette narrowly escaping rape (1968 edn, p. 164). This scene is contrasted with Burke's famous eulogy of the French Queen in which she is remembered as a virtuous beauty surrounded by the flower of French chivalry (pp. 169–70). The difference in the way the Queen is treated is supposed to mark the difference between the *ancien régime* and revolutionary France – showing that 'the age of chivalry is gone. . . . That of sophisters, oeconomists, and calculators, has succeeded; and the glory of Europe is extinguished for ever' (p. 170). Such a contrast embodies Burke's political argument in a brilliant emotive prose whose techniques had been theorised in the *Philosophical Enquiry* over thirty years earlier. If Burke's argument is that society needs its illusions, its rites and traditions, in order to conceal and muffle the 'natural' destructive tendencies of human beings, these passages on the French Queen's fate seem to embody this position through their rhetorical power. Marie-Antoinette's treatment thus seems a concrete symbol for the fact that, in radical societies, 'All the decent drapery of life is to be rudely torn off. All the super-added ideas, furnished from the wardrobe of a moral imagination . . . necessary to cover the defects of our naked shivering nature . . . are to be exploded as a ridiculous, absurd, and antiquated fashion' (p. 171).

The *Reflections* was hugely popular amongst the classes it was aimed at – those in power who were flirting with revolutionary ideas – selling 30,000 copies and earning approval, ironically, from George III, the king whose overweening power Burke had always sought to limit. But for radicals such as Thomas Paine, and for old Whig colleagues and friends such as Charles James Fox, Burke had betrayed a lifetime's commitment to reform. For his part, Burke tried to demonstrate – in *An Appeal from the New to the Old Whigs* (1791) – that his position over the French Revolution remained as consistent with the principles of 1688 as the rest of his works were. Whether that is so or not remains a point of discussion, but what counts is that *Reflections* had an immediate and indelible impact on political and literary writing:

> Its phrasing passed immediately into English political discourse, with notions of chivalry and of rich, organic nationhood duly emerging in the parliamentary rhetoric of Pitt, Windham, and also Fox, . . . [its] arguments made their way into the configurations of European ideological warfare [and] are felt in imaginative literature too . . . [since] many of the penetrative features of the *Reflections* – its irrationalism, its hatred of system and argument, its reverence for the family, neighbourhood, and native country – recur in English and German literature around 1800, at the point when "Romanticism" first makes a distinct appearance.   (Marilyn Butler, *Burke, Paine, Godwin, and the Revolution Controversy*, 1984)

So powerful were Burke's formulations that they dictated the terms and the language of radical discourse in the flurry of writing that followed the publication of *Reflections*: Thomas Paine, Mary Wollstonecraft and William Godwin produced their finest writing in this 'Revolution controversy' (1790–3). As James K. Chandler argues, Burke paradoxically 'contributed much to the formation of an articulate radical ideology in the England of the 1790's'. For although Burke's 'intention was to nip the English Jacobin movement in the bud, his book became the occasion of its greatest flowering' (*Wordsworth's Second Nature*, 1984). The centrality of this 'pamphlet war' can be seen not only in that it helped articulate the polarisation of British society during the war with France (Burke successfully persuading Pitt's government of the need to fear British reformers as well as revolutionary France, Paine constructing a political language for the disenfranchised) but also in that it provides the terms in which the 'first generation' of Romantic poets expressed both their early radical allegiances and their subsequent betrayal. For Burke's *Reflections* – drawing as it does on the aesthetic theory he had developed in 1757 – politicised aesthetics (and language) as well as aestheticising politics. Burke's impact on Wordsworth, for example, is still being thought out. Chandler finds it central to an understanding of both Wordsworth's politics and his poetics; for, while Wordsworth's unpublished 'Letter to the Bishop of Llandaff' (1793) attacks Bishop Watson in the same terms as Paine's response to Burke, at some point between 1820 and 1828 Wordsworth could insert into *The Prelude* the passage calling on the 'Genius of Burke' to 'forgive the pen seduced' by radical philosophy into being slow to understand Burke's political wisdom (*The Prelude*, 1850 text, VII, 512–43). But, if Wordsworth's admirers see this as a late aberration whose conservatism is almost analogous to a failure of poetic power (see E. P. Thompson, 'Disenchantment or Default? A Lay Sermon,' in *Power and Consciousness*, ed. Conor Cruise O'Brien, 1969), Chandler tries to show that even the Preface to *Lyrical Ballads* (1800) – which is often seen as an extension of Paine's political thought into poetic theory – is grounded in assumptions which derive from Burke's *Reflections*.

TOM FURNISS

Burns, Robert (1759–96)

Robert Burns, the eldest son of a tenant farmer in Ayrshire, Scotland, grew up to a life of hard physical work, poverty, and acute awareness of social disadvantage. It was to find 'some kind of counterpoise' to this harsh set of circumstances, and to amuse himself by transcribing 'the various feelings, the loves, the griefs, the hopes, the fears, in his own breast', that he began to write poetry (Preface to *Poems, Chiefly in the Scottish Dialect*). By his mid-twenties he displayed exceptional mastery of both satire and lyric in Lowland Scots. In the summer of 1786, when he was on the point of abandoning farming in Scotland and emigrating to the West Indies, he published his first collection of poems, in an edition of 612 copies printed in the country town of Kilmarnock. *Poems, Chiefly in the Scottish Dialect* met with such success that he changed all his plans, and journeyed to Edinburgh, where he was enthusiastically welcomed by a number of leading literary figures, partly because the quality of his work appeared to confirm current primitivist theories of genius. Among them was Henry Mackenzie, whose sentimental novel *The Man of Feeling* Burns had long admired and claimed to prize 'next to the Bible'. In an influential essay in his periodical *The Lounger*, Mackenzie praised the 'power of genius' of 'this Heaven-taught ploughman', and he helped Burns arrange publication of an expanded edition of his *Poems* in the spring of 1788.

With the money he earned from publication, Burns toured the Scottish Borders and Highlands, and spent a second winter in Edinburgh. Eventually he returned, somewhat reluctantly, to tenant farming in south-west Scotland. He combined excise work with farming for a time, then became a full-time excise officer in Dumfries. His most famous poem, 'Tam o' Shanter', was written in 1790, but for the most part he devoted the leisure hours of his later years to the writing and collecting of Scottish songs, in which he was passionately interested. From 1788 until his death he was the principal contributor to and virtual editor of the greatest of all Scottish song collections, James Johnson's *Scots Musical Museum* (6 vols, 1787–1803). He also supplied the words of many songs for George Thomson's *Select Collection of Original Scottish Airs* (5 vols, 1793–1818), which boasted among its musical contributors Haydn and Beethoven. In all, he wrote some 200 songs.

Burns admired the poetry of Gray, Goldsmith and Shenstone, as well as the sentimental prose of his fellow countrymen Henry Mackenzie and James Macpherson. His reading in the literature of sentiment profoundly influenced him. Its effects can be seen in such poems as 'To a Mountain Daisy' and 'The Cotter's Saturday Night', both of which greatly appealed to his contemporaries; in many of his songs; and in his rather stilted letters. He was at his best, however, in satires such as 'The Holy Fair' and 'Address to the Deil', or when the expression of sentimental ideas was controlled and balanced by his fund of common sense and down-to-earth humour, as for instance in his Scots verse epistles and mock elegies. 'What an antithetical mind!' exclaimed Byron, 'tenderness, roughness – delicacy, coarseness – sentiment, sensuality – soaring and grovelling, dirt and deity – all mixed up in that one compound of inspired clay!' (Journal, 13 Dec 1813).

Burns was indebted both to 'polite' literature in English and to a vernacular oral tradition which is often homely and as often racy. What makes him exceptionally interesting is that, like all major writers, he makes up his own rules and defies facile categorising. Thus in one mood he harks back to the robust values and belief in clarity of Pope, his greatest eighteenth-century predecessor in poetic satire; in another he writes sentimentally, almost as if cast in the role of Harley, Mackenzie's Man of Feeling; and every so often – arguably while obeying his deepest artistic instinct – he anticipates the energy and cutting edge of full-blown Romantic protest. It is, above all, in song that his combination of simplicity, real personal feeling and memorable phrasing places him with the Romantics. 'A Man's a Man for A' That', for example, bears the stamp of his mind no less clearly than *The Marriage of Heaven and Hell* has that of Blake. Like nearly all his songs, it was given to the world casually and anonymously; but its powerfully direct assertion of Burns's vision of shared humanity and anticipated social and political change has carried it round the world, along with love songs such as 'Ae Fond Kiss' and the now traditional song of parting, 'Auld Lang Syne'.

Wordsworth more than once acknowledged a debt to Burns, writing in one poem of the particular sadness he felt when the Scottish poet died at the age of thirty-seven:

> I mourned with thousands, but as one
> More deeply grieved, for He was gone
> Whose light I hailed when first it shone,
>   And showed my youth
> How Verse may build a princely throne
>   On humble truth.

('At the Grave of Burns', ll. 31–6)

DONALD A. LOW

Byron, George Gordon, sixth Lord (1788–1824)

Byron was born in London in 1788, inherited his title at the age of ten, moved from Scotland to the Byrons' dilapidated estate near Nottingham (Newstead Abbey, which is the basis for 'Norman Abbey' in *Don Juan*, XIII–XVII), and was educated at Harrow and Cambridge. In 1809–11 he made a tour of Portugal, Spain, Greece and Turkey which he described in the first two cantos of *Childe Harold's Pilgrimage*. With the publication of these in 1812 he found instant and enduring fame. He wrote next a series of tales mainly set in the Near East which fascinated early readers by their exotic locations and the psychology of their villain heroes (all versions of the 'Byronic hero'). These are *The Giaour* (1813), *The Bride of Abydos* (1813), *The Corsair* (1814), *Lara* (1814), *The Siege of Corinth* (1816) and *Parisina* (1816). In 1815 he married Anne Isabella Milbanke, whom he described in a letter to Lady Melbourne (20 Apr 1814) as 'a very superior woman, a little encumbered with Virtue'. Byron later satirised her as Donna Inez in canto I of *Don Juan*. They separated after fifteen months amidst rumour, largely instigated by Byron's own compulsive indiscretion, of incest with his half-sister Augusta Leigh and homosexuality. Leaving England permanently in 1816, he travelled through Belgium (visiting the scene of Waterloo) and Germany to Switzerland, where he met Shelley for the first time and wrote *Manfred* (1817), and then settled in Venice, from whence he visited Rome (1817). These peregrinations were turned into cantos III and IV of *Childe Harold* (1816, 1818). In Venice, after some picturesque dissolution, he became the established lover or *cavalier servente* of the Countess Teresa Guiccioli. The role of *cavalier servente* is mocked in his Venetian comic poem *Beppo* (1818). A different Venice, cruel but magnificent, is examined in two political tragedies, *Marino Faliero* (1820) and *The Two Foscari* (1821). Byron's sojourn in Italy (Venice, Ravenna, Pisa, Genoa) from 1816 to 1823 produced his greatest work, and his growing acquaintance with Italian art and familiarity with the Italian language profoundly influenced both his style and his attitude to life. In particular, Italian culture helped him to sketch and sustain a comic view in *Don Juan* and much of his later poetry from *Mazeppa* (1819) to *The Vision of Judgment (1822)* and *The Island* (1823). This complements that insistence upon suffering which is so central to *Childe Harold*, the early tales, *The Prisoner of Chillon* (1816), *Manfred*, *Sardanapalus* (1821), the Venetian tragedies, *Cain* (1821) and *The Deformed Transformed* (1824). Byron's political understanding was complex but he remained always sympathetic to liberty though as distrustful of revolutionary and democratic 'cant' as of any other. In Italy he was involved with early movements for Italian independence, and in 1823 he went to Greece in order to assist and organise the movement for Greek independence from the Turks. He died there in 1824, as much, it would seem, of exhaustion as of a specific medical ailment. His death, in such circumstances, intensified the international fame which was already his. His body was returned to England, refused burial in Westminster Abbey, and finally laid to rest in his ancestral vaults near Newstead Abbey.

Byron's fame in the nineteenth century rested largely on his sympathy for liberty and national cultures, and the titanic pessimism of *Childe Harold*, the tales, *Manfred* and *Cain*. This image persists. In England, a reaction occurred soon after his death. Byron was considered to be less moral and wise than Wordsworth and less 'pure' a poet than Keats and Shelley. In the early twentieth century, only his 'anti-Romantic' poems (*Beppo*, *The Vision of Judgment*, and *Don Juan*) were praised, and that with qualification. More recently, there has been a substantial revaluation led by American critics and renewed interest in the whole range of Byron's work. In particular, his remarkable narrative skills, his position inside and ironically outside Romantic attitudes, his sympathetic grasp of familiar life, and his almost existentialist insights into life's 'nothingness' (seen as nihilistic or as profoundly religious) have been examined with ever-increasing respect.

Childe Harold's Pilgrimage

The poem appeared in three stages and draws directly on Byron's travels in Europe and the Near East. The first two cantos are based on his Grand Tour of Portugal, Spain, Malta, Greece and Turkey, begun in July 1809. Byron began writing these cantos in Janina (31 Oct 1809) and completed them in Smyrna (28 Mar 1810), though, as was his usual habit, he continued to interpolate stanzas after 'completion'. Canto III was begun in the English Channel (Apr 1816); the first draft was finished by 8 June 1816 but additions were made until 4–10 July 1816 at Geneva. The canto is based on

Byron's departure from England after leaving his wife and child (alluded to at the beginning and end of the canto), and his trip through Belgium (Waterloo) and Germany (along the Rhine) to Switzerland (the Alps). The first draft of canto IV was written in Venice (26 June – 19 July 1817) but Byron added at various times another sixty stanzas to the original 126 before its publication on 28 April 1818. The canto is based on Byron's first acquaintance with Italy, especially Venice and Rome. The whole poem in four cantos was published the following year. In a sense it remains unfinished because, *en route* between Italy and Greece in 1823, Byron thought of beginning a fifth canto.

Early reviewers praised the originality of *Childe Harold*. This originality consists in the successful joining together of different traditions. The poem is in part a travelogue, a series of topographical meditations, and it is the dramatisation through the figure of Harold of a central part of Byron's tormented personality. Greece, for instance, is presented simultaneously as a classical landscape where 'Apollo still thy long, long summer gilds' (II.lxxxvii) and as a wild romantic terrain inhabited by brigands who could appear in a Walter Scott novel. Essentially the poem uses landscape, art, architecture and history to symbolise the poet's suffering, but also explores and represents them as possible appeasements of that suffering. Canto III, set in Northern Europe and by the Swiss Alps, was influenced by Shelley, whom Byron had just met and by whom he was persuaded to reread Wordsworth's poems. In canto IV, however, set in Venice and Rome rather than romantic nature, Byron clearly parts company with Shelley and Wordsworth. It is in the art and architecture of Rome that Byron finds represented an endurance and serenity in suffering which, in some measure, provides a resting-point for Harold's pilgrimage.

Childe Harold was the most famous poem that Byron produced. It was immediately translated, imitated, and even (by Lamartine and others) extended into another canto or (by Berlioz) made the notional basis of a symphonic work. It is a poem about Europe, and Europe was delighted to recognise itself in this passionate, elegiac, conservative yet liberal and revolutionary masterpiece. The public, rhetorical manner, however, in which even the deepest self-exposure in the poem is conducted became increasingly alien to nineteenth-century (lyrical/sincere) and twentieth-century (imagistic/ironic) taste. *Don Juan* has displaced it as the long Byron poem that everyone reads. Recently, however, Anglo-American criticism has attended to the elusive structure and resilient idiom of *Childe Harold* with renewed interest and praise.

Don Juan

Canto I was begun in Venice (3 July 1818) but the poem was put aside on Teresa Guiccioli's insistence after Canto V (finished 10 Nov 1820 at Ravenna). The poem was resumed in Pisa (Apr 1822). The last cantos of the poem (X–XVI) were written rapidly in Genoa (Sep 1822–May 1823). Byron began canto XVII in May 1823 but he had only completed fourteen stanzas when he left for Greece in July. It seems clear that he could not have continued such a poem whilst in the role of hero and military commander. Byron's physical condition, too, is bizarrely relevant to his poem's composition. He checked his tendency to obesity by vigorous dieting. The first nine cantos were written by a man whom visitors found noticeably corpulent. The last cantos were written by a virtual ascetic. This division roughly corresponds to a perceptible change in the poem itself.

Cantos I–II were published in July 1819, III–V in August 1820, VI–VIII in July 1823, IX–XI in August 1823, XII–XIV in December 1823, XV–XVI in March 1824 and the fourteen stanzas of XVII in 1903. Though some stanzas were added or omitted and many emended in the course of composition or before publication, it seems as if the digressive character and intermittent publication of the poem allowed Byron to leave the poem alone and not interpolate considerable extra material as he did in *The Giaour* or *Childe Harold*. The first five cantos were published by Byron's habitual publisher and friend John Murray, but the remaining cantos, declined by Murray because of their scandalous nature, were published by John Hunt. The poem was widely read in cheaper pirated editions. It is this 'scandalous' nature of *Don Juan* which determined early reactions to the poem. Even Keats was appalled by the shipwreck scene in canto II. Readers disliked two things. First, the subject matter of the poem was presented in such a way as to disclose unmistakably the instinctual basis (survival, sexuality, power) of most individual and social behaviour. Secondly, readers disliked Byron's practice of apparently undermining the effect of elevated poetic passages by always juxtaposing them with low behaviour and idiom. This is the substance, for instance, of Jeffrey's charges against *Don Juan* in the *Edinburgh Review* (Feb 1822). Despite the praise lavished on it by Scott and Shelley ('Nothing has ever been written like it in English' – letter to Byron 21 Oct 1821), despite Goethe's admiration and Pushkin's imitation of its manner in *Eugene Onegin*, *Don Juan* took second place in nineteenth-century estimation to *Childe Harold*. Twentieth-century criticism has elevated the poem far beyond all Byron's other works, and its status, alongside Wordsworth's *The Prelude*, as one of the greatest works of English Romantic literature seems assured. Initially the revaluation of *Don Juan* was accomplished by those who saw it as an anti-Romantic riposte to nineteenth-century poetry in general. It authorised therefore the characteristic respect from the 1920s to the 1950s for satire and for dryness in poetry. More recently, it has been read as a quasi-existentialist masterpiece in which the boundaries between fiction and reality, nihilism and comic trust, are expertly delineated and criss-crossed. Its apparently careless but exactly calculated juxtapositions of idioms and events, its unremitting energy and final integrity are more widely acknowledged and admired than ever.

The poem is based on the traditional Don Juan story but inverts that fable in two essentials. Byron's Juan is seduced rather than seducing and he is still alive rather than damned at the end of an inherently unfinishable narrative. Juan's love adventures with some eight assorted heroines (especially with Haidée in cantos II–IV), his shipwreck (II), and his experience of war in the siege of Ismail (VII–VIII) form one half of our interest. The other is the narrator of the poem, who digresses constantly and normally presents a viewpoint antithetical to that of the narrative. Beyond these options of trusting to life as Juan does and reflecting upon it in the sceptical manner of the narrator, the poem itself seems committed, increasingly, to some third containing perspective. This seems especially true of the last cantos of the poem, set in Norman Abbey (XIII–XVII). These juxtapose the satirically observed Lady Adeline Amundeville, the religious positive of the 'Radiant and grave' Aurora Raby, and the anarchic sexual energies of the Duchess of Fitz-Fulke. These three figures are made even more complex by their interraction with the two appearances of the ghost of the Black Friar (canto XVI). The unmasking of the ghost leaves the poem finally suspended between religious myth and farce despite the satirical emphasis of the last cantos as a whole.

The dramas

Byron wrote eight plays of different kinds, none of which, with the possible exception of *Werner*, was meant to be performed.

The first two acts of *Manfred* were begun in Geneva in August 1816 and completed by October. Act III was written in January–February 1817 but Byron, taking the advice of William Gifford, decided that the act was too weak to publish. He rewrote it substantially whilst in Rome and the play was published 16 June 1817. Byron frequently added lines to his poems but did not substantially revise them. The rewriting of Act III of *Manfred* is the most considerable revision that he undertook. The new act is a vast improvement on its predecessor. The composition of the play was directly influenced by the Alpine scenery surrounding Byron in autumn 1816, by Byron's conversations with Shelley about poetry, nature and Wordsworth, and by hearing 'Monk' Lewis reading his own translation of scenes from Goethe's *Faust*. It doubtless carries references to Byron's probable incest with his half-sister. Byron's play, like Goethe's, presents its hero's pursuit of knowledge but is more concerned with the intense vacancy of Manfred's suffering and guilt and his quest for mental, natural and supernatural agencies which will bring him release from this strange pain. The revised last act grants him release from, yet still vindicates, his agonised self-assertion. The play is one of the most powerful and influential, albeit perhaps the least lucid, of Byron's major poems. It is the basis for Schumann's *Manfred* overture and incidental music and for Tchaikovsky's *Manfred* Symphony.

Byron wrote three classical tragedies on historical themes. Two of these, *Marino Faliero* (written Apr–July 1820, published 21 Apr 1821) and *The Two Foscari* (written June–July 1821, published 19 Dec 1821) are set in and largely concerned with Venice. *Sardanapalus* (written 13 Jan – 27 May 1821, published 19 December 1821) is set in ancient Nineveh. All three are political plays. They dramatise the ways in which royal or noble figures attempt to initiate or sustain some revolutionary actions that will displace the dark cycle of cruelty which makes up Venetian, Assyrian or any human history. They do so with mixed motives and, always, without final success.

Werner is a Gothic intrigue based on repeated patterns of murder, deception and guilt. It is the least interesting of his dramas but was quite often performed in the nineteenth century. Byron derived the story from 'Kruitzner, or the German's Tale' in Harriet Lee's *The Canterbury Tales* (1797–1805). He attempted a narrative version of this when he was thirteen years old. In 1815, when he was a member of the sub-committee of management of Drury Lane Theatre, he wrote the first act of a play based on the same story, with numerous stage directions. He rewrote it *ab initio* in December 1821 and January 1822. It was published on 22 November 1822.

Cain (written July–Sep 1821, published 19 Dec 1821) is a remarkable version of the biblical story. It dramatises the connection and the gap between the will's obscure involvement with murderous intent and reason's horror both at death and at any moral justification for it as punishment. These paradoxes, which haunted Byron, were not grasped by contemporary readers. The publication of *Cain* created a sensation and the play was almost invariably seen as endorsing Lucifer's attack on orthodox belief.

Heaven and Earth (written Oct 1821, published 1 Jan 1823 in *The Liberal*, no. 2) is something of a sequel to *Cain*. It dramatises the intermarriage of fallen angels and human beings just before the Flood. Like *Cain*, it juxtaposes the inner psychology and external fact of destruction.

The Deformed Transformed (probably written early in 1822, published 20 Feb 1824) is unfinished. It is based on Goethe's *Faust* and a little-known play by Joshua Pickersgill called *The Three Brothers* (1803). It interweaves the methods and concerns of *Manfred*, *Cain* and (via the dramatised version of the sack of Rome in 1527) Byron's historical tragedies. It is both lyrical and almost cinematic in character. It was reviewed badly by Byron's contemporaries but Goethe praised it as 'new and original, close, genuine, and spirited' (*Conversations*, 1874). Arnold, the play's hunchback hero, is enabled by a compact with Lucifer ('the Stranger') to gain the shape of Achilles, whilst the Stranger assumes that of the hunchback and the name of 'Caesar'. In this guise, Arnold rescues the beautiful Olimpia during the sack of Rome. A memorandum by Byron on the manuscript suggests that Arnold will become jealous of Olimpia's attraction to the Stranger's personality in Arnold's original shape. The play dramatises the destructive mismatch of desire, intellectual understanding and inherent human limitation in a way which both mocks and is respectful to orthodox belief.

Byron's *Cain* and *Manfred* achieved fame, even notoriety, in Byron's lifetime, but his other plays were little regarded. Readers had become so accustomed to detecting projections of Byron himself in his poems that they failed to see Byron's capacity to dramatise situations in deliberately different ways, underestimated his concern with history, and did not recognise the almost-allegorical presentation of his ideas. The plays remain little read, but all of them have been discussed with considerable respect by recent criticism. G. Wilson Knight largely inaugurated this concern and wanted them performed in a specially designed theatre like that at Bayreuth. Nearly all modern critics concur in admiring the originality and evident seriousness of Byron's dramatic concern with politics, religion and gender.

The oriental tales

Byron's oriental tales were composed in England between his return from the Orient and publication of the first two cantos of *Childe Harold* (1812) and his return to the Continent in 1816 after his separation from his wife. They coincide with and are partly responsible for Byron's astonishing notoriety in upper-class English society at that time. They retail to that society an exotic other world, Muslim or Catholic, consumed by dangerous passions and engaged in subversive activities.

The Giaour (pronounced 'jowr') was written probably in late 1812 and published on 5 June 1813. The first edition had 684 lines, but Byron kept adding sections up to the seventh edition (Dec 1813). The final version has 1334 lines. This suggests incoherence, but the narrative is presented as a sequence of fragments from different points of view. *The Bride of Abydos* was written between 1 and 8 November 1813 and published, with some additions, on 29 November. *The Corsair* was begun on 18 December 1813 and Byron finished correcting proofs (a process that usually involved augmentation) by 16 or 17 January 1814. The poem was published on 1 February 1814, together with the sixth edition of *The Bride of Abydos*, the seventh edition of *Childe Harold*, I and II, and the ninth edition of *The Giaour*. It surpassed even these in instant popularity: 25,000 copies were sold by March 1814. *Lara* was written between 15 May and 12 June 1814 and published with Byron's customary additions shortly after 5 August. The advertisement to the poem declares that 'the reader . . . may probably regard it as a sequel to the *Corsair*' but not all readers have found this possible. *The Siege of Corinth* may have been started in autumn 1812; it was certainly restarted in October 1813; and it was eventually put together between January and November 1815. The same material, originally intended for one tale, seems to have been separated into *The Siege of Corinth* and *Parisina*. Both poems were published on 6 February 1816.

The oriental tales differ in circumstance but utilise the same basic ingredients. They are all set in the Near East except for *Lara*, which is set nominally in Spain but in fact in a notional Gothic setting. They use these exotic but incidentally detailed locations to make their vivid narratives and extreme psychological portraits credible. Heroes are dark, guilty, aloof, but good leaders of men and tender to women. They are, with Harold and Manfred, versions of the original 'Byronic hero'. Settings are fixed and characters trapped within themselves but the verse itself moves with careless rapidity. These features have been interpreted in diverse ways. For some, and especially for English readers from the mid nineteenth century onwards, the tales are too stereotyped and slapdash to be taken seriously. For others, the simultaneous intensity and off-handedness of their narrative form and the stark antitheses of their structures excite admiration. Byron's insight into the psychology of guilt, his insistence on punishment, and his sensitivity to the ways in which gender differences promote value differences are extensively commented upon. So too is the seriousness with which Byron explores religious and political questions in these tales directly and by implication. Byron's interest in different modes of verse narrative, originating in the oriental tales, persisted in *Parisina* (1816), *The Prisoner of Chillon* (1816), the brilliant comic tale *Beppo* (1818), *Mazeppa* (1819) and, most substantial of all, *The Island* (1823). The concern and something of the method of the oriental tales re-emerge in *Don Juan* (especially cantos II–IV).

The Vision of Judgment

This poem was begun in May 1821, left, then taken up and completed in September at Ravenna. It was a reply to and a satire on Robert Southey's tribute to the lately dead George III (*A Vision of Judgement*) published in April 1821. In the preface to this poem, Southey had attacked 'men of diseased hearts and depraved imaginations' who had set up 'what may properly be called the Satanic School'. Byron, quite rightly, saw this as an attack on himself and associated it with an early rumoured slur by Southey (though he denied it) on Byron's and Shelley's sexual morals. Byron had other grounds for his riposte. Southey's earlier political opinions had been radical. He had written a play, *Wat Tyler*, in 1794 (reprinted maliciously in a pirated edition 1817) in which he praised revolutionary attack on monarchic government. Southey subsequently became a convinced Tory and, from 1813, Poet Laureate. Byron thus attacked him as a political renegade and, of course, could not accept Southey's eulogy of George III's life and the triumphs of illiberal policies at home and abroad. In particular, Byron was outraged by Southey's confidence in placing George III and Louis XVI in heaven and their opponents in hell. Byron found Southey's poem 'ludicrous and blasphemous' (appendix to *The Two Foscari*). Finally, Southey's poem was written in a version of classical hexameters quite alien to English metrics. It was thus a reactionary poem written in revolutionary idiom, whereas Byron's beliefs and poetic practice were the opposite of this.

Byron sent the manuscript to his usual publishers, John Murray, in October 1822, but Murray was reluctant to publish it because of its political character. Byron accordingly asked Murray to send it to John Hunt, who published a new journal, *The Liberal*, edited by his brother Leigh Hunt. Murray sent on the manuscript, not the corrected proofs nor the preface, and it was published in this form in October 1822 in the first edition of *The Liberal*. It was the occasion for a private prosecution of the publishers which led eventually to the firing of Hunt. A corrected version of the poem appeared in the second edition of *The Liberal* (1 Jan 1823) with an errata slip correcting some misprints and substituting two mitigating versions of contemptuous phrases about George III and his queen. Only very recent editions have restored these.

The poem describes the arrival of George III at the gates of heaven, the dispute between Michael and Lucifer over whether George belongs to heaven or hell, and it ends farcically with the King slipping into heaven under cover of a heavenly riot caused by the reading of a very boring poem. This poem is Southey's original *A Vision of Judgement* and Southey is more the object of Byron's satire than the King himself. The most impressive section of Byron's *Vision* is the first meeting of Satan and Michael. Byron here presents them simultaneously as convincing antithetical moral archetypes and as aristocratic friends who regret that they now belong to opposite factions.

The Vision of Judgment outraged most of Byron's English contemporaries because of its subject matter. Charles Lamb, however, noted that 'Lord Byron's satire, is one of the most good-natured description – no malevolence' (recorded in Henry Crabb Robinson's Diary 8 Jan 1823). Goethe admired it immensely, but Victorian taste generally (Swinburne is the most notable exception) disliked the poem because of its apparent debunking of the visionary imagination. Twentieth-century criticism has often admired it for precisely this reason. Both reactions simplify the poem, for it adroitly combines reverence and irreverence, gentle whimsy and firm satiric purpose. Swinburne's advice in the Preface to his selection of Byron's poems (published 1866) remains sound: 'Those who read it with due delight, not too gravely and not too lightly, will understand more than can be set down; those who read it otherwise will not understand anything.'

BERNARD BEATTY

Campbell, Thomas (1777–1844)

Thomas Campbell was born in Glasgow, and educated at the university there. After a series of posts as a tutor to families in the Scottish Highlands, he attracted attention with his poem *The Pleasures of Hope* (1799). He subsequently settled in London, and became a prominent member of the London literary scene. His *Gertrude of Wyoming* appeared in 1809, and he edited a seven-volume anthology, *Specimens of the British Poets*, published in 1819. In 1820 he became editor of the *New Monthly Magazine*, and was active in the movement to found a university in London. He was regarded as an eminent figure in his own day, and was buried in Westminster Abbey.

The Pleasures of Hope is characteristic of its type of 'Pleasures' poem (see under ROGERS), written in heroic couplets and reflective in a straightforward and undemanding way. Campbell has the ability to write pithy lines ("Tis distance lends enchantment to the view') and occasionally skilful passages, such as the description of Hope and Imagination in part II. *Theodric* is partly set in the Swiss Alps, and concerns the love of Theodric for Julia, the Swiss girl who dies for love of him, and Constance, the English lady whom he also loves. *Gertrude of Wyoming*, written in Spenserian stanzas, is also about love and death, set against a background of conflict between American Indians and white settlers (Wyoming is a village in Pennsylvania, not the state): the killing of Gertrude and her father during their retreat to the fort is characteristic of Campbell's strong subject matter. This father–daughter relationship is a recurring theme in Campbell's work: it is found again in 'Lord Ullin's Daughter', 'The Maid of Neidpath' and 'The Ritter Bann'. Love and battle are found in 'The Wounded Hussar' and the song 'When Napoleon Was Flying'. Campbell is perhaps best remembered for his energetic and patriotic poems (although their sentiments are not fashionable in the late twentieth century) such as 'Battle of the Baltic' and 'Ye Mariners of England', subtitled 'A Naval Ode'; more appealing to modern taste is the sombre 'Hohenlinden'.

J. R. WATSON

Cartoons

The political cartoon had been developed in England in the seventeenth century, for instance in the portrayal of Charles I as a martyr in *Eikon Basilike*; in the eighteenth century Hogarth carried the art further, satirising modern manners, fanatical Methodism, political jobbery and city wickedness. He was succeeded in the later eighteenth century by two great artists in political and social caricature.

Thomas Rowlandson (1756–1827) began as a conventional artist, but during the 1780s he developed as a caricaturist and as an acute observer of people and manners. He was influenced by his wide travels on the Continent, and by the Dutch School, and began to produce modern comic subjects. He was particularly adept at crowded scenes with people tumbling over one another, or grotesque figures contrasted with beautiful ones. His men often have long noses, pock-marked faces, and pot bellies: the contrast between human beings and nature, or old and young, is made sharper when the old men appear as voyeurs, as in *The Exhibition Stare-case*, in which the men are taking the opportunity to ogle the ladies who have fallen down with their legs in the air. Rowlandson, like Hogarth, is sharp on drunkenness, lack of self-control, and human pride: his pictures are a telling indictment of late-eighteenth-century society. His *Tour of Dr Syntax in Search of the Picturesque* (1812) is a humorous satire on a fashion for picturesque travel; *The English Dance of Death* (1815–16) is a brilliant series of engravings which emphasise human folly, pride and mortality. His contemporary and friend James Gillray (1756–1815) was more explicitly political, and may be called, as the twentieth-century cartoonist David Low called him, 'the father of the political caricature'.

Gillray began as an illustrator, although by 1782 he was producing engravings satirising the Rockingham ministry, with cheeky titles such as *The Kettle Hooting the Porridge Pot* (on the resignation of Fox from the foreign-secretaryship). He continued to produce satires on political subjects and on the clergy and their tithes (*Balaam*, 1783), but it was the French Revolution which really brought out his full power. His reactionary attitude may be seen in his cartoon on Paine's *Rights of Man*, which was entitled *'The Rights of Man' – or – Tommy Paine, the Little American Taylor, Taking the Measure of the Crown, for a New Pair of Revolution Breeches* (1791).

Subsequently he portrayed the French as absurd fanatics and savage cut-throats, in a whole series of cartoons from the September massacres (1792) onwards. *The Zenith of French Glory* (1793) is an appalling depiction of the Terror, showing a man being guillotined before a large crowd, the bodies of two priests and a judge hanging, and a savage-looking *sans-culotte* sitting on a street lamp and playing a violin. At the foot of the cartoon is written 'Religion, Justice, Loyalty, & all the Bugbears of Unenlighten'd Minds, Farewell!' Gillray's support of Pitt's government, possibly for financial reasons, is shown in *The Giant-Factotum Amusing Himself* (1797), in which Pitt bestrides the House of Commons like a colossus, crushing the opposition with his foot.

Gillray was a master of the telling stroke or emblem, which he used in the anti-revolutionary cause of the government. In *The Apotheosis of Hoche* (1798), the revolutionary general Hoche is seen ascending out of his boots towards a revolutionary heaven peopled with Jacobins, playing a guillotine instead of a harp. Later Gillray directed his satire against Napoleon, although Napoleon was said to have been amused by Gillray's depiction of the Peace of Amiens, *The First Kiss This Ten Years! – or – the Meeting of Britannia & Citizen François* (1803), in which a thin French officer just manages to kiss the cheek of an extremely buxom Britannia.

Gillray's drawing has a tremendous energy and zest, but he has sometimes been seen as reactionary, unprincipled and lacking in charity. Osbert Lancaster said that 'one thing alone Gillray lacked – compassion; had he possessed it, he would be, what he so nearly is, the equal of Daumier'. Nevertheless, Gillray's caricatures provide a sharp and ironic commentary on the events and people of the time, and are useful evidence of a British reaction against the Revolution in France.

J. R. WATSON

Chatterton, Thomas (1752–70)

Thomas Chatterton was born in the parish of St Mary Redcliffe, Bristol, and died in London. As an orphaned boy, he attended

Colston's charity school, which he left after learning the 'three Rs', to become an indented apprentice to an attorney (to 'be educated a scrivener'). There, and in other circumstances, he had leisure to examine old legal documents and parchment manuscripts and to become acquainted with both legal jargon and older forms of the English language. He also discovered the names of some officials of Bristol in the fifteenth century. All this set a vivid imagination to work.

Chatterton's work may be divided into two categories: what he wrote in eighteenth-century English, and what he composed when conjuring up the vision he had of Bristol in the fifteenth century or earlier, written in a language he had borrowed or imagined. The latter sort was first brought to the notice of his fellow-citizens when he gave a narrative of the opening of the bridge by the mayor in 1247, on the occasion of a new bridge being opened in 1768. This was the first of a series of prose texts, and of drawings: coats of arms, monuments or parts of them (see *The Complete Works of Thomas Chatterton*, ed. D. S. Taylor and B. Hoover, 1970). He composed prose pieces relating aspects of Bristol life as he partly imagined it, and partly reconstructed it, round the central character (and genuine figure) of William Canynge, five times mayor, and Thomas Rowley, whom he made a priest. Rowley is most often to be seen as Chatterton himself telling stories. Much more important is his poetry in pseudo-Middle English, not so much for its story-telling as for the poetic talent that it reveals – its facility of rhyme and rhythm and its happy handling of different metres. This verse and the related prose pieces together constitute the 'Rowley cycle'. Among the best poems are 'Bristowe Tragedye', 'Ynn auncient Days', 'The Tournament', 'The Battle of Hastynges', I and II, 'Song toe Ælla', 'Parlyamente of Sprytes', *Ælla* and three eclogues. *Ælla* contains the poetical creed of Chatterton, that of the poet's right to unbounded imagination.

Obviously various influences can be felt in Chatterton's poetry, from Chaucer to William Collins, and James Macpherson in some prose pieces. But there are reasons for appreciating his poems as individual and original: the wealth and precision of imagery in 'A Bristowe Tragedye', the melody of the 'Mynstrelles Songe' in *Ælla*; the clear-sightedness of the evolution of poetry outlined in 'Letter to the Dygne Mastre Canynge'; the attempt to re-create the mediaeval atmosphere of the city, with its public feasts and processions of all the religious orders and its faith. By May 1769 the Rowley cycle was nearly over, and Chatterton embarked on different poetry and prose: he yearned after the literary fame which his talents could have brought him, but which famous men such as Horace Walpole and Thomas Gray denied him, when they saw through his forgery 'The Ryse of Peyncteynge, yn Englande'.

This fame he could only find in London, or with the London reading public. He therefore shifted to more conventional poetry: love verse written on behalf of a school friend, descriptive lines ('To Mr Holland', 'Clifton', mixing past and present), a 'burletta', an elegy. But the main characteristic of his inspiration remained the same: he wrote of Bristol and Bristolians, whether praising or satirising them. Satire was the consequence when he did not receive the expected reward of his genius. It began with local subjects: 'The Constabiliad', 'Hervenis'. In 1770 he mixed poetical and prose satire on larger subjects. He found himself in the aftermath of radical agitation: the cry was 'Wilkes and Liberty'; the *Junius Letters* had appeared. What he had started with Bristol subjects, he thought he would now practise on a larger scale, with wider social and political themes. His prose includes letters to London newspapers, in the style of Junius, inveighing against the Duke of Grafton, or George III's mother; 'The Whore of Babylon' is a 575-line poem that violently attacks politicians, churchmen devoted to the Crown, and miserly Bristol aldermen. The last months of Chatterton's life (Apr–July 1770) were spent in London, where he wrote further lengthy satirical poems, which sometimes repeated whole passages already published. 'Resignation', 'Kew Gardens' and 'The Exhibition' lash out at all the strata of the ruling classes. 'A Letter to the Lord Mayor' brought him into contact with the addressee, William Beckford, but the young poet's death intervened to end all hopes of patronage. His last prose fragments criticise the manners of the time, and a few lines sing of Africa, or go back for the last time to the world of Rowley ('An Excelente Balade of Charitie'). Several letters to his sister and friends give hints as to his occupations, but the reason for his death is not known.

Chatterton's work remains highly original, although he borrowed from his predecessors and followed some models. His reconstruction of the past is rich and colourful, and displays feelings of Christian charity. His attitude to the Church is very mixed: for the Church of his own time he had nothing but contempt, but he glorified the mediaeval faith as the leading force that united society. His view of contemporary society is often bitter, because he felt rejected and could not rise above his condition, despite his deep poetic aspirations. That is why he became a symbol of neglected genius, and the subject of de Vigny's drama in 1835. The tribute paid to him by later English poets, such as Wordsworth, Keats and D. G. Rossetti, shows that he ranks among the most interesting pre-Romantic poets, even if he is still but little read. His Rowley poetry was first edited by Thomas Tyrwhitt in 1777, and the first important edition was published by Robert Southey and Joseph Cottle, his fellow citizens of Bristol, in 1803.

GEORGES LAMOINE

Clare, John (1793–1864)

John Clare was born on 13 July 1793 in the village of Helpstone, Northamptonshire; his father was a thresher, and after brief schooling Clare found himself working in the fields as thresher, limeburner and gardener. He started to write for his own amusement, and by 1818 had enough material to contemplate publication; he met a local bookseller, Edward Drury, who encouraged him. Drury and his cousin, the London publisher John Taylor, published *Poems Descriptive of Rural Life and Scenery* in January 1820. This was a success, and Clare was soon mixing with all the *London Magazine* circle as the famous 'Northamptonshire Peasant Poet'. *The Village Minstrel* followed in 1821 (by which time the first volume had gone into four editions); a long period of ill health, and wrangling between poet and publisher, preceded *The Shepherd's Calendar* (1827), and more ill health lay behind the production of *The Rural Muse* (1835). In 1837 Clare was removed to Dr Matthew Allen's asylum in Epping Forest, from which he escaped in July 1841, hoping to return to his childhood love, Mary Joyce (she had died three years earlier); after six months at home in Northborough, Clare was taken to the General Lunatic Asylum at Northampton, where he stayed until his death on 20 May 1864. When his body was taken back to his native village, there was nobody to receive it.

It has become impossible to read Clare's work without these grim biographical facts in our minds: so much of his work is an attempt to come to terms with his life, with the 'contrarietys' and the 'chain of contradictions' that dog him. It is no accident that one of his surviving prose pieces is to do with 'Self Identity'. Many people were only too ready to dub him the 'peasant poet', thereby allotting him a safe niche, where the flames of his ambition could not be fanned. But Clare knew that he did not fit so cosily into such literary and social conveniences, that he was not just another Stephen Duck or Robert Bloomfield. His whole writing life (and it was a long and abundant one – a complete works would easily run to ten volumes) was devoted to establishing a distinctive voice that had its roots both in the long literary tradition he absorbed (his impressive library survives in Northampton Public Library as a testament to his wide reading), but also in the oral tradition of ballads (which he collected avidly, along with the tunes that he could play on his beloved violin). *The Shepherd's Calendar* represents one aspect of his poetic maturity, in its unblinkered celebration of a way of life that no poet before him had understood so well or so closely; the poems contained in his last volume, *The Rural Muse*, show his ability to connect such a vision with his elegiac, lyric sense of pain and loss, operating at both a personal and a communal level. In finding his own voice, Clare finds the voice of his community (hence the anger that spills out in *The Parish*, dating from 1823, a satire never published in his lifetime). In the asylum, apart from some extraordinary 'Byronic' poems – 'Don Juan' and 'Child Harold' – Clare tends to revert to the songs and ballads he had known when young; some of his most remarkable poems represent the distillation of the lyric impulse which runs through his life. There is obviously a danger of romanticising the visionary poems of the asylum, to make of Clare a second Blake; but it is clear that he himself saw his work in the asylum as the painful summation of a life devoted to the Muse in all her forms.

Early poems

Clare's second volume of poems, *The Village Minstrel* (1821), was an important attempt to build on the achievement of his first book, *Poems Descriptive of Rural Life and Scenery* (1820), which went into its fourth edition at the beginning of 1821. In his introduction to the first volume, John Taylor (Clare's publisher) had emphasised the circumstances under which the poems were written, and much of the contemporary response had concentrated on this phenomenon of the 'Northamptonshire Peasant Poet', as much as on the intrinsic merit of the verse. Although this attention ensured the volume's success, it was not a fair reflection of what Clare was attempting; in spite of his self-consciousness in the face of so many pressures (from patrons, advisers, editors, friends) – reflected in poems such as 'Helpstone' and 'Dawnings of Genius' – he was trying to find a poetic voice true to his own vision of the country-side. Initially, his immersion in Thomson, Beattie, Goldsmith and Bloomfield had at least provided him with some kind of model; but there was always a danger of his verse seeming merely pastiche. Even in his first volume, however, this ability to write with clarity and precision shines through; if a common criticism was that he did not know how to stop, his reply came in his cultivation of the sonnet, which provided him with the very structure he needed, at this stage, to contain the drama of the world as he saw it. Nor was he prepared to limit himself to the conventional utterances about the joys of the countryside: several poems are about the hardships of poverty, the corresponding damage done by the wealthy, or about the unpastoralised people he knew from his own experi-ence. Significantly, some of these poems were forced out of later editions. But, when it came to the second volume, Clare was determined to push forward this particular line of anti-pastoral. The title poem – originally written at the end of 1819 as 'The Peasant Boy' – was Clare's audacious attempt to contain his own responses to the natural world in a wider context that embraced the whole village community. It was an answer to those who had asked for a long poem, and a rebuke to those who wanted reassur-ance that he would not indulge in the 'radical lines' deplored by Lord Radstock. The poem allows Clare to focus on the hero Lubin – clearly an image of himself, and yet sufficiently distanced for him to be objective – with humour and compassion. Clare gets away from the awkwardness of much of his earlier verse, and, appar-ently rejoicing in the freedom of the stanzaic form (comparable both to Beattie's *Minstrel* and Reynolds' *Romance of Youth*), com-bines infectious joy at the pleasures of the country with rage at their loss as a result of enclosure. It is quite clear that he is preparing the ground for the much more powerful dramatisation of social consciousness in *The Shepherd's Calendar* (1827).

While Clare was working on *The Village Minstrel*, John Taylor suggested a rather more planned poem, which might be called either 'Ways of a Village' or 'Week in a Village'. As soon as *The Village Minstrel* was out of the way, Clare settled down to something even more ambitious: he was particularly anxious to combine his talents for so-called descriptive writing with his interest in 'village tales' (partly encouraged by his reading of Crabbe). The genesis of *The Shepherd's Calendar* was a protracted and tortured affair, partly because of Clare's own uncertainties, partly because of his increasing ill health, and partly because he was writing several different things at once, whilst at the same time receiving conflicting advice from his publishers, Taylor and Hessey. By January 1823 he had finished a long satirical poem called *The Parish*, which even his admirer Mrs Emmerson (not fond of satire) had to admit was 'powerfully written'; Clare thought it 'the best thing that I have ever written'. But it was never published in his lifetime. It has enormous historical importance, as an attack on the corruption of the day; it is also an essential indication of the cast of Clare's mind, which was always ready to turn to the harshly satirical. In this respect *The Parish* is the foil to the more lyrical, more celebratory, but none the less open-eyed *Shepherd's Calendar*. By the end of 1823 it was agreed that Clare would write a month-by-month account of the year, with a narrative poem for each section. But his publishers were far from satisfied with the quality of Clare's manuscripts, let alone their legibility. When the book finally appeared in 1827, Clare had virtually lost interest, and Taylor acknowledged that 'the time has passed away in which Poetry will answer'. The book, one of Clare's finest achievements, caused hardly a ripple in literary waters. The original plan had not been followed, in that only a handful of narrative poems, along with some others such as 'Superstition's Dream', were included at the end, after the monthly poems. But Clare demonstrates conclusively here his ability to write a long poem, personal in tone and yet full of acknowledgements to the literary tradition; each section has its own distinctive shape and voice, as Clare explores the various possibilities of different metrical structures, while managing to give the poem a sense of architectural wholeness.

Clare's last published volume was to be *The Rural Muse* (1835) – likewise the result of a long process of wrangling and disagreement. He had initially hoped to publish in 1832 *The Midsummer*

Cushion, a fat manuscript volume which survives as one of his most beautiful fair copies; but he had to settle for a selection from this cornucopia, edited by Mrs Emmerson. By this stage in his life, in spite of ill health, Clare had become remarkably self-assured, as his experiments with the sonnet form, with songs and ballads and with a series of poems on birds and birds' nests testify. He was able to rise poetically to one of the worst crises of his life up to this point, when he moved from Helpstone to Northborough, three miles away. 'Remembrances', 'Decay' and 'The Flitting' are powerful expressions of loss and isolation, showing the fragility of the very act of imaginative re-creation. But, by the time the volume was published, Clare was too ill to feel much sense of achievement. Within two years he was to enter an asylum in Epping Forest.

Later poems

In Dr Matthew Allen's asylum at High Beech, Epping Forest, Clare wrote two long poems, 'Don Juan' and 'Child Harold'; he referred to them in letters of May 1841, and when he escaped in July he continued to work at them, in the small, cramped notebook he took with him (Northampton MS 8). At home in Northborough, until the end of December 1841, he made what amounted to fair copies of both poems (Northampton MS 6), but it is clear from what he said later in the 1840s that he came to regard anything he wrote as a continuation of 'Child Harold', or as he called it (after James Montgomery's poem of the same name) 'Prison Amusements'. Since neither poem was published in his lifetime, we have to view them ('Child Harold' especially) as poems 'in process' as much as finished works of art. Clare's Byronic obsession began in the 1820s and lasted all his life. His attempt at 'Don Juan' is coarse, though amusing, continuing the vein of satire of *The Parish* (1823). Clare's desire to assert his poetic identity ('Though laurel wreaths my brows did ne'er environ / I think myself as great a bard as Byron') comes out forcefully in 'Child Harold', which is a lament for the loss of his childhood love, Mary Joyce; the poem consists of nine-line Spenserian stanzas, interspersed with songs and ballads (but also with a variety of disparate material, including biblical paraphrases). Just as different kinds of material jostle for our attention on the page (even in the 'fair copy'), so different poetic voices clamour in the text, as affirmation and denial vie with each other.

In December 1841 Clare was taken to the Northampton General Lunatic Asylum, where he remained until his death in 1864. He continued to write incessantly; apart from Northampton MS 19, most of the poems survive only in the transcripts made by the steward, William Knight. Much of the verse during these years is repetitive, love songs addressed to a range of different women, remembered from his past or encountered on his trips into Northampton (before these were curtailed in the mid-1840s). But Clare also rises to some of his greatest heights, in poems such as 'I am', 'It is the Evening Hour', 'The Sleep of Spring', 'An Invite to Eternity', 'Secret Love' and 'A Vision'. The last can almost be taken as a culmination of all his poetry, with its defiant emphasis on the freedom of the spirit that grows with the genius of poetry, even when imprisoned in an asylum. But Clare was too honest to remain content with that proud affirmation; and he confronted despair in verse of dreadful poignancy. Yet he continued writing: his last known poem, 'Birds Nests' (which survives in his own hand), appears to have been written in the winter of 1863.

MARK STOREY

Cobbett, William (1763–1835)

William Cobbett was born at Farnham, Surrey, then a picturesque part of rural England. He was educated by his father, a small farmer, and to the end of his life retained a scepticism about formal education, especially if it was divorced from experience. At the age of fourteen he read Swift's *Tale of a Tub*, which produced, he said, 'a sort of birth of the intellect'. After a spell in London as a lawyers' clerk, he joined the army in 1784, where he rapidly achieved promotion and became a sergeant-major, continuing his habit of self-improvement through study. After serving in Canada he left the army in 1791, having unsuccessfully attempted to expose a scandal in the quartermaster's stores. He lived for a short time in northern France, but was alarmed by the course of the Revolution, and after the deposition of the King in August 1792 he left for the United States. He lived in Pennsylvania for some years, supporting himself at first by teaching English to French settlers; he then began to publish a series of anti-radical pamphlets under the pseudonym of 'Peter Porcupine', followed by a daily news-sheet, entitled *Porcupine's Gazette*, attacking what he saw as corruption and folly. His headstrong articles and forthright opinions brought libel actions upon his head, and after unsuccessfully contesting one he left for England in 1799.

He was welcomed by the government, which needed good political writers; and after an unsuccessful venture called *The Porcupine* (1800–1) he launched the weekly *Political Register*, which continued to appear until his death in 1835. He attacked the Peace of Amiens, defended the slave trade against the reforming Wilberforce (whom he detested), and defended bear-baiting and other field sports; his robust economics (against paper money, Jews and the City of London) and his growing disappointment with the policies of Pitt and Addington turned him towards a position of extreme radicalism which made him impatient with both Whigs and Tories. He argued for a complete reform of the abuses of society, which he saw as benefiting the financiers and people in power at the expense of the ordinary people. He argued for the abolition of sinecures, the reform of Parliament, and a dismantling of the whole system of financial and political privilege. He came to detest London, which he called 'the Wen', seeing it as opposed to all the qualities of English life that he admired; he also abhorred religious movements such as evangelicalism and Methodism, which he saw as trying 'to teach the poor to starve without making a noise'. His combative energy and his passionate advocacy of causes he embraced led him to publish an article in July 1809 on the brutal flogging of some soldiers. He was tried for sedition, and sentenced to two years' imprisonment in Newgate, from which he continued to edit the *Political Register*.

The end of the Napoleonic wars was followed by a period of poverty and depression, especially among the rural poor. Cobbett's repeated attacks on the war and its economic consequences ('before you have paid for the play, you will find that there is no money left for the supper') again made him unpopular in official circles. It was during this period that he became convinced of the need for parliamentary reform. He spent a further period in America (from May 1817 to November 1819) and continued his campaign from there, resuming it with more effectiveness on his return. His 'Rural Rides' began in October 1821, and were collected and published in 1830 (extended and revised edn, 1853). He stood for Parliament unsuccessfully in 1820 and 1826, but was elected as independent member for Oldham in 1832, in the first parliament to sit after the Reform Bill.

Carlyle saw Cobbett as healthy and honest: 'strong as a rhinoceros, and with singular humanities and genialities shining through his thick skin' ('Essay on Scott', 1838). Cobbett's impulsive and volatile nature made him appear inconsistent, and he certainly held different views at different times of his life. Hazlitt saw him as 'made up of mere antipathies', and described him as a wrestler or boxer whose blows lacked direction and purpose:

> If his blows were straightforward and steadily directed to the same object, no unpopular minister could live before him; instead of which he lays about right and left impartially and remorselessly, makes a clear stage, has all the ring to himself, and then runs out of it, just when he should stand his ground. . . .   (*The Spirit of the Age*, 1825 edn)

But Hazlitt, exasperated as he was by Cobbett's lack of direction, recognised his power as a political writer: 'he is a kind of *fourth estate* in the politics of the country'. Cobbett's strength, and his weakness, was that he wrote from experience: it gave his writing great force, but it also allowed him to air his prejudices. At its best, in the descriptions of England in *Rural Rides*, it is full of acute observation by one who knows what he is talking about. He notices things that other writers would miss: he sees, for example, that the people of the Wiltshire Downs are unfortunate because they have no wood for fires, as the Sussex labourers have (Oct 1825); he mocks the 'sickly, childish taste' of Gothic follies in a gentleman's park in Hampshire (Oct 1821); he remarks with pleasure on the absence of fir-trees in Hertfordshire (June 1822) and on the neatness of the houses in Suffolk (Mar 1830); he laments the absence of singing birds in Lincolnshire (Apr 1830). His sharp condemnation of abuses or folly is made more powerful by his readiness to appreciate things that he likes: good farming, clean towns, healthy people, decent work. His descriptions of neat and attractive English towns (Guildford, Bristol, Ipswich, Hull, Nottingham) contrast nicely with his view of the colleges of Oxford: 'I could not help reflecting on the drones they contain and the wasps they send forth!' His attacks on abuses in the army were those of an old soldier; his defence of the rural poor was that of one who had been brought up in the countryside. His knowledge of the working poor, both in cities and in the country, led him to attack those whom he saw as responsible; and his vigorous, manly style makes him an enjoyable as well as an effective writer. The emblem on his commemorative medal shows a plough and a pen, and the two things, separately and in combination, are useful indicators of Cobbett's individual excellence as a political writer and polemicist.

J. R. WATSON

Coleridge, Samuel Taylor (1772–1834)

Samuel Taylor Coleridge was born at Ottery St Mary, Devon, the youngest child of the Revd John and Ann Coleridge. In 1782 he entered Christ's Hospital School in London, where at the age of fifteen he was, it was said, 'converted' to free thought by reading, among other things, Voltaire's *Dictionnaire philosophique*. In his essay 'Christ's Hospital, Five and Thirty Years Ago' (*Essays of Elia*, 1823) his fellow pupil Charles Lamb recalls Coleridge's reading of Homer, Pindar, Iamblichus and Plotinus, 'for even in those years thou waxedst not pale at such philosophic draughts'.

In 1791, Coleridge entered Jesus College, Cambridge. With men such as William Frend, a fellow of the College, he was drawn to political radicalism, an admiration for the principles of the French Revolution, and Unitarianism. He left Cambridge without a degree, and on a visit to Oxford in June 1794 met Robert Southey, with whom he planned the ill-fated scheme of Pantisocracy – the establishment of a republican paradise on the banks of the Susquehanna river in the United States. In pursuit of this scheme he married, ill-advisedly, Southey's sister-in-law, Sara Fricker. During 1794–5 Coleridge published his first poetry in the *Morning Chronicle*, his 'Sonnets on Eminent Characters' (including Burke, Priestley, Godwin and La Fayette), and delivered his lectures on politics and religion in Bristol, to finance Pantisocracy. They include the Six Lectures on Revealed Religion (May–June 1795), an anti-establishment view of Christianity with Christ as patriot and revolutionary.

Cooling of relations with Southey and the abandonment of Pantisocracy led to a period in 1795–6 when Coleridge considered entering the Unitarian ministry and briefly edited the journal *The Watchman* (Mar–May 1796), a radical publication which proclaimed in its prospectus that 'in the present perilous state of our Constitution the Friends of Freedom, of Reason, and of Human Nature, must feel it their duty by every mean in their power to supply or circulate political information'. In 1796 he also published *Poems on Various Subjects*, which included 'The Eolian Harp'. In 1797 he met William Wordsworth at Racedown, Dorset, and from their discussions 'The Rime of the Ancient Mariner' was begun and *Lyrical Ballads* (1798) planned, 'written chiefly with a view to ascertain how far the language of conversation in the middle and lower classes of society is adapted to the purposes of poetic pleasure' (Advertisement). Initially, the work was received unenthusiastically.

During 1797–8 Coleridge remained at Nether Stowey, Somerset, in close harmony with Wordsworth, and moving away from the revolutionary politics of his younger days, though he continued to preach Unitarian sermons (at Shrewsbury in January 1798). 1798 was his poetic *annus mirabilis*, when he completed 'The Ancient Mariner', and wrote 'Frost at Midnight', 'Fears in Solitude' and 'Kubla Khan'. In September 1798 he travelled to Germany with William and Dorothy Wordsworth, met Klopstock in Hamburg, and studied German at Göttingen. He read Lessing (particularly important for Coleridge's work on biblical interpretation, especially in *Confessions of an Inquiring Spirit*, published in 1840), Leibniz, and above all Kant and Schelling. Coleridge became one of the most important channels for the reception of Kant and German idealism in England.

In October and November 1799 he first visited the Lake District with Wordsworth, and he moved to Greta Hall, Keswick, in July 1800. The early months of 1801 were a period of intense intellectual activity when Kant took hold of him 'as with a giant's hands'. He moved essentially from the eighteenth-century tradition of Locke, Hartley and necessitarianism to the Kantian notion of the mind active in perception – 'we receive but what we give, / And in our life alone does Nature live' ('Dejection: An Ode'). It was also a period of domestic strife, increasing addiction to opium, and his hopeless love for Sara Hutchinson, from which emerged the 'Dejection' ode (1802). In an attempt to restore his health and equanimity, Coleridge went to Malta in April 1804, becoming secretary to the Governor. He did not return to England until August 1806, after a period of intense religious reassessment recorded in his notebooks for these years. 'Historical Christianity' continued to be a stumbling-block, and his abandonment of Unitarianism was reached through a metaphysical rather than a historical approach (see Kathleen Coburn, *The Self Conscious Imagination*, 1974). On returning to England he became finally estranged from his wife and lived with the Wordsworths at Coleorton, Leicestershire (Dec 1806–Feb 1807), where Wordsworth read to him his 'Poem to Coleridge', later to become *The Prelude*. Between June 1809 and March 1810 he edited the journal *The Friend*, republished in 1818 in a three-volume *rifacimento*. He described *The Friend* as 'a series of essays . . . to aid in the formation of fixed principles in politics, morals, and religion, with literary amusements interspersed'. It includes the definitive version of the 'Essays on Method'. Between 1811 and 1814 came several series of lectures on Shakespeare in London and Bristol, and in the same period a longstanding quarrel with Wordsworth (dating from October 1810) was patched up.

In 1813–14 Coleridge was openly professing a return to Christian beliefs, and submitting to medical care for his opium addiction. He was also working on his *Biographia Literaria*, which was finally published in 1817. Originally envisaged as a prelude to his great, never-completed work, the 'Opus Maximum', and a preface to his poems, it became a longer discourse on poetry and a critique of Wordsworth. Often criticised for large-scale plagiarisms of Schelling, the *Biographia* reveals Coleridge's close and critical readings of German thinkers, particularly Kant and Fichte, and chapter 13 contains his crucial distinction between fancy on the one hand and the primary and the secondary imagination on the other. It pursues the unification of philosophy and aesthetic feeling, since only from such a synthesis can true knowledge come.

By 1817 Coleridge was lodging with Dr Gillman in Highgate, where he remained for the rest of his life. In the early 1820s he was engaged on the 'Logic' manuscript (vol. XIII in the *Collected Works*), wishing to provide a system of logic 'applied to the purposes of real life'. Much of his later work, including *Aids to Reflection* (1826) – his most sustained writing on Christianity – and *Confessions of an*

Inquiring Spirit, is intended to encourage readers to think for themselves in a manner both consistent and self-aware. Also important are the two *Lay Sermons* (1816, 1817), dealing with questions of reform and moral responsibility, and *On the Constitution of the Church and State* (1830), dealing with national culture and the role of the 'clerisy' ('the learned of all denominations, the professors of all those arts and sciences, the possession and application of which constitute the civilization of a country'). This latter work was an influence on Matthew Arnold and J. H. Newman.

The 'Sage of Highgate' (Carlyle's term) was an important figure for many of the younger intellectuals, including Carlyle, Sterling, J. S. Mill and F. D. Maurice. His extensive and important letters have been published under the editorship of E. L. Griggs, and his notebooks continue to appear, edited by Kathleen Coburn. The later notebooks are particularly important as evidence of his final philosophic and theological reflections.

Biographia Literaria or Biographical Sketches of my Literary Life and Opinions

The first edition of 1817 was the only English edition in Coleridge's lifetime. No manuscript has ever been discovered. In April 1815, Wordsworth published his *Poems* with a new preface, prompting Coleridge to collect his poems with a 'preface'. The writing began in May, and the 'preface', having become a book in its own right, was dispatched to the printer on 19 September. Difficulties of printing and completion delayed publication until July 1817. Coleridge regarded *Biographia* as 'an important Pioneer to the great Work on the *Logos*, Divine and Human [the never-completed "Opus Maximum"], on which I have set my Heart and hope to ground my ultimate reputation' (letter to Gutch, 17 Sep 1815).

Part I is broadly autobiographical, describing his education and youthful literary experience, and his friendships with Southey and Wordsworth. The narrative is gradually overwhelmed by a series of 'philosophical chapters' (I, chs 5–13), formidable metaphysical discussions which are heavily dependent on his reading of Kant, Fichte and Schelling (from whom it is often argued that he heavily plagiarised). Chapter 13 contains the crucial discussion of fancy, and primary and secondary imagination. Part II is an extended discussion of poetics, an attack on Wordsworth's Preface to *Lyrical Ballads* and a vindication of the poems themselves.

Kathleen Wheeler has shown how metaphor and irony constitute both the structure and point of engagement with the reader (*Biographia Literaria, Sources, Processes and Methods*, 1980), as a work seeking to unify philosophy and aesthetic feeling. William Hazlitt, in an extravagant notice in the *Edinburgh Review* (Aug 1817), describing Kant as a 'most wilful and monstrous absurdity', attacked Coleridge for 'indulging his maudlin egotism and his mawkish spleen in fulsome eulogies of his own virtues, and nauseous abuse of his contemporaries'. The *Monthly Review* (Sep 1817) complained that 'he now imposes on us the pain of seeing him exhibit the decrepitude of genius.'

'Christabel'

According to Coleridge's own preface to the poem, the first part was written in 1797, at Stowey, Somerset; the second part in 1800, at Keswick. It was never completed, although a third part was planned, and the year before his death he claimed to have the plan of 'Christabel' 'entire from beginning to end in my mind', but the idea was 'an extremely difficult and subtle one' (*Table Talk*, 6 July 1833). He was unusually reticent about the origins and purposes of his poem. It is not mentioned in his letters before October 1799 (to Southey), and he is clearly uncomfortable with its progress. 'Every line has been produced by me with labor-pangs', he wrote to James Tobin (Sep 1800). He admitted to Humphry Davy that he disliked the poem (*Collected Letters*, ed. E. L. Griggs, 1956–71, I, 631–2), while to the publisher of *Lyrical Ballads* Wordsworth wrote that it should be excluded, for 'the style of this Poem was so discordant from my own' (letter, 18 Dec 1800).

The precise nature of the action and relationship between Christabel, the beautiful and apparently malignant Geraldine, and Sir Leoline is obscure in the two cantos which were eventually published in 1816. The imagery is strongly sexual. In James Gillman's *The Life of Samuel Taylor Coleridge* (1838) it is suggested that 'the pious and good Christabel suffers and prays for "the weal of her lover" . . . and she thus defeats the power of evil represented in the person of Geraldine'. Derwent Coleridge disputes this account of the evil Geraldine, who is 'no witch or goblin, or malignant being of any kind' (*Poems of Samuel Taylor Coleridge*, 1870). Byron admired 'Christabel' greatly (letter to Coleridge, 18 Oct 1815), but reviews were generally unenthusiastic. William Hazlitt in *The Examiner* (June 1816) wrote that 'there is something disgusting at the bottom of his subject'; Josiah Conder in the *Eclectic Review* (June 1816) said, 'We cannot conceal that the effect of the present publication upon readers in general, will be that of disappointment'.

Coleridge's sources are explored in A. Nethercot, *The Road to Tryermaine* (1939).

The conversation poems

It was only 'The Nightingale', published in *Lyrical Ballads* (1798), which Coleridge prefaced with the epithet 'a Conversational Poem'. But, following the suggestion of G. M. Harper in his essay 'Coleridge's Conversation Poems' (*Quarterly Review*, CCXLIV, 1925), six poems of Coleridge's Somerset years have become known as the 'conversation poems'. These, composed between 1795 and 1798, are 'The Eolian Harp', 'Reflections on Having Left a Place of Retirement', 'This Lime-Tree Bower my Prison', 'Frost at Midnight', 'Fears in Solitude' and 'The Nightingale'.

Influences were both personal and literary. Charles Lamb urged Coleridge to abandon his early elaborate Miltonic style for a simplicity which 'springs spontaneous from the heart', and described 'Reflections' as a 'very beautiful poem' (*Letters of Charles and Mary Anne Lamb*, ed. E. W. Marrs, Jr, 1975, I, 65). The eighteenth-century tradition of loco-descriptive poetry was important: described by Dr Johnson as '*local poetry*, of which the fundamental subject is some particular landscape' (*Life of Denham*), the genre originated in Sir John Denham's *Cooper's Hill* (1642). It was lyricised in the sonnets of William Lisle Bowles (1789), mentioned in *Biographia Literaria* (I, ch. 1) as allowing the expansion of 'my fancy, and the love of nature, and the sense of beauty in forms and sounds'. Coleridge also refers to William Crowe's *Lewesdon Hill* (1788), a probable influence on 'Reflections'.

The conversation poems are all in blank-verse measure, possibly owing something to William Cowper's *The Task* (1784). They have a tripartite structure, beginning with the establishment of a physical setting, continuing with a reverie inspired by the place, and concluding with a return to the original setting. All are dominated by what Humphry House called 'the seeing, remembering, projecting mind' of Coleridge, who in a dramatic monologue registers an encounter with other minds – either Lamb, the Wordsworths, Hartley or Sara Coleridge.

Early reviewers were quietly enthusiastic. C. L. Moody in the *Monthly Review* (May 1799) noted that Coleridge 'seems solicitous to consecrate his lyre to truth, virtue, and humanity'. The most comprehensive modern study is Kelvin Everest, *Coleridge's Secret Ministry: The Context of the Conversation Poems 1795–8* (1979).

'Dejection: An Ode'

The earliest draft of the poem is a letter addressed to Sara Hutchinson (Asra), Wordsworth's sister-in-law, whom Coleridge loved hopelessly. It was dated 4 April 1802, and first published in the *Morning Post* on 4 October 1802, Coleridge's seventh wedding-anniversary and Wordsworth's wedding-day. It was later published in *Sibylline Leaves* (1817). There is another version in a letter to William Sotheby, 19 July 1802. The published version has a greater unity, and omits the more personal and self-pitying passages of the original letter.

'Dejection' was written at a time of great intimacy with Wordsworth in the Lake District, when the two men were discussing, in March 1802, Wordsworth's partly composed 'Ode: Intimations of Immortality', and the poet's recognition that 'there hath passed away a glory from the earth'. In his ode, Coleridge describes the loss of his poetical powers, his broken marriage and love for Sara Hutchinson, the paralysing effects of metaphysical speculation (or opium?), and his dulled response to nature. In the opening lines from the 'Ballad of Sir Patrick Spence' the Bard is 'weather-wise' because he is, unlike Coleridge, in touch with the primitive powers of earth and air.

'Dejection' draws heavily upon biblical imagery, notably Acts 17:24–8 (in ll. 47–8) and Revelation 21:1 (l. 69), and upon theological associations of the word 'joy' as a quality in divine creativity, now lost to the poet. A detailed discussion of the poem's background can be found in George Dekker, *Coleridge and the Literature of Sensibility* (1978).

Beginning in the context of a dialogue with Wordsworth,

'Dejection' may be regarded as Coleridge's last great poem, but remained almost unmentioned in the nineteenth century. In the *Quarterly Review* for August 1834, H. N. Coleridge described it as 'one of the most characteristic and beautiful of [Coleridge's] lyric poems'. In our own time, Stephen Prickett has called it 'one of his greatest insights into the workings of his own creativity' (*Coleridge and Wordsworth: The Poetry of Growth*, 1970).

'Kubla Khan, or A Vision in a Dream. A Fragment'

Although not published until 1816, when Coleridge wrote the prose preface, 'Kubla Khan' was probably written in summer 1797 (according to the preface) or autumn 1797 (according to the Crewe manuscript, and a letter to John Thelwall, 14 Oct 1797). Other dates between 1798 and 1800 are possible. It was written in a farmhouse between Porlock and Lynton in Exmoor, supposedly under the influence of opium, and broken in upon by the intrusion of 'a person on business from Porlock'. Coleridge records that it was published at the request of Byron.

The preface should be read as critically as the poem, not least because its version of the poem's origin differs from that in the Crewe manuscript, which was only discovered in 1934. The sources of 'Kubla Khan' are explored in J. L. Lowes's *The Road to Xanadu* (1927), but in his preface Coleridge alludes to one major influence, the *Pilgrimage* of Samuel Purchas (1617), which is the origin of Kubla's palace. Another is probably Milton's *Paradise Lost*, which gives the clue to Mount Abora, untraced by Lowes, but in the Crewe manuscript written as 'Mount Amora' – a Freudian slip, or a reference to Mount Amara (*Paradise Lost*, IV. 280–2)?

Norman Fruman in *Coleridge, the Damaged Archangel* (1972) reads the poem in terms of powerful erotic imagery. The opium dream may not be unrelated to Piranesi's *Carceri d'invenzione* (1745, 1761). Probably, however, there is deliberate irony in the preface, the person from Porlock enacting a typically Romantic fragmentation, the broken form of the conscious intellect. Coleridge claims to offer the poem 'rather as a psychological curiosity, than on the grounds of any supposed *poetic* merits'. It is not known why Coleridge delayed so long before publishing (there are two references to 'Kublai Khan' in the notebooks of 1802 and 1804). Eventual reviews were uniformly dismissive, despite Byron's admiration. Thomas

Moore wrote in the *Edinburgh Review* (Sep 1816), 'there is literally not one couplet in the publication before us which would be reckoned poetry, or even sense'. 'The poem itself is below criticism', reported the *Monthly Review* (Jan 1817).

'The Rime of the Ancient Mariner'

It was first mentioned in a letter by Dorothy Wordsworth of 20 November 1797 that Wordsworth and Coleridge were 'employing themselves in laying the plan of a ballad'. In notes dictated by Wordsworth to Isabella Fenwick in 1843, he suggests that the poem was planned in the spring of 1798 on a walk from Alfoxden, Somerset, to Lynton and the Valley of Rocks, but 13 November 1797 is a more probable date. On 18 February 1798 Coleridge wrote to Joseph Cottle, 'I have finished my ballad – it is 340 lines' (*Collected Letters* I, 387). In the first edition of *Lyrical Ballads* (1798) it was 658 lines. For the second edition of 1800, Wordsworth moved the poem from its initial position in the collection, claiming that it 'has upon the whole been an injury to the volume' (letter to Cottle, 24 June 1799). The subtitle 'A Poet's Reverie' was added in 1800. The final, less archaic form with its prose gloss was published in *Sibylline Leaves* (1817).

Wordsworth records that he had been reading George Shelvocke's *A Voyage round the World* (1726), which describes the albatrosses at Cape Horn. Coleridge was widely read in sea-voyage stories, and of particular importance are the *Pilgrimage* of Samuel Purchas (1617) and Richard Hakluyt's *Principal Navigations, Voyages, and Discoveries of the English Nation* (1589) (see Lowes, *The Road to Xanadu*). The poem almost certainly owes something to the contemporary incident of the mutiny on HMS *Bounty* (see *Collected Notebooks*, I, 174), and to a dream of Coleridge's friend John Cruikshank about 'a skeleton ship with figures on it'.

Early criticism was almost uniformly hostile. A notice (attributed to Southey) in the *Critical Review* for October 1798 complained that 'Many of the stanzas are laboriously beautiful; but in connection they are absurd or unintelligible.' The *Monthly Review* for June 1799 stated, 'it is the strangest story of a cock and bull that we ever saw on paper', but 'there are in it poetical touches of an exquisite kind'.

DAVID JASPER

Collins, William (1721–59)

William Collins is considered one of the finest lyric poets of the mid-eighteenth century. Sensitive, passionate, gifted with a vivid imagination, he aspired to join the great tradition of English poetry represented by Spenser, Shakespeare and Milton.

The son of a prosperous Chichester hatter, Collins was educated at Winchester and Oxford, where he struck up a lasting friendship with Joseph Warton. In January 1742, while an undergraduate, he published his rather tame *Persian Eclogues* (reprinted as *Oriental Eclogues* in 1757). The couplets often echo Pope's *Pastorals*, but some emotional lines indulging in 'silent Horrors', 'wild'ring Fear and desp'rate Sorrow', foreshadow Collins' later poetry. In December 1743 he addressed *Verses* to Sir Thomas Hanmer (*An Epistle* in the second, revised edition of May 1744), where he traces the progress of drama from ancient Greece to Shakespeare's England and deals with the connection between poetry and painting. The 'Drafts and Fragments' of poems probably written in 1744–5 (and edited in 1956) evince the same interest in the 'Sister Arts'.

In 1744 Collins came to London to try his fortune and met Dr Armstrong, Dr Johnson, the Wartons and James Thomson. He was 'a literary adventurer, with many projects in his head' (Johnson, *Lives of the Poets*: 'Collins') including several tragedies, a *History of the Revival of Learning* and a translation of Aristotle's *Poetics*. None of his schemes came to fruition, except a thin volume of twelve original *Odes on Several Descriptive and Allegoric Subjects* published in December 1746 (dated 1747). They failed to attract attention at the time, in spite of their strange beauty. Sadly disappointed, the poet himself bought and destroyed the unsold copies. The volume contains the famous 'Ode to Evening' and 'How sleep the Brave' as well as odes to Pity, Fear, Simplicity, Liberty, and so on. Following the example of the Wartons and Akenside, Collins chose to write principally in the lyric form. In 1749, he published a touching ode on the death of James Thomson, 'In yonder Grave a Druid lies', and in 1750 he presented to the Scottish dramatist John Home a draft of his 'Ode to a Friend', published in 1788 as 'An Ode on the Popular Superstitions of the Highlands of Scotland'.

Collins left London in the early 1750s, was seriously ill, travelled to France to regain health, returned to England and gradually fell into a state of acute melancholia. He died in Chichester, tended by his sister, in utter obscurity.

If Collins' contemporaries ascribed his misfortunes to indolence, 'irresolution' and 'long habits of dissipation' (Johnson), modern criticism has drawn the reader's attention to the socio-cultural context that led so many eighteenth-century poets to depression, insanity and suicide. Poverty and frustrated ambition increased Collins' physical and mental distress. But he also suffered from feelings of inadequacy resulting from the pessimistic view that true poetic power is a 'God-like Gift' granted to few.

Collins projected his anxiety into his work, which is haunted by a death wish signalled by the recurrence of 'rest' and 'tomb'. Peaceful images of the 'sylvan Grave' or 'Earthly Bed' alternate with terrifying figures, 'hideous' forms or 'gliding Ghosts' that suggest the poet's intense awareness of the demonic forces at work in the universe. Collins combines the natural and the supernatural, like Coleridge.

A close study of the 'Ode to Fear' and the 'Ode on the Poetical Character' reveals Collins' interest in the nature of poetry. It is inspired and prophetic, founded on the power of vision and connected with the divine. The poet is a seer who transcends reality to achieve a deep insight into the heart of things. Johnson himself noted that Collins was 'delighted with those flights of imagination which pass the bounds of nature'; and Coleridge admired his visionary boldness as well as his gift to feel emotion and to arouse it in others. Collins agreed with Joseph Warton's view that it was necessary 'to bring back poetry into its right channel' (Advertisement to the Odes on Several . . . Subjects), meaning invention and imagination. He appealed to terror, culti-vated the sublime and finally turned to Scottish superstitions, beliefs and legends to revive poetic inspiration. The evocation of the water-kelpie drowning the 'luckless Swain' has the force of dream imagery.

The poet's vision is 'self-justifying' and images often are 'the source of illumination' (P. M. Spacks, The Poetry of Vision, 1967). Collins' imagery is never merely decorative, for it provides both the substance and the structure of the poem, as in the 'Ode to Fear'. Collins prefers significant or suggestive details to visual accuracy and physical descriptions. Far from being too abstract or generalised, his personifications acquire a symbolic value as in the evanescent 'Ode to Evening', where the invoked figure merges into twilight and silence. His metrical variety offers a new departure from the couplet. He ranges from simple stanzaic forms to intricate and irregular Pindaric odes which convey stronger emotions.

Collins' odes were to have considerable influence in the 1760s, when Richard Hurd's *Dissertation on the Idea of Universal Poetry* (1766) developed a new approach to poetry. The Romantics paid full tribute to Collins' originality.

MICHÈLE PLAISANT

Constable, John (1776–1837)

John Constable was born at East Bergholt, Suffolk, the son of a miller. He showed early promise as a painter of local landscape, and was encouraged as a young man by Sir George Beaumont, the wealthy patron and connoisseur. He was enrolled as a student in the Royal Academy in February 1799, and first exhibited in 1802; it was in this year that he told his friend John Dunthorne, 'there is room enough for a natural painter' (C. R. Leslie, *Memoirs of the Life of John Constable*, 1845, ed. J. Mayne, 1951, p. 15). It was part of every painter's education, in the days of the picturesque, to experience a mountainous countryside: Constable's uncle, David Pike Watts, therefore arranged for him to visit the Lake District in 1806. Although he did some painting there, he did not feel at home in the landscape: he told his biographer, C. R. Leslie, that 'the solitude of mountains oppressed his spirits' (*Memoirs*, p. 18). In succeeding years Constable continued to persevere in his own particular branch of landscape art, finding encouragement from friends and from the occasional patron such as John Fisher, later Archdeacon of Salisbury. Constable survived during those years by painting portraits and other occasional commissions, but his landscape work did not receive much general recognition. During the summer he would revisit Suffolk, recovering his health and spirits, and painting in the open air: one of the first major pictures resulting from this was *Boat-building*, painted during the summer of 1814, and exhibited at the Royal Academy in 1815.

After a long courtship, and many objections from her relatives, Constable married Maria Bicknell in 1816. He was elected an Associate of the Royal Academy in 1819, but his distinctive subject matter and his disinclination to paint to the popular taste prevented his election to full membership until 1829, three months after his wife's death in November 1828. This election, the accolade of a successful painter in his day, was a belated tribute to his art; in a bitter allusion to his long wait and his recent bereavement, Constable quoted Dr Johnson to the effect that the honour had been delayed 'until I am solitary, and cannot impart it' (*Memoirs*, p. 171). He was also upset by the patronising attitude of Sir Thomas Lawrence, President of the Royal Academy, who suggested to Constable that he had (as a landscape-painter) been lucky to be elected.

Many of his great pictures had been completed during the 1820s, and some were first generally appreciated at an exhibition of British painting in Paris in 1824. His major canvases included *The Hay Wain* (1821), *The Leaping Horse* (1825), *The Cornfield* (1826), *Dedham Vale* (1828 – the version now in the National Gallery of Scotland), and *Hadleigh Castle* (1829). After his wife's death, Constable spent much time in a publishing-project for a series of reproductions from his pictures by the engraver David Lucas, to be known as *English Landscape* (5 parts, 1830–2); at other times he undertook teaching and committee work at the Royal Academy. In his painting he continued to experiment with different techniques, adopting a broader and more impressionistic style in his sketches, and using his palette knife to startling effect in such original pictures as *Waterloo Bridge* (exhibited 1832). He died suddenly in London in 1837.

During his lifetime, Constable's art was overshadowed by that of his great contemporary Turner, and after his death Ruskin's praise of Turner in *Modern Painters* ensured that Constable remained in Turner's shadow. It was only in the last part of the nineteenth century that Constable began to be appreciated in his own right as a great Romantic landscape-painter. One reason for his long neglect was that, compared with Turner, Constable had a narrow range of subject matter and technique. He never went abroad, and he painted only some parts of England, which he returned to again and again: the Stour valley on the border between Essex and Suffolk, Salisbury and Dorset, Hampstead Heath, Brighton, and a few other places. By declining to paint large-scale historical pictures, Constable was making a statement about what he considered important: the relationship between human beings and nature, in an environment which he saw as benevolent and healthful (in *The Dark Side of the Landscape*, 1980, John Barrell has suggested that this was because Constable distanced himself from his human subjects). He therefore turned back to his beloved Suffolk again and again, and painted the people there in their daily tasks. His love for that part of the country was in part owing to his upbringing: as Wordsworth's childhood in the Lake District produced impressions upon him that never faded, so Constable's childhood influenced him deeply. Those scenes, he said, 'made him a painter' (*Memoirs*, p. 1). His chief instruction to students was to study nature and not art, and he is sometimes thought of (using his own words to Dunthorne) as 'the natural painter'. This does not mean that he painted nature naturally, an impossible thing to do; indeed his art was (as he said) as rigorous as science, and it also owed much to the observation of poets such as Milton and James Thomson, author of *The Seasons*. One of his favourite poets was Cowper, whom he thought of as the poet of religion and nature (a taste which he shared with his wife), and his favourite piece of verse was from Shakespeare's *The Winter's Tale*:

> daffodils
> That come before the swallow dares, and take
> The winds of March with beauty.

He admired verse which drew his attention to hitherto unobserved aspects of natural beauty, and (like Coleridge in poetry) saw his own painting as helping to restore neglected insights into natural life and beauty. He wrote to C. R. Leslie in 1833 of 'my "light" – my "dews" – my "breezes" – my *bloom* – and my *freshness* – no one of which qualities has yet been perfected on the canvas of any painter in the world' (*Correspondence of John Constable*, ed. R. B. Beckett, 1965, III, 96). By contrast, much art was 'used only to blind our eyes, and to prevent us from seeing the sun shine – the fields bloom – the trees blossom – and from hearing the foliage rustle; while old – black – rubbed-out and dirty canvases take the place of God's own works' (*Memoirs*, p. 218). Other painters attempted to persuade Constable to varnish his pictures, a procedure he resisted, saying on one occasion, 'there goes all my dew' (*Memoirs*,

p. 177). The result was that to his contemporaries his pictures seemed startlingly bright and vividly coloured: the flecks of white paint which Constable put on with his palette knife often gave the impression of a bright and restless scene. In this respect he was one of those artists who have changed people's ways of looking at the world, and at art: his paintings now seem entirely natural and representational. To his contemporaries, such as Sir George Beaumont (who asked Constable how he knew where to place his brown tree), and the hanging-committee of the Royal Academy (which dismissed one of Constable's pictures as a 'nasty green thing'), the greenness of Constable's pictures, and their painterly techniques of white flecks of paint, were surprising and even shocking. Constable was often told that his pictures, even the finished versions exhibited at the Royal Academy, seemed unfinished (twentieth-century critics have sometimes, by contrast, preferred the impressionistic power of the sketches to the more deliberate composition of the final versions). For his part, Constable stuck to his methods and principles with a commendable integrity, and his art has long been accepted as one of the most effective examples of the pictorial representation of nature.

J. R. WATSON

Cowper, William (1731–1800)

William Cowper was born at Great Berkhamsted in Hertfordshire, the son of the rector, John Cowper, who was chaplain to George II, and of Ann, née Donne, whose family claimed distant royal descent. He was educated at private schools and at Westminster. He went on to the Middle Temple, and was eventually called to the Bar in 1754. His mother's death when he was six, and persecution by an older boy at Dr Pittman's, one of his early schools, had given rise to serious psychological problems, which were severely aggravated in 1753 by the breakdown of his love affair with his cousin, Theadora Cowper. She was the 'Delia' of a remarkable series of personal lyrics (published posthumously) charting a descent from happy expectations to the nightmare world of separation and feverish longing. The melancholia subsided, and in the mid-1750s Cowper was moving confidently in fashionable circles in the metropolis, contributing papers to the literary periodical *The Connoisseur*. He transferred to the Inner Temple in 1757. Theadora's father, Ashley Cowper, had opposed his daughter's alliance with the young lawyer, but in 1763 he used his patronage to gain him an important clerkship of journals in the House of Lords. Controversy over his uncle's right of nomination, however, and the prospect of a formal examination for the post plunged Cowper into the vortex of suicidal derangement. His profound terror at this time is vividly set out in his autobiographical *Memoir* (written around 1766, published 1816), which also looks back to the child's perilous growth to self-consciousness and describes his subsequent recovery and religious conversion at Dr Cotton's asylum at St Albans in 1765. He lived thenceforth in retirement, first at Huntingdon with the Revd Morley Unwin and his wife, Mary, and then, after Unwin's death in 1767, at Olney in Buckinghamshire, where he was accompanied by Mrs Unwin and came under the influence of the prominent evangelical clergyman John Newton. His engagement to Mrs Unwin was terminated in 1773 during another bout of despondency, which now involved a conviction of his own eternal damnation. His life thereafter was a mixture of calm, contemplative seclusion and recurrent dereliction: his creative achievements emerged in part as a sane reaction to and transcending of inner turmoil and in part as an expression of that darker realm of emotion and imaginative threat.

Olney Hymns, written in collaboration with Newton, appeared in 1779. Cowper's sixty-seven items represent a major contribution to English hymnody. They are often as much subjective lyrics as hymns for congregational use, however, and at their most powerful, as in 'Oh! for a closer walk with God' or 'God moves in a mysterious way', speak, significantly, less of joyful assurance than of desire and the challenge of uncertainty. The volume of *Poems* published in 1782 consists chiefly of eight long reflective essays in rhymed couplets: 'Table Talk', 'The Progress of Error', 'Truth', 'Expostulation', 'Hope', 'Charity', 'Conversation' and 'Retirement'. These poems established Cowper as a leading advocate of evangelical concerns and attitudes, but they are notable also for their trenchant, 'muscular' style, lively satiric sketches, astute comments on topical issues (the loss of the American colonies, the Gordon Riots, the slave trade, anti-semitism), and, especially in

'Table Talk', an assessment of contemporary poetry that under-lines (well before Wordsworth's Preface to *Lyrical Ballads*) the need to rediscover worthy purposes in the face of a prevalent taste for mere 'push-pin play' and 'clockwork tintinabulum'. The upvaluing of the sequestered life and the contemplation of nature in 'Retire-ment' foreshadows *The Task: A Poem in Six Books* (1785), a master-piece *sui generis* whose subject matter and meditative blank verse were an important influence on Wordsworth and Coleridge.

Cowper was at the same time an accomplished writer of 'oc-casional', humorous and other verses, as can be seen from the famous comic ballad *John Gilpin* (1782), the sonorous lines of 'The Loss of the Royal George', and the superb mock-elegy 'On the Death of Mrs Throckmorton's Bulfinch' (1789). These shorter poems are on one level triumphs of form, but what gives so many of them their peculiar edge is the presence of complex currents and configurations of inward experience: Gilpin's ride, for example, is a spirited yet precarious version of Cowper's preoccupation with the individual's subjection to fate and adversity, while Mrs Throck-morton's caged bird is a double of the captive poet, and its destruc-tion by a foraging rat his dream of how death will suddenly come. Coleridge called Cowper the 'best of modern poets', the first to combine 'natural thoughts' with 'natural diction', 'the heart with the head'. His originality is in fact subtler and deeper than this implies, not only in the meditations of *The Task* (which Coleridge has in mind) but also in a foregrounding of the self in a variety of contexts, whether implicitly or with the stunning explicitness of a statement such as 'Lines Written during a Period of Insanity' (1774, published 1816). He claims ordinary subjects and incidents for poetic treatment, yet frequently transforms them into the materials of extraordinary, even surreal, psychodrama.

Cowper's years at Weston Underwood, where he took up resi-dence with Mrs Unwin in 1786, produced a range of important poems, including those against slavery (1788) and several not published until after his death, most notably 'Yardley Oak' (com-posed in 1791), the sonnet 'To Mrs Unwin' (1793), and the poised but relentlessly intimate lines 'To Mary' (1793), provoked by Mrs Unwin's inexorable decline following a paralytic stroke in 1791. 'On the Receipt of my Mother's Picture', written in 1790 and published in 1798, makes successful trial of memory as a means of stabilising a present fraught with a sense of tragic destiny, and is an outstanding early example of the extended Romantic lyric of feeling, confession and therapeutic introspection, carrying Puritan habits of autodidactic self-examination into the most common emotional circumstances. Cowper was befriended during this period by Lady Hesketh; the author William Hayley, who helped to secure him a pension from the King in 1794; and his kinsman John Johnson, with whom he retired to Norfolk in 1795. His health, both mental and physical, deteriorated rapidly after Mrs Unwin's death in 1796, though he continued to revise the blank-verse translation of Homer which he had published, not unsuccessfully, in 1791 as a rival to Pope's standard version, and in 1799 he composed his great poem of isolation and despair, 'The Castaway'. He died at East Dereham in 1800.

Cowper is no longer viewed simply as the 'stricken deer' or the humane poet of religion, rural scenes, and the routine of daily life, but is recognised as a potent and driven innovator whose art and vision, his 'journey within' and revaluation of the world without, have firm and enduring status both in their own right and for their bearing on Romantic and modern trends. He was also a brilliant exponent of the familiar letter, leaving in his considerable correspondence a rich account of the times, as well as insights into his own existence and relationships. His translations into English verse, apart from the Homer, include the Latin and Italian poems of Milton and the devotional poems of the French quietist Madame Guion.

'The Castaway'

The poem was written at East Dereham in Norfolk over the period 20–5 March 1799, a brief creative interval during the progress of Cowper's last mental and physical decline, which brought his death in 1800. It was prompted by a passage in Richard Walter's *A Voyage round the World by . . . George Anson* (1748) describing the 'unhappy fate' of a seaman 'canted overboard' in a violent storm. Cowper's kinsman John Johnson was a subscriber to the 1798 edition of Walter's book and read it aloud to the poet. Another source is Odysseus's fight against 'the swelling surge' in Pope's *Homer's Odyssey* (V.405ff.), a work Cowper would have had in mind from revising his own rival translation at this time. 'The Castaway' was first published in 1803 in William Hayley's *Life and Letters of William Cowper* (1803–6).

The poem operates on three levels: narrative, authorial psycho-drama, and the communication of a whole view of human life and destiny. Cowper immediately identifies with the protagonist–victim, whom he calls 'such a destin'd wretch as I', but then with-holds exact definition of the interrelationship while depicting the situation and struggle of the outcast mariner in eight stanzas which subtly universalise their significance. The setting with its discreet yet sustained personifications ('Obscurest night involved the sky . . .') situates the individual in a field of hostile forces, which he opposes courageously but ultimately without hope, 'supported by despair of life'. The events are fraught with irony: for example, the crew try to help their drowning comrade but, 'pitiless perforce', must yield to circumstance and save themselves by racing away on the very wind that carries his cries towards them; and, in another characteristic blend of psychological precision and emblematic effect, the mariner understands their haste but also knows how 'bitter felt it still to die / Deserted, and his friends so nigh' – in the final analysis each man is alone with his fate. The climax is a good example of the figurative depth that Cowper brings to a chaste, unelaborate style: 'For then, by toil subdued, he drank / The stifling wave, and then he sank.' We usually drink for refreshment or to stay alive, but here the act of imbibing is a terrible consumma-tion as the protagonist, exhausted by his efforts to survive, co-operates in his own destruction. It is a dreadful communion with an adverse and unyielding power – a viaticum for the descent to death and nothingness.

This is the culmination of Cowper's lifelong tragic vision, clearly influenced by his Calvinistic belief in a world of rigid predestina-tion and by his conviction of personal damnation. His habits of self-projection have been evident throughout the poem – in, for instance, the image of an unstable existence and illusions of se-curity in the designation of the mariner's ship as a 'floating home' (which recalls his comparison of himself to a vessel 'tempest-toss'd' in 'On the Receipt of my Mother's Picture'), or in the configuration of aloneness in the midst of apparent community (which had surfaced in several confessional letters, and in precari-ously comic form in *John Gilpin*). The castaway's predicament is a mirror of the poet's own. Death, however, is not only horror but the end of suffering, and it is noticeable that the word 'sank' signals a relaxation in the pulsating rise and fall of the rhythms of 'The Castaway', which have borne the stress of a mind encountering and striving to make sense of its own dark preoccupations (the verse form is a personalisation of the stanza of 'eights' and 'sixes' used by Cowper for many of his hymns, but with an additional couplet allowing meaning to be summarised or stressed). The moment of going under is at once Cowper's confirmation of his painful destiny and a dream of release. In the concluding three stanzas he then stands back to assess and announce the purposes of the poem. It has not been his intention simply to recount 'a narrative sincere', or to 'immortalize the dead' in the manner of elegy. Rather, his motives have been urgently subjective: 'misery still delights to trace / Its 'semblance in another's case'. This 'delight' may refer to the therapeutic gains of contemplating the finality of death, as well as of finding an anguish equal to one's own; but it is in the end much more the aggressive delight of declaring a *distinction* between himself and Anson's mariner, for 'We perish'd, each alone: / But I beneath a rougher sea, / And whelm'd in deeper gulphs than he' – the 'gulphs' of persistent inward and spiritual turmoil. Cowper thus draws stature from the extremity of his lot. When he says, just previously, 'No voice divine the storm allay'd', he not only points to God's positive absence in his hour of need but also defiantly accepts that for him the universe is indeed Godless, as similarly in a poem written on 12 March 1799, 'Montes Glaciales', he had bidden farewell to all poetic consolations and plunged into the sphere of the daemonic. In 'The Castaway' he transcends his fate by embracing it, upvaluing its singularity, snatching affirmation out of the jaws of denial.

'The Castaway' has always been recognised as a striking expression of private experience. The idea that it is somehow shot through with self-pity can now be safely discounted together with the whole sentimental misreading of Cowper as the gentle 'stricken deer'. It is a poem of great emotional force and intellectual control, a complex mental topography in which thought manages but does not diminish the thrust of feeling. Within the context of Romanticism, it signifies, like many of Cowper's works, the rise of subjectivism and a concern with identity – that urge both to stabilise the self and to write it uniquely into existence which yields, in Northrop Frye's terminology (see 'Towards Defining an Age of Sensibility', *English Literary History*, 1956), a poetry of 'process' rather than 'product' and often involves oracular identification with a perceived image or the authoring of a role. Cowper takes us explicitly into the post-Enlightenment realm of existential need and heroism; and therein lies a measure, too, of his broader modernity. A full revaluation of 'The Castaway' and its place in Cowper's achievement appears in Vincent Newey, *Cowper's Poetry: A Critical Study and Reassessment* (1982), and there is relevant commentary in Bill Hutchings, *The Poetry of William Cowper* (London, 1983).

The Task

The poem was begun in October 1783, when Lady Austen playfully suggested to Cowper that he should write in blank verse on the subject of 'the sofa'. By the time of its completion in the autumn of 1784 it had grown to a work of six books and around 5000 lines. It was published in July 1785, and acclaimed as a masterpiece. Though stemming from the tradition of didactic–descriptive verse given impetus by Thomson's *The Seasons* (1726–30), it manifests a wholly original bent, texture and range of concerns.

The Task is a poem on man, nature and society, but also distinctively about the self. Cowper explained its unifying principle as a tendency 'to discountenance the modern enthusiasm after a London life, and to recommend rural ease and leisure as friendly to piety and virtue'. 'God made the country, and man made the town' (book I: 'The Sofa'): the moral design is apparent from the first, and includes not only generalised attacks on metropolitan corruptness but specific reactions to topical issues such as the slave trade and contemporary education (book II: 'The Time-Piece') or the loss of the American colonies (book IV: 'The Winter Evening'), where Cowper, in sentiment and style, is the advocate of values blending Evangelicalism and the classical ideals of *humanitas* and *gravitas*. This thread, together with the exploration of rural and domestic life, makes *The Task*, in a sense, a truly national poem. The public dimension and objectives, however, are constantly interwoven with or replaced by radically personal ones. There is overt autobiography and confession, as in the famous passage in book III ('The Garden') where Cowper describes himself as 'a stricken deer, that left the herd', but self-revelation also surfaces in the form of mental currents and preoccupations, involving such repeated image patterns as that of confinement and release. The account of the death-in-life of the 'paralytic' in book I, for example, projects the poet's own psychological fears, which are themselves countered in the Ouse Valley section (which lies behind Wordsworth's 'Tintern Abbey') through an affirmation of his unfailing capacity for being renovated by the influence of nature. These episodes represent the same oscillation of mind as is reported in a letter to John Newton in 1788 where Cowper talks of swinging restlessly between thoughts of 'a happy eternity' and being 'cast down to hell'.

Books III and IV focus more particularly upon the poet's sequestered life. The activities of the gardener, though the occasion for subtle humour, are developed as parables of how man should live in the world he inherits. Cowper is powerfully aware of a 'civilization' beyond, which is re-forming itself on the principles of acquisitive commercialism and manufacture, so that men 'incorporated . . . Build factories with blood' and the 'Midas finger of the state' reaches ruinously even into the countryside. Thus alienated from the collective present, in a posture that strikingly foretells the divorce of virtue from the practical sphere which is characteristic of the Romantic and post-Romantic aesthetic, he formulates in his vision of the innocent yet creative labours of 'the happy man' a sustaining myth of optimum existence and harmony with the given energies of the creation. Like the poem itself – a 'task' inscribed in that of the gardener – this myth is both a vehicle of outgoing idealism and a source of stability for Cowper himself, the means of framing a positive identity out of the materials of his enforced seclusion.

Books V ('The Winter Morning Walk') and VI ('The Winter Walk at Noon') are notable on one level for their forceful anti-deist arguments from the design of the creation up to the God of revealed religion, unrivalled as Christian apologia at least until the appearance of William Paley's *Natural Theology* in 1802. There is 'A soul in all things, and that soul is God'. Yet Cowper sees God as the *source* as well as the end of heightened perception – 'Acquaint thyself with God, if thou would'st taste / His works' – and so makes the very ability to appreciate nature a sign of grace and spiritual well-being. His dynamic interactions between self and a living organic reality, covering an amazing range of moods and perspectives, from the set-piece descriptions of book I through the interiorised 'variegated' landscape amid the frost in book IV (which facilitated the meditation of Coleridge's 'Frost at Midnight') to the kaleidoscope of various close engagements with sentient and non-sentient nature in the later books, are at once an immediate emotional fulfilment and trials of inner strength and wholeness, an experiential route to 'salvation'. They are also the medium of intuitive wisdom: 'the heart / May give an useful lesson to the head', and we see into the life of things. Cowper's *Task* then draws to a close, however, with further examples of discord in the world (man's treatment of animals, for instance) and the perverse 'religious' rituals of the Handel and Garrick festivals ('Man praises man'), followed by a wishful prophecy of the Last Day. Cowper may look forward to this future time when all will be restored to prelapsarian beauty, but in the present he must use his resources, above all his liberty to create and go on creating, in order to survive beneath the shadow of an overarching human chaos. In a final return to the theme of the sequestered life he emerges at last as the enigmatic solitary, useless in the eyes of society but the exemplar of patient privacy, the individual for whom 'contemplation is his bliss' and 'his warfare is within'. *The Task,* and its psychological journey, are open-ended: uncertain whether his life is a privilege or a 'doom', faced always with the 'warfare . . . within' though armed with the gifts of active 'contemplation', he is not so much 'the happy man' of ancient tradition as the modern existentialist hero.

The early reviewers, who praised Cowper for his piety and descriptive realism, and for the interest of his character, set the tone for subsequent approaches to *The Task.* Recent criticism, however, has seriously reassessed its importance as a poem of ontological quest and a philosophic statement. Among other things, it established a 'natural faith' and model of being-in-the-world central to the Romantic project of rediscovering value in a world rendered potentially void of meaning by Newtonian science and other mechanistic trends in Enlightenment thought. It made the self, though cast out to the periphery of society, the centre of creative opportunity and attention; it celebrated God but made the mind of man, whose trials and triumphs are everywhere richly constellated in the writing, the ultimate location of truth. In upvaluing experience and the miracles of the commonplace, it both democratised and redeemed imagination.

VINCENT NEWEY

Crabbe, George (1754–1832)

George Crabbe was born on 24 December 1754 in Aldeburgh, Suffolk; he set up in practice as a surgeon in 1775, the year that his poem *Inebriety* was published anonymously. By 1780 he had decided that he wanted to be a writer, and went to London, where he eventually secured the patronage of Edmund Burke and the publication of *The Library* in 1781. He returned to Aldeburgh as its curate, and in 1782 became chaplain to the Duke of Rutland at Belvoir Castle, Leicestershire. His eleven-year engagement to Sarah Elmy culminated in marriage in 1783, the year *The Village* was published. *The Newspaper* followed in 1785. No more poetry was published until *The Parish Register* of 1807, when Crabbe became an established poetic figure: *The Register* followed in 1810, *Tales* in 1812, and *Tales of the Hall* in 1819. In 1832 Crabbe died at Trowbridge, Wiltshire, where he had held a living since 1813.

Although Crabbe achieved considerable popularity in his own day, there were dissenting voices, particularly amongst those who felt he was an anachronism, challenging the new emphasis on the powers of the imagination. As the *Quarterly Review* put it, 'He is the poet of reality, of reality in low life', and Wordsworth and Coleridge found their own unease expressed by Hazlitt in *The Spirit of the Age*: 'the adept in Dutch interiors, hovels , and pig-styes must find in Mr Crabbe a man after his own heart'. As Wordsworth put it, 'Crabbe's Pictures are mere matters of fact, with which the Muses have just about as much to do as they have with a collection of medical reports, or law cases'. Amongst the Romantics it was Byron who celebrated Crabbe's particular virtues: 'Though nature's sternest painter, yet the best' (*English Bards and Scotch Reviewers*, l. 858). Another admirer was Jane Austen, and several recent critics have pointed to an affinity between Johnson, Jane Austen, Byron and Crabbe that challenges the dismissal of him as 'Pope in worsted stockings'.

The apparently even tenor of Crabbe's verse has led many readers to assume that all of his poetry is the same, that it shows no development. Although the heroic couplet remained his basic form, he played many variations on the theme of truth, and how it is to be perceived in verse (in this he foreshadowed Byron, without attempting to match him in bravura). *The Village* deals in generalities, using the apparently easy appeal of the pastoral form to undermine its very ethos. 'I paint the Cot', he declares, 'As Truth will paint it, and as Bards will not.' People do not exist in the poem as named individuals; the view is panoramic, deliberately held at arm's length. One well-known passage, however, hints at the power of detail that will distinguish his later verse. In an inversion of the Miltonic Eden, he catches the way in which the natural world becomes an image of the sterile life of those who live and work there (ll. 65–70). In 'Peter Grimes', one of the most famous tales in *The Borough*, Crabbe is able to relate such details to the complex psychology of his central character. It is perhaps in *The Borough* that we hear most insistently the echo of Crabbe's hidden life as a novelist – three novels written and burnt on a bonfire in the early years of the century. For *The Borough* operates on the vast and generalised scale implied by the title: the opening 'General Description' is deceptive, for the apparent calm and placidity is offset by the turbulence of the sea, which is always there as a threat, 'Various and vast, sublime in all its forms'. And, just as no settled perspective is possible there, so the individual sections of the poem explore the ambiguities of feeling and judgement that lurk, whether the subject be 'Clubs and Social Meetings', 'Inns', or 'Peter Grimes'. In the last, what might have been a simple narrative is presented in an extraordinarily complex fashion, as one perspective gives way to another: as in one of Wordsworth's Lyrical Ballads, judgement is as much against the people as against the pitiful yet heroic Grimes. Significantly, this tale is followed by an episode entitled 'Prisons', where again Crabbe explores that realm of terror which seems to be peculiarly his; we might at times hear anticipations of Byron or echoes of Scott, but the difference is that Crabbe's awareness of that dread world of nightmare is filtered through a verse of remarkable control and decorum. He writes, in 'Peter Grimes', of the 'small stream, confined in narrow bound' that 'Ran with a dull, unvaried, sadd'ning sound; / Where all, presented to the eye or ear, / Oppress'd the soul with misery, grief, and fear.' That could stand as an epigraph to much of his own work. He is honest enough to recognise the soul's oppression – which is not to say that he ends up the turgid moralist he is sometimes painted. There is in his best work a deep humanity, and an abiding sense of humour. These characteristics come to the fore in the *Tales* of 1812, which have an almost Shakespearean richness (each tale is prefaced by several Shakespearean quotations, as though Crabbe wants us to think in those terms). Crabbe dramatises particular situations, relying on idiosyncratic details to bolster his insights into the self-deceptions which fascinate him. If there is a feeling that he cannot after 1812 sustain the quality of his best work, then there is still a large amount on which his reputation can rest. Francis Jeffrey spoke in the *Edinburgh Review* of Crabbe as 'the most original writer who has ever come before us'. Any just assessment of Crabbe needs to understand how Jeffrey felt able to say that.

MARK STOREY

De Quincey, Thomas (1785–1859)

Thomas De Quincey was born in Manchester, the son of a well-to-do merchant. His father suffered from ill-health, and spent much time abroad, so that his children saw little of him; he died when Thomas was 7 years old. During his early years, he was brought up in the company of his mother and sisters, and the death of his sister Elizabeth, also when he was 7, profoundly affected him. He was educated at Manchester, Bath, in Wiltshire, and again in Manchester, at the Grammar School, from which he ran away. He lived in Cheshire and Wales for a short time, and then went to London, where he existed in poor lodgings and made friends with the outcast women and prostitutes. His description of the kindness of one of them, 'Ann', is memorably described in *Confessions of an English Opium Eater* (1822). He was rescued by friends, and went to Worcester College, Oxford, which he left without taking a degree. It was during this period, on a visit to London probably in 1804, that he first had recourse to opium, in the first instance as a cure for the pain of a bad tooth with a dental abscess.

He had long admired Wordsworth and Coleridge, and finally met Coleridge in 1807. He escorted Mrs Coleridge and the children, at Samuel Taylor's request, to the Lake District, and thus met Wordsworth, whom he had been too timid to visit earlier. After a further period in London, during which he saw Wordsworth's pamphlet on the Convention of Cintra through the press, he returned to live in Grasmere; in 1816 he married Margaret Simpson, daughter of a local 'statesman' (or 'estatesman', one who farmed his own land). He made a living by writing, chiefly for the *Quarterly* and *Blackwood's*, but his addiction to opium continued: the 'Confessions of an English Opium Eater' appeared in the *London Magazine* for October and November 1821, and was published in 1822. His many contributions to periodicals based in Edinburgh, such as the *Edinburgh Literary Gazette* and *Blackwood's*, led to his removal to Edinburgh, and he settled there permanently in 1828. His financial problems, his generosity to those even poorer than himself, and his dependence on opium, made his life difficult, and he spent periods in the debtors' prison, but he also continued to write articles and essays of remarkable value. His 'Recollections of the Lake Poets' appeared between 1834 and 1839, causing offence to Coleridge's family by its suggestions of plagiarism; his 'Sketches from the Autobiography of an English Opium Eater' between 1834 and 1841, later entitled *Autobiographic Sketches*. He became, in Alethea Hayter's phrase, 'the prophet of opium' (*Opium and the Romantic Imagination*, 1968, p. 104); he intended to write a master-piece about the effects of addiction, entitled *Suspiria de Profundis*, but much of the material was subsumed into *Autobiographic Sketches* and other essays. Perhaps the most notable of the essays on this subject was 'The English Mail Coach' (*Blackwood's*, 1849), in which he recounted a memorable dream and showed an awareness of its origins in childhood and early experience (see under DREAMS). His other work as a critic is best represented by the brilliant 'On the Knocking on the Gate in Macbeth' and 'Murder Considered as one of the Fine Arts'. He died in reduced circumstances in Edinburgh in 1859. His *Collected Writings* were edited by David Masson (Edinburgh, 1889–90); the most recent biography, by Grevel Lindop, is entitled *The Opium Eater* (1981).

J. R. WATSON

Drama

The Romantic period was not an age of great stage plays, although there were some remarkable dramas written by major poets; but it was a great age of individual actors. The period was characterised by the dominance upon the stage of one or two individuals, while the ensemble playing was neglected (there is evidence that re-hearsals by the company were few, or were not taken seriously). The most famous eighteenth-century practitioner of the individual style was David Garrick (1717–79), who set new standards of excitement, energy and histrionic style. When he retired in 1776, he was succeeded as the great actor of the day by John Philip Kemble (1757–1823), together with Kemble's sister, Sarah Siddons (1755–1831). In 1788 Kemble became the actor–manager of Drury Lane Theatre, and there and at Covent Garden he dominated the English stage for some thirty years; his style was more intense and controlled than that of Garrick. At Drury Lane and Covent Garden he specialised in Shakespearean roles, notably those of the great individual tragic heroes, such as Hamlet, Othello and Coriolanus.

After Kemble, the volatile Edmund Kean (1787–1833) became the most celebrated actor of his day. Coleridge memorably described his acting as 'like reading Shakespeare by flashes of lightning', and Hazlitt (who disliked Kemble's style) was also a great admirer. Kean in his turn was succeeded by William Charles Macready (1793–1873), who – among his other achievements – was the first actor in 150 years to perform Shakespeare's original *King Lear*, as opposed to Nahum Tate's altered version.

The most notable stage plays of the second half of the eighteenth century had been Goldsmith's *She Stoops to Conquer* (1773), and Sheridan's *The Rivals* (1775) and *The School for Scandal* (1777); these plays looked back to the comedies of manners and intrigue found in the Restoration theatre, and before that to Ben Jonson. Another kind of drama had begun in 1768 with Horace Walpole's never-acted *The Mysterious Mother* (a play which deals with incest): the Gothic revival was represented by M. G. ('Monk') Lewis's *The Castle Spectre* (1797), and many other plays.

In addition to the Gothic, other Romantic themes appeared. The Jacobin novelist Thomas Holcroft wrote *The Road to Ruin* (1792), about greed and extravagance, and Elizabeth Inchbald, another Jacobin novelist and a practising actress herself, wrote a number of plays, including *Every One Has His Faults* (1793) and *Lovers' Vows* (1798), adapted from a German play by August Friedrich von Kotzebue, whose plays enjoyed a considerable vogue at the turn of the century. *Lovers' Vows*, with its mixture of melodramatic plot and political undertones (portraying the wicked Baron Wildenhaim), is interesting in its own right, but is chiefly remembered for being the play which was rehearsed (though never acted) in Jane Austen's *Mansfield Park*. Other notable writers of plays during this period were Richard Cumberland, whose *The Mysterious Husband* (1783) was influenced by Horace Walpole's play, and George Colman the Younger (1762–1836), whose *The Iron Chest* (1796) is a dramatisation of Godwin's *Caleb Williams*, but who also wrote comedies.

The period saw the rise of melodrama, begun by M. G. Lewis and Cumberland, and continued in Thomas Holcroft's *A Tale of Mystery* (1802) and James Robinson Planché's *The Vampire* (1820); the beginnings of costume drama, found in W. T. Moncrieff's *Rochester, or Charles the Second's Merry Days* (1818); and the beginnings of farce, as in George Colman's *Love Laughs at Locksmiths* (1802). It was also notable for the adaptation of Scott's novels and poems for the stage; many were thus adapted, *Rob Roy* proving extremely popular. Scott himself was persuaded to write a melo-drama, *The Doom of Devorgoyl*, but he did not like the theatre, regarding badly behaved spectators as 'a national nuisance' (Allar-dyce Nicoll, *A History of English Drama, 1660–1900*, 1955).

Scott's dislike of theatre-goers was symptomatic of a certain distance between major authors and the stage. Most of the prin-cipal Romantic poets wrote plays, but with the exception of Cole-ridge's *Remorse*, acted in 1813, and Byron's *Marino Faliero*, acted in 1821, they were not put on the stage. Wordsworth thought of staging *The Borderers* (1796–7), but the plan came to nothing, probably because the plot was too complicated; and Shelley's *The Cenci* (1819) was unsuitable to the taste of the time. Byron's other dramas, such as *Manfred* (1817), were not well fitted for the stage (indeed, Byron did not wish *Manfred* to be performed); and Shel-ley's *Prometheus Unbound*, though written in acts and scenes, is more a dramatic poem than a drama. It would seem that the Romantic poets preferred to work in their own private and inspi-rational ways, and were on the whole unwilling to accommodate their work to the practical needs of the stage or to the tastes of the theatre-going public of the time.

J. R. WATSON

Dreams

Literature has in all ages occasionally referred to dreams, some-times even – as in some Shakespearean plays – used them as dramatic elements. But never before Romanticism had dream as a psychical reality experienced by the artist been raised into the theme of a poem or of a work of art. Beyond the external universe and the world of waking consciousness, the dream world is ap-proached by the Romantic artist as a fascinating virgin territory whose exploration or representation involves the creation of hitherto unattempted modes of expression by the Promethean creator. The meditative prologue of *The Fall of Hyperion, a Dream* – which we may be tempted to read as Keats's ultimate message on poetic art – goes so far as to define the creative poet as he who can 'trace' on paper the figures and visions of his dream: 'For Poesy alone can tell her dreams, / With the fine spell of words alone can save / Imagination from the sable charm / And dumb enchantment' (I.8–11). In such works the artist's achievement is to be measured in terms of the 'fidelity' (to use Coleridge's word in his preface to 'Kubla Khan') with which the work reproduces or re-creates the dream experience.

Coleridge himself exemplifies this in the two contrasted dream poems 'Kubla Khan' and 'The Pains of Sleep' (published together in 1816); so does Keats in his sonnet 'On a Dream' and implicitly in the mysterious ballad 'La Belle Dame sans Merci', which records a smaller dream within its narrative (stanzas ix–xi) but reads actually, as a whole, like a narrative within a larger dream; and De Quincey's autobiography, *The Confessions of an English Opium Eater* (1822), culminates in a final section which describes the dream processes and transcribes the opium dreams. These are examples of the Romantic attempt to create rhythms, syntactic forms and images fit to render the specific quality of the dream experience, with its inexpressible blend of unintelligibility and emotional poignancy. The literary dream text thus tends to free itself from all logical structure and presents itself as a sheer flow of intense images ('Kubla Khan') or as a sequence of strange, anxiety-generating events ('La Belle Dame sans Merci'; De Quincey's opium dreams).

The Romantic's fascination with dreams is easy to understand in view of the general context of the Romantic quest for new frontiers of consciousness and for an expansion of human experience into 'unknown modes of being'. But, as soon as the nature or the field of this trans-consciousness has to be conceptually articulated, Romantic theory fluctuates – sometimes, confusingly enough, within the same discourse. Thus De Quincey's essay 'Dreaming' (written years after the first edition of the *Confessions*, as a sort of postscript) offers two different theories of the dreaming-faculty: one mystically and the other psychologically oriented. In one passage the sleeper's consciousness is described as a 'mirror' receiving in a dream 'dark reflections from eternities below all life' and thus as a unique intuitive faculty of metaphysical knowledge: 'the magnificent apparatus which forces the infinite into the chambers of a human brain'. Such a concept of dream as mystical synecdoche, where in a privileged moment an infinite Mind is brought to coincide with a finite consciousness, underlies indeed the dramatic scenarios of some major Romantic myths. In *The Fall of Hyperion* initiation into the vision of the primal gods' 'high tragedy' – the theme of the epic – passes through two successive dream avenues: that of the poet's private dream and that of Moneta, an archetypal, universal mind and repository of primordial dreams. In Shelley's *Prometheus Unbound*, Act II, the dynamism of the soul's initiatory journey down to the world's ultimate power (Demogorgon) is provided by the twofold dream of Asia (Prometheus's anima), which first epiphanises the soul's unsuspected depths, then summons her to explore those depths ('Follow!') and leads her to a final vision of the unthinkable – but dreamable – 'rays of gloom . . . Ungazed upon and shapeless'.

Elsewhere in 'Dreaming', however, De Quincey defines dream as the 'reproductive faculty', by which he means that dream, especially when 'assisted' by opium – which does not radically alter, but intensifies the faculty – can resurrect the dreamer's remotest past experiences in all their native intensity, thus illuminating 'the whole depths of the theatre' of infancy which had long lain dormant in the memory: 'now [under the influence of opium] the agitations of my childhood reopened in strength; now first they swept in upon the brain with power and the grandeur of recovered life'. The key metaphor here is no longer that of the mystical 'mirror', but that of 'The Palimpsest of the Human Brain' (the title of another essay of the time), where the memories of primal experiences, most of all 'the deep, deep tragedies of infancy', subsequently covered in the course of time by numberless layers of new experience and seemingly obliterated, remain as 'secret inscriptions' in the mind and rise to the surface again under the action of the 'fierce chemistry' of dream. Such a theory indeed underlies the design of the *Confessions*, with the final nightmares presented as 'reverberations', variations and majestic amplifications, of the writer's primal tragedies (the loss of Elizabeth, the passionately loved elder sister, called up in an autobiographical sketch, and its implicit repetition in the loss of Ann, the sister soul, included in the first part of the *Confessions*). One cannot fail to be struck by the resemblances between this perspective and that of later psychoanalytic theory: the decisive importance of infantile (particularly traumatic) experiences in the generation of the personality; the persistence of those experiences as unconscious memories or impressions – a notion undoubtedly implied (although De

Quincey does not use the term 'unconscious') by the image of the mind's palimpsest and by the notion of memories that 'remain lurking' or 'sleeping' in or below the mind; and the free access of those memories to dream life, their occasional re-enactment as dream scenes, and, when they reappear, their astonishing freshness and intensity, as if no time had elapsed since the events that occasioned them, and as if the impressions recalled 'were virtually immortal' (in Freud's own phrase). All these De Quinceyan intuitions are remarkably congruent with Freud's *Traumdeutung* (interpretation of dreams).

It would be a mistake, however, to press the analogy further, for De Quincey's dream theory lacks what Freud calls the 'true child of night': namely, the repressed desire which is the true generator of the dream fantasy, in which it expresses and fulfils itself in a hallucinatory way (possibly reactualising an experience in which it was fulfilled, or retrospectively modifying an experience to achieve fulfilment). In this respect De Quincey's dream theory differs fundamentally from Freud's, being unaware of the wish-fulfilling function of dream, and therefore ignorant of the complex techniques of distortion brought into play by dream to make the repressed (because offensive) wish unrecognisable to the ego's consciousness: this 'dream work', far more complex than De Quincey's dream 'chemistry', combines such techniques as displacement, condensation and symbolic disguise, which transform the 'latent dream thoughts' into the 'manifest content' of the dream. De Quincey would no doubt have rejected the very notion of repression and of unavowable wishes (notably, in his case, of incestuous wishes), though, to a modern reader acquainted with Freudian psychoanalysis, these might well appear to form the core of the dreams transcribed in the *Confessions*, constituting the true 'unconscious of the text', the true 'secret inscriptions'.

There were, however, some penetrating thinkers of the Romantic period who were intuitively acquainted with the psychoanalytic concept of 'repression', and the correlative notion of the 'return of the repressed' (the irruption into the dreamer's consciousness of unacceptable impulses or thoughts usually held down in the unconscious but able to escape control during sleep): 'unconscious impressions', Hazlitt observed, 'give a colour to and react upon our conscious ones . . . we check these rising thoughts and fancy we have them not. In dreams when we are off our guard, they return securely and unbidden.' Such a view of the dreaming (or daydreaming) psyche is no doubt behind Shelley's imaginative creation of the Furies that confront the self in Act I of *Prometheus Unbound*: 'Phantasms foul' that Prometheus, for all his horror, yet beholds 'in loathsome sympathy', and that describe themselves, addressing the hero, as 'dread thought beneath thy brain, / And foul desire round thine astonished heart' (I.447–51, 488–9). The same view underlies, more generally, the Romantic reinterpretation of the supernatural (or the creation of a new 'fantastic' literature) in terms of *revenance* – of the ghost's 'return' from the 'undiscovered country' and uncanny manifestation of itself to the ego. Such returning dream figures include Shelley's Furies; Coleridge's Geraldine, met by the heroine in the opening night scene of 'Christabel', 'A sight to dream of, not to tell'; and Gil-Martin, the fiendish figure of James Hogg's *Confessions of a Justified Sinner*, of whom the hero observes that 'All my dreams corresponded exactly with his suggestions'. In each case, the poem's or the novel's basic drama is the transcription of a dream mechanism.

The Romantic dream scene is probably at its best in the expression of this type of dream, where the sleeper passively witnesses the manifestation of his own 'returned' unconscious desire, in what the Romantic artist used to call the 'nightmare' and what depth psychology defines as 'anxiety dream': namely, that sort of dream which fulfils in a barely disguised form an unconscious impulse which the ego experiences as repellent or monstrous, so that its fulfilment generates intense anxiety. The strength of the Romantic 'nightmare' – particularly apparent to the modern reader – lies in the artist's capacity to objectify the monstrous in the dream scene while gradually blurring the boundaries between the monstrous and the self, which becomes disseminated, subject and object, and finally omnipresent as the text works towards a latent epiphany that shows forth the monstrous within the dreaming 'I'.

Three examples, from three different genres – poetry, prose and painting – will serve to illustrate this pattern. First, in Coleridge's 'The Pains of Sleep' the poem obliquely internalises the brawling 'fiendish crowd / Of shapes' which beset the dreamer by showing the 'I' finally wakened by his 'own loud scream'. Secondly, in the narrative of De Quincey's oriental nightmares in the *Confessions*, an unnamed 'deed' of the dreamer's own doing is metonymically reflected in the monstrousness of the primitive animal gods appointed to judge it, and when the 'I' suddenly comes upon Isis and Osiris (the mother–son or sister–brother divine couple) it is hinted that the deed is incest. The same deed is disguised in the night-marish experience of the 'cancerous kisses [of] crocodiles', an anxiety-dream image of the loved mother made repulsive by the dread of incest, which acts as a counter to desire and turns the love scene into a horror scene which the dreamer witnesses 'loathing and fascinated'.

The third example is furnished by the celebrated *Nightmare* painted around 1790 by Füssli (Fuseli). In this plastic (and thereby static) representation of the anxiety dream, the internalisation of the monstrous cannot, of course, be achieved through any developing discourse as in verbal art. But it is effected, more subtly, in the shifting interpretation which the enigmatic painting generates in us as we look at it. To begin with, we see a beautiful female sleeper lying paralysed by two monsters: the incubus that weighs down on her breast, and the petrifying night-mare (female horse) that stares through the curtains. Looking more deeply, we see that the beautiful sleeper, in the lower left-hand corner of the picture, is no longer the dreaming subject, but the object of a dreamer's desire, and that the Medusa-like mare in the upper right-hand corner is the negative image of the same sleeper, repelling all erotic desire and placed under taboo through the fear of incest. These two figures represent conflicting facets of one and the same mother figure, between which – in De Quincey's words – the 'I' stands 'loathing and fascinated'. In this deeper reading of the pictorial dream text, the dreaming 'I' is found to be represented in the figure that hovers *between* the two female images, the sleeper and the mare: that is, the whole picture hinges upon the small demonic incubus, in which we (and the artist's desire) locate the dreaming self.

CHRISTIAN LA CASSAGNÈRE

The Fantastic

Edmund Burke's *Philosophical Enquiry into the Origin of our Ideas of the Sublime and Beautiful* (1757) emphasised both the suggestive and the disturbing quality of art, for in Burke's 'sublime' mankind is upset or threatened by that which lies beyond human control or understanding. In Shelley's words, 'sorrow, terror, anguish, despair itself, are often the chosen expression of an approximation to the higher good'. Sublimity is seen in terms of exaggeration, art as frightening through distortion and violence of light.

Giovanni Battista Piranesi (1720–78) was born in Venice and settled in Rome in 1740. Horace Walpole, English art connoisseur and author of the Gothic novel *The Castle of Otranto* (1764), wrote of Piranesi in the Advertisement to volume IV of *Anecdotes of Painting in England* (1771) that he 'seems to have conceived visions of Rome beyond what is boasted even in the meridian of its splendour. Savage as Salvator Rosa – fierce as Michael Angelo, and exuberant as Rubens, he has imagined scenes that would startle geometry, and exhaust the Indies to realise. He piles palaces on bridges, and temples on palaces, and scales Heaven with mountains of edifices.' First widely known for his *Vedute di Roma*, begun in 1745 and continued through his lifetime to the number of 137 etchings of ancient and modern Rome, Piranesi became most famous for his *Carceri d'Invenzione*, which were begun about 1745 and reworked in 1761. It is these etchings of imaginary prisons which characterise most startlingly the art of the fantastic. In his *Confessions of an English Opium Eater* (1822), Thomas De Quincey records how

> many years ago, when I was looking over Piranesi's *Antiquities of Rome* [the *Vedute*], Mr Coleridge, who was standing by, described to me a set of plates by that artist, called his *Dreams* [the *Carceri*], and which record the scenery of his own visions during the delirium of a fever. Some of them . . . represented vast Gothic halls: on the floor of which stood all sorts of engines and machinery, wheels, cables, pulleys, levers, catapults, etc. etc. expressive of enormous power put forth and resistance overcome . . . elevate your eye, and a still more aerial flight of stairs is beheld: and again is poor Piranesi busy on his aspiring labours: and so on, until the unfinished stairs and Piranesi both are lost in the upper gloom of the hall. With the same power of

endless growth and self-reproduction did my architecture pro-
ceed in dreams.

Not only did De Quincey and Coleridge find in Piranesi's art the
wild expression of their opium nightmares, but it also reflected the
Romantic obsessions with the 'modalities of fragmentation' (see
Thomas McFarland, *Romanticism and the Forms of Ruin*, 1981) and
with dreams – obsessions also explored in Coleridge's 'Kubla
Khan', for example.

Piranesi's architecture answers, too, to Burke's analysis of the
effect of 'infinity and things multiplied without end'. Through
Robert and James Adam, Piranesi's engravings came to exercise a
widespread influence on the internal decoration of English houses.
Through Walpole and William Beckford, they provided inspiration
in the tradition of the Gothic novel. Beckford refers to Piranesi
more than once in his *Dreams, Waking Thoughts and Incidents* (1783),
recording particularly how in Venice 'I could not dine in peace, so
strongly was my imagination affected; but snatching my pencil, I
drew chasms, and subterranean hollows, the domain of fear and
torture, with chains, racks, wheels, and dreadful engines in the
style of Piranesi.'

Johann Heinrich Füssli (1741–1825) was born in Zürich. He was
forced by his father to become a Zwinglian minister in 1761, but
failed both as a clergyman and later as a translator and hack writer.
His literary connections, however, brought him into contact with
J. G. Herder and the German writers of the movement labelled *Sturm
und Drang* (Storm and Stress). He settled in England in 1763, as
'Henry Fuseli', and was encouraged by Sir Joshua Reynolds to
become a painter. Like Piranesi he was deeply affected by the
antiquities of Rome, where he studied from 1770 to 1778; the
experience is reflected in the drawing known as *The Artist Moved by
Magnitude of Antique Fragments* (1778), a depiction of a monstrous,
fragmentary and ruinous hand and foot from classical statuary.
Fuseli's first great success was *The Nightmare*, exhibited at the
Royal Academy in 1781, which, in the words of Nicholas Powell,
does not 'set out to illustrate a dream, so much as to depict the
sensations of terror and stifling oppression experienced in a nightmare' (William Vaughan, *Romantic Art*, 1978, p. 50). The painting,
of an abandoned female form oppressed by a devil and the staring
eyes of a ghostly mare, explores the subconscious, irrational drives
of eroticism and is typical of the artist in its distortion and exagger-

ation of form, particularly the female (see also under DREAMS).

Fuseli's art exploits the horror in the human psyche. He ab-horred vague speculation and complained in 1798 that 'the expec-tations of romantic fancy, like those of ignorance, are indefinite'. He was a pragmatist, and the *New Monthly Magazine* remarked in 1831 that 'it was he who made real and visible to us the vague and insubstantial phantoms which haunt like dim dreams the op-pressed imagination'. His visionary quality inspired Blake, whom he met about 1787, and in 1790 he began a series of forty-six huge paintings illustrating Milton. Professor of Painting at the Royal Academy from 1799, he attracted with his extravagant and fantas-tic art the attention of many of the key figures in English Roman-ticism. Mary Wollstonecraft fell in love with him before her marriage to Godwin, even to the point of begging Fuseli's wife to allow her to live as a member of their family. Impulsive and eccentric, Fuseli exemplifies the Romantics' predilection for the unfinished and fragmentary; their preoccupation with the work-ings of the creative mind, with emotional encounter and sensi-tivity; and their obsession with the heroic and the vast – an obsession fed in Fuseli by his admiration of ancient Rome and the powerful figures of Michelangelo.

DAVID JASPER

The French Revolution

Preceded by the American Revolution, and to some extent inspired by it, the French Revolution was an epoch-making upheaval by which, in the last decade of the eighteenth century, the old order in France was overthrown and such forces as democracy, national-ism and socialism were set in motion throughout Europe. Rousseau's doctrine of the sovereignty of the people as the only legitimate foundation of government directly challenged the right to exist of every monarchy in Europe. The French Revolution was the conse-quence of a moral, financial, economic and social crisis. The auth-ority of the French King, an absolute monarch of divine right, was questioned by the eighteenth-century French *philosophes*

Montesquieu, Voltaire, Rousseau and Diderot, whose *Encyclopédie* (published between 1751 and 1772) was a war machine pointed at the *ancien régime*. Masonic lodges were instrumental in propagating the new ideas of the age of Enlightenment among the middle classes. The French monarchy was criticised all the more because of the enormous deficit sustained by the royal finances due to inflation and ever-increasing expenditure. The year 1788 marked the beginning of a severe economic crisis for France: the bad harvest of 1788, followed by a bitter winter, led to a tremendous rise in prices, and resuscitated the spectre of famine; French textile industries, unable to meet competition from Great Britain, were hit by unemployment. The economic crisis could not but aggravate latent social antagonisms, making the oppression of the lower classes by the nobles, higher clergy and upper middle classes even more intolerable.

The meeting of the States General at Versailles (5 May 1789) is traditionally regarded as opening the first phase of the French Revolution. The only institution in France that could represent the nation had not met since 1614: Louis XVI agreed (Aug 1788) to summon the States General because such a meeting was demanded – though not for the same reasons – simultaneously by the Clergy, the Nobility, and the Third Estate. As a concession to popular opinion, Necker, the King's chief minister and financier, had decided (Dec 1788) that the Third Estate should have as many representatives as the Clergy and the Nobility put together. In his pamphlet *Qu'est-ce que le tiers état?* (Jan 1789) the abbé Sieyès claimed that the will of the nation should be expressed by the representatives of nine-tenths of its population, as the Third Estate numbered 26 million, the Nobility 400,000, and the Clergy 135,000. For six weeks, each order of the States General, namely the 300 deputies of the Clergy, the 300 deputies of the Nobility, and the 600 deputies of the Third Estate, met separately. On 17 June 1789, the representatives of the Third Estate, who occupied the great Salle des Menus Plaisirs, and wanted the Clergy and Nobility to join them with a view to constituting a single body, declared themselves to be the National Assembly. This was a direct challenge to the King, who insisted on the three orders' sitting separately. Because the Third Estate refused to disperse 'except at the point of the bayonet', Louis XVI, whose troops were not trustworthy, finally ordered the Clergy and Nobility to meet with the Third Estate as a single assembly, in which the deputies were to vote individually. This was a great victory for the Third Estate, which, by winning a few supporters from the other orders, especially the lower clergy, would be able to control the States General.

On 14 July 1789 the Parisians, hearing that Necker was dismissed and that royal troops threatened the capital, stormed the Bastille, an old fort prison viewed as a symbol of feudalism and tyranny. Louis XVI recalled Necker, and appointed Lafayette commandant of a newly created militia, the National Guard. Judging from the *cahiers de doléances*, in which people wrote down their grievances and asked for reforms, the main claims were for the abolition of class privileges, more equitable taxes, and a definition of subjects' rights as opposed to the prerogatives of the crown. A few hours' debate during the night of 4 August gave rise to the legislation of 11 August, which abolished the remnants of serfdom, made all feudal rights extinguishable by purchase, did away with seigniorial privileges and monopolies, put an end to inequality of taxation, and opened military and civil promotion to all classes. The Declaration of the Rights of Man and of the Citizen, voted by the National Constituent Assembly on 26 August, owed not a little to the American Declaration of Independence (1776), in so far as it laid down the 'natural and imprescriptible' right of every citizen, whose duty it was to share in the armed defence of the community, to liberty, equality, property and security. According to the doctrine of the sovereignty of the people and Montesquieu's principle of the separation of powers, the making of laws was delegated to a Legislative Assembly which was elected for two years. The royal prerogatives were limited (11 Sep) to a suspensive veto.

On 5 October, while the capital suffered from food shortage, market women set out for Versailles to petition the King for bread and compel him to accept the measures of 4 August. After breaking into the royal apartments, they obliged Louis XVI and the unpopular Marie-Antoinette to follow them back to Paris and take residence in the Tuileries. Besides the opposition of the King, the National Assembly, acting through a great many committees, had to face the passive resistance of local authorities as well as the refusal of the people to pay taxes. The Fête de la Fédération (14 July 1790), held in Paris to celebrate the anniversary of the storming of the Bastille, was an attempt to create a feeling of national unity. Discontent grew among the workers in the big towns. Wages failed to keep pace with the rise in prices, which reached a peak in 1789 and 1790. Many domestics and workmen were thrown out of employment by the emigration of wealthy people. The *loi Le Chapelier* (14 June 1791), by prohibiting all professional associations, workers' meetings and strikes, deprived the proletariat of any hope of improving their lot. But the worst problem that the National Assembly had to meet was the bankruptcy of the treasury. To overcome the financial crisis, it decreed (2 Nov 1789) that 'all church property' was 'at the disposal of the nation'. The lands and buildings of the French Church (representing one fifth of all such wealth in the country) were to be sold by auction, clerical salaries being paid by the treasury thereafter. When Pope Pius VI condemned the civil constitution of the clergy, to which French priests were expected to take an oath of allegiance, the Church was split between 'constitutional' clergy, receiving state support, and 'refractory' or 'non-juror' clergy, who were exposed to heavy penalties. The sale of the nationalised Church property was encouraged by the issue of *assignats*, government certificates entitling their holders to buy houses and land. The *assignats*, which were soon circulated as a paper currency, enabled France to avoid national bankruptcy (though their purchasing-power never stopped decreasing) until the loot of the countries conquered by the republican armies could replenish the treasury.

Louis XVI was drawn by the Queen into plans for a flight towards the Rhine frontier, from where the *émigrés* and the Emperor's troops would march on Paris, dismiss the Assembly and restore him to his throne. On the night of 20 June 1791, the King, disguised as a valet, and his family escaped from the Tuileries, to which they were ignominiously brought back five days later, after being recognised and stopped at Varennes. The flight to Varennes, together with the massacre of the Champ de Mars on 17 July, was a decisive factor in the fall of monarchy and the development of republican ideas. The Legislative Assembly, which replaced the Constituent in pursuance of the Constitution of 1791, contained a majority of new deputies, 'the Girondists' (so called because their leaders came from the Gironde), who, though republican idealists, championed the rights of property – significantly, the framers of the Constitution of 1791, for whom equality meant at most equality before the law, had limited the franchise to the property-owning classes. These deputies soon came into conflict with the 'Montagne' ('Mountain'), a group of Jacobins whose republicanism verged more and more on socialism, and who relied for their power on the Parisian mob. Because the crown was now known to be in sympathy with the *émigrés* and their foreign protectors, an armed attack on the Tuileries was mounted by the Parisian Jacobins (10 Aug 1792), and the royal family were immured in the old Temple as prisoners of the Paris Commune.

The next six weeks brought to the foreground an unprincipled demagogue of unusual powers, Danton: his call to arms sent thousands of volunteers to the front, as members of the new army which stopped the Prussian army at Valmy (20 Sep 1792) and pushed back the French frontiers to the Netherlands and the Rhine. It was also Danton who originated the famous Committee of Public Safety, which he invested with dictatorial powers. Since 10 August the revolutionaries had been demanding exemplary punishment for all the 'conspirators'. More than a thousand prisoners supposedly favourable to the King were sadistically butchered. Initiated by the Comité de Surveillance of the Commune, the September massacres were inspired by Marat, a former medical man, now a ruthless terrorist of the Left. To the leaders of the Mountain, men such as Robespierre and Saint-Just, Louis XVI was a traitor who from the first had been plotting against the Revolution. After a parody of a trial before the National Convention – which in its opening meeting on 21 September had proclaimed the Republic – the King was sentenced to death and guillotined in the Place de la Révolution (21 Jan 1793). Ten days later (1 Feb) the Convention declared war on the King of England and the Stathouder of Holland.

The adoption of the Jacobin Constitution of 1793 (24 June 1793) proved fatal to the Girondists, whose leaders, regarded as too liberal and suspected of betraying the country, were arrested. If some of them survived, many, such as Brissot and Vergniaud, were the victims of the proscription following Charlotte Corday's murder of Marat (13 July) and were guillotined (31 Oct). The dominating figures of the Committee of Public Safety, whose dictatorship soon became formidable, were successively Danton (until 10 July) and Robespierre (from 27 July). The authority of the Committee, which succeeded in controlling the administrative bodies, the generals and *commissaires* of the Republic, was based on its ruthless use of force: its local courts were allowed to inflict the death penalty; its travelling agents were empowered to dismiss or send to trial in Paris all those who were suspected of obstructing the government's policy; the revolutionary tribunal in Paris, whose public prosecutor Fouquier-Tinville stuck to a merciless interpretation of the laws against *incivisme* and *conspiration*, was a most efficient instrument of terror. Powerful as it was, the Jacobin regime had to fight constantly against the opposition of royalist agents, non-juror clergy, and Girondists intent on revenge. Even more than devotion to the monarchy, the attacks by the Convention on the priesthood and the enforcement of conscription caused the counter-revolutionary insurrection of the 'Chouans' in the west of France. Brutal ferocity on both sides characterised the wars of the Vendée (Mar 1793 – June 1796). Robespierre and his followers (Lazare Carnot, Billaud-Varenne, Collot d'Herbois) proved incapable of coping with the economic difficulties: bread and meat had to be rationed in Paris, and a scheme to ensure supplies by enforcing the sale of consumer goods at fixed prices (*lois du maximum*, May–Sept 1793) proved a failure. Between September 1793, when the Law of Suspects was passed, and March 1794, the number of executions rose every month. The 'Reign of Terror', during which many thousands (including Marie-Antoinette, Hébert and Danton) were put to death, reached a peak with the law of 22 Prairial (calculated to speed up the work of the revolutionary tribunal). Saint-Just's decrees of Ventôse (Feb–Mar 1794) aimed at distributing the confiscated property of suspects to poor citizens could not soothe the wave of popular discontent caused by the enforcement of the *maximum*. The moderates in the Convention, horrified by the excesses of the Terror, organised a movement against the Committee of Public Safety. On 26 July 1794, Robespierre delivered a fiery speech before the Convention, in which he attacked head-on all those who, according to him, discredited the Revolution. The next day (9 Thermidor) 'the Incorruptible' (as he had been nicknamed) was arrested, together with Couthon and Saint-Just. He was guillotined a few hours later.

The fall of Robespierre (28 July 1794) was followed by the partial suppression of the Committee of Public Safety and of the revolutionary tribunal, the emptying of the prisons, and the purging of the *sociétés populaires*, which had been hotbeds of Jacobinism. The Thermidorian reaction saw the emergence in the provinces of a counter-revolutionary terror. While Jacobins were being massacred, the royalists landed an *émigré* force in Quiberon Bay, which was annihilated by Hoche (21–22 July 1795). This republican victory led the Convention, which disbanded on 26 October 1795, to introduce the Constitution of the Year III, designed to prevent either dictatorship or democracy.

The government of the Directory, with its executive of five directors and its two chambers (the Anciens and the Cinq-Cents), lasted from November 1795 to November 1799. Babeuf, an eccentric Jacobin who had long dreamt of an equal distribution of wealth, was executed on 27 May 1797 for heading the *conspiration des égaux*. This was followed by an anti-Jacobin reaction. But the 'balance of power' policy did not survive. General Bonaparte, whose appointment as commander of the army of Italy (2 Mar 1796) was followed by spectacular victories, was soon convinced that the Revolution could not but end in a military dictatorship. With the *coup d'état* of 18 Brumaire (9 Nov 1799) ten years of Revolution gave way to Caesarism.

The appeal of the French Revolution, despite its disappointing conclusion, was enormous all over Europe. Goethe, who was present at Valmy, significantly noted in his diary, 'From this place and this day dates a new epoch in the history of the world.' In Britain the Whigs were neutral or sympathetic to the Revolution while the Tories were hostile to it. To Burke's denunciation of revolutionary change in his *Reflections on the Revolution in France* (1790) Thomas Paine replied with his *Rights of Man* (1791), in which he pleaded for universal suffrage and the abolition of monarchy. Paine's book was very popular and largely contributed to the growth of a lower-class radicalism whose tone and language were those of revolutionary France. It is not surprising that Pitt's government had several radical leaders arrested and asked Parliament to pass more stringent treason and sedition acts (1795, 1799). In literary circles, the French Revolution was at first immensely popular. 'Bliss was it in that dawn to be alive', Wordsworth later exclaimed in *The Prelude*. Blake gave vent to his revolutionary feelings in such 'prophetic books' as *The French Revolution* and *The Marriage of Heaven and Hell*. Southey and Coleridge wrote a crude drama entitled *The Fall of Robespierre*, and planned to establish in America an egalitarian community called Pantisocracy. Unlike the Romantic poets of the first generation (Wordsworth, Coleridge, Southey), whose early enthusiasm for the French Revolution was later superseded by a return to Toryism, Shelley and Byron remained radicals from first to last. Godwin's *Political Justice* (1793), which influenced Shelley, developed notions in harmony with the more extreme currents of feeling stemming from revolutionary France. *Prometheus Unbound* (1820), a lyrical drama, was the best expression of Shelley's revolutionary idealism. The French Revolution, of which Carlyle gave a partial but poetic view (1837), was a cataclysmic drama whose impact on English Romantic poets and men of letters proved considerable.

JEAN RAIMOND

German Literature

British interest in German literature rose steeply in the years around 1780, and remained high throughout the Romantic period. The *Scots Magazine* for 1799 declared that the 'Literature of Germany seems for some time to have taken the lead among the nations of Europe'. The response was by no means universally favourable; indeed the reputation of the leading German authors was not secure until the 1820s or 1830s. But the controversy surrounding them in the earlier period was perhaps a more genuine tribute to their power than their later Victorian respectability.

The British reception of German literature in the Romantic period has two distinct phases. In the first, lasting until 1813, the impact was made by writers of the so-called *Sturm und Drang* (Storm and Stress) movement: by the Goethe of the 1770s, the Schiller of the 1780s, by Bürger and his ballads in the popular manner. Following in their wake, but achieving a much more widespread success, both in Germany and England, were the best-sellers of the 1790s and early 1800s: August von Kotzebue's dramas, August Lafontaine's novels of domestic life, and the various tales of wonder and terror which rapidly made German literature synonymous with the sublime and Romantic. The second phase began in 1813 with *De l'Allemagne* by Madame de Staël (see GERMAN PHILOSOPHY AND CRITICISM), which shifted critical attention to the mature works of Goethe and Schiller. Of their younger contemporaries, only the Schlegel brothers (see GERMAN PHILOSOPHY AND CRITICISM), Johann Paul Friedrich Richter ('Jean Paul') and Friedrich de la Motte Fouqué (*Undine*, 1811) received much acclaim in Britain before 1830.

The seminal works of the first phase were the historical drama *Götz von Berlichingen* (1773) and the epistolary novel *Die Leiden des jungen Werther* (1774), both by Johann Wolfgang Goethe; Friedrich Schiller's play *Die Räuber* (1781) and his unfinished novel *Der Geisterseher* (1787–9); and Gottfried August Bürger's ballad 'Lenore' (1773). Each of these carried the anti-classical tendencies of the age to a new pitch. The 'refined sentiment and sensibility' of *Werther*, 'which has so much captivated the young and the romantic of every country it has reached' (*Scots Magazine*, 1790); the historical colour and derring-do of *Götz*; the extreme situations and the eloquence, 'impassioned and sublime', of *Die Räuber* (Henry Mackenzie,

Account of the German Theatre, 1790); the uncanny terrors of *Der Geisterseher*, of the ride with the ghostly lover in 'Lenore' – each in its own way helped to crystallise Romantic forms and feelings across Europe, both through their own continuing influence and through the numerous imitations to which they gave rise.

Hazlitt's reaction was typical of that of the British Romantics in the first flush of enthusiasm: 'How eagerly I slaked my thirst of German sentiment . . .; how I bathed and revelled, and added my floods of tears to Goethe's The Sorrows of Werter, and to Schiller's Robbers' (*Complete Works*, ed. P. P. Howe, 1930–4, XII, 226). The young Southey declared he would 'hardly be satisfied till I have got a ballad as good as *Lenora*' (letter to C. W. W. Wynn, 1799). The young Coleridge asked, 'My God . . ., who is this Schiller, this convulser of the heart?' (letter to Southey, Nov 1794) and addressed a sonnet 'To the Author of the "Robbers"' (published 1796); he later (1800) translated the second and third parts of Schiller's *Wallenstein* trilogy. *Werther* was translated seven times between 1779 and 1807; Walter Scott's was one of two translations of *Götz* in 1799; *Die Räuber* had four translations between 1792 and 1800, and *Der Geisterseher* two, in 1795 and 1800; 'Lenore' was translated six times between 1795 and 1800, once by Scott (see Violet A. A. Stockley, *German Literature as Known in England 1750–1830*, 1929; and E. Purdie, 'German Influence on the Literary Ballad in England during the Romantic Period', *Proceedings of the English Goethe Society*, N. S., 3, 1926). Many ballads, novels and plays in the 'wild, irregular, and romantic' German taste (*Monthly Mirror*, 1800) were inspired by these works, directly or indirectly. The two authors of note most deeply under the German influence were Walter Scott and Matthew ('Monk') Lewis. Lewis's works from *The Monk* (1796) onward were imbued with the 'extravagant horror of the German school' (*European Magazine and London Review*, 1803), which descends from the most 'Gothic' elements of *Götz*, *Die Räuber* and *Der Geisterseher*; in later years, he was one of the first and most ardent champions of Goethe's *Faust* (Part I, 1808). Scott's interest in German literature was more short-lived than Lewis's, but it was intense at an early and crucial stage of his development. Between 1796 and 1801 he translated not only *Götz* and 'Lenore', but also further ballads by Bürger and Goethe, and four other German plays, including Schiller's historical drama *Die Verschwörung des Fiesco zu Genua* (1783), which, Scott says, he would read 'to sobbing and weeping audiences' (G. H. Needler, *Goethe and Scott*, 1950). The immediate impression can be seen in confessedly imitative works such as *Glenfinlas* or the 'half-mad German tragedy' *The House of Aspen* (Scott to Lady Abercorn, 17 May 1811). Greater things are adumbrated in Scott's praise of *Götz* as a work 'in which the ancient manners of the country are faithfully and forcibly painted' (preface to his translation, quoted in Needler, *Goethe and Scott*), a remark which seems to point the way to the Waverley novels. Both *Götz* and *Waverley* are elegies for 'the ancient manners' of a more heroic age.

Until 1813, Goethe was known in England chiefly for his *Götz* and *Werther*, and Schiller was still the author of *Die Räuber* and *Der Geisterseher*. Both had the allure of somewhat dangerous, possibly immoral genius. Their transformation into monuments of respectability began with Madame de Staël, who introduced her public to the works of their maturity – Schiller's later dramas and their lofty idealism, Goethe's plays in the classical style, his *Faust*, and his novel *Wilhelm Meisters Lehrjahre* (1795–6). What de Staël began was completed by Carlyle, who in the 1820s became the most influential British propagandist of the German classics. He translated *Wilhelm Meisters Lehrjahre* in 1824, and its sequel, *Wilhelm Meisters Wanderjahre*, in 1827. An even more important landmark was his hero-worshipping *Life of Schiller* (1823–4). By these efforts, British critical opinion was converted to a view of Goethe and Schiller as classic authors and as apostles of the noblest human ideals. The marmoreal calm which now descended was lightened only by the popularity of Jean Paul's quirky fictions, which found champions and translators in De Quincey and Carlyle; the latter's *Sartor Resartus* (1833–4), heavily influenced by Jean Paul, can be considered a late offshoot of German rather than English Romanticism.

KEVIN HILLIARD

German Philosophy and Criticism

'We suspect that German philosophy is at present the noblest in Europe; and we are sure that German criticism is at present the best.' The reviewer in the *London Magazine* (1820, p. 66) gave voice to a sentiment which had been on the increase since the beginning of the century. It was through and in the writings of Immanuel Kant, Johann Gottlieb Fichte, Friedrich Wilhelm Joseph Schelling, Friedrich Schlegel, August Wilhelm Schlegel and others that the Romantic movement became most fully conscious of itself, of its religious, metaphysical, aesthetic and literary principles. English Romanticism, too, incurred a debt to German theory.

The principal relevant works are as follows (in each case, usual English title after colon).

Kant, *Kritik der reinen Vernunft* (1st edn 1781; 2nd edn 1787): *Critique of Pure Reason*

Kant, *Kritik der praktischen Vernunft* (1788): *Critique of Practical Reason*

Kant, *Kritik der Urteilskraft* (1790): *Critique of Judgement*

Fichte, *Über den Begriff der Wissenschaftslehre* (1794): *On the Concept of the Science of Knowledge*

Fichte, *Grundlage der gesamten Wissenschaftslehre* (1794–5): *Foundations of the Science of Knowledge*

Schelling, *System des transzendentalen Idealismus* (1800): *System of Transcendental Idealism*

Schelling, *Über das Verhältnis der bildenden Künste zu der Natur* (1807): *On the Relation of the Fine Arts to Nature*

F. Schlegel, *Geschichte der alten und neuen Literatur* (1815, tr. 1818): *History of Ancient and Modern Literature*

A. W. Schlegel, *Über dramatische Kunst und Literatur* (1809–11, tr. 1815): *On Dramatic Art and Literature*

To this list should be added Mme de Staël's *De l'Allemagne*, which, published in London in 1813 and translated into English in the same year, made the ideas of the German Romantics accessible to a wider public for the first time, and popularised the distinction between classical and Romantic art; book III was entirely devoted to German philosophy. The main British mediators were Coleridge (see below), Henry Crabb Robinson (who studied in Jena in 1802–5 and published articles in the *Monthly Register*, 1802–3), Carlyle (see below), and, to a lesser extent, Hazlitt (review of de Staël in the *Morning Chronicle*, 1814) and De Quincey (essays in various periodicals in the 1820s and 1830s).

'There is no doubt that Coleridge's mind is much more German than English' (Henry Crabb Robinson, letter to Mrs Clarkson,

29 Nov 1811). Much the same has been said about his writings; even in his own lifetime, Coleridge had to defend himself against the charge of plagiarism (see *Biographia Literaria*, ch. 9; also *Anima Poetae*, 1895, p. 89). He had studied Kant thoroughly by 1801, and put his studies to use especially in *The Friend* (1808), *Essays on the Principles of Genial Criticism* (1814), the 'Logic' manuscript (1822) and *Aids to Reflection* (1825) (see René Wellek, *Immanuel Kant in England 1793–1838*, 1931, pp. 65–135). The influence of Schelling (especially the *System*) is particularly marked in *Biographia Literaria*, chs 12–13; 'On Poesy or Art' (1818) is 'little more than a paraphrase' of Schelling's *Über das Verhältnis der bildenden Künste zu der Natur* (René Wellek, *A History of Modern Criticism, 1750–1950*, II: *The Romantic Age*, 1955). Portions of A. W. Schlegel's lectures *Über dramatische Kunst* were absorbed into Coleridge's own lectures on Shakespeare (1811–12). He thus became the only British critic to espouse and share the view of poetry developed by the German Romantics.

> If the artist copies mere nature, the *natura naturata*, what idle rivalry! . . . Believe me, you must master the essence, the *natura naturans*, which presupposes a bond between nature in the higher sense and the soul of man. . . . In the objects of nature are presented, as in a mirror, all the possible elements, steps, and processes of intellect antecedent to consciousness . . .; and man's mind is the very focus of all the rays of intellect which are scattered throughout the images of nature. ('On Poesy or Art', in *Biographia Literaria*, ed. J. Shawcross, 1907, II, 257–8)

This is quintessential philosophical Romanticism. The lesson the younger Germans drew from their (mis-)reading of Kant was that the given, immediate world of sense perception was not the ultimate reality, but the product of prior operations of *Geist* (mind or spirit). Nature, apparently alien to us, was in reality a form of spirit, a symbolic universe, which would reveal its more than human wisdom only to philosophical speculation or imaginative poetic divination. This philosophy, which placed great stress on truths inaccessible to empirical reasoning, seemed for a while to lend new credence to religious belief, and Coleridge eagerly seized upon this aspect of German thought. In aesthetics, the German Romantics helped him to articulate his unease about Wordsworth's (theoretically) naïve naturalism and develop his own theory of the poetic imagination (*Biographia Literaria,* chs 1–12). The figure of Mr Flosky in Peacock's *Nightmare Abbey* is a satire on Coleridge's leanings towards German philosophy. To the strictly philosophical influences should be added his early study (in Göttingen, 1799) of the German, historicist school of biblical scholarship (Lessing, Herder, Michaelis, Semler, Eichhorn; see E. S. Shaffer, *'Kubla Khan' and the Fall of Jerusalem,* 1975).

Carlyle's reception of German thought was narrower, more eclectic and (on a popular level) more influential than Coleridge's. His early views were conditioned by de Staël's book. Not until 1826 did he turn to Kant himself (giving up, however, after 150 pages of the *Kritik der reinen Vernunft*); Fichte he studied only in the popular work *Das Wesen des Gelehrten* (1805). The fruit of this reading is found chiefly in *Wotton Reinfred* (1827), the essay on Novalis (*Foreign Review,* 1829), and the 'philosophy of clothes' of *Sartor Resartus* (1833–4). Carlyle followed Coleridge in using German philosophy to justify religious belief and to combat materialism: the world of the senses was an insubstantial garment through which, at times, a higher reality could be apprehended. Man participated in this higher realm through moral action, undertaken for its own sake, not for utilitarian reasons (a doctrine derived from Kant's ethics).

Whatever its actual influence, German Romantic thought has often seemed to hold out the promise of a deeper, theoretical understanding of Romantic literature in general. In this sense, Henry Crabb Robinson was able to speak of the 'German bent' of Wordsworth's mind (*Diary, Reminiscences and Correspondence,* ed. Thomas Sadler, 1869, I. 482), even though Wordsworth knew very little about German philosophy and criticism. Similarly, A. C. Bradley argued that 'in many of the thoughts' of the Germans 'we feel the presence of the same spirit that spoke in English [Romantic] poetry' (*English Poetry and German Philosophy in the Age of Wordsworth,* 1909). Arguments for the national idiosyncrasy of the various Romantic literatures (A. O. Lovejoy, 'On the Discrimination of Romanticisms', 1924) have checked this synoptic tendency; but in recent decades the juxtaposition of German philosophers and English poets has become commonplace in histories of the period.

KEVIN HILLIARD

Godwin, William (1756–1836)

William Godwin was born in Wisbech, in Cambridgeshire, of dissenting stock. His father attended the dissenters' academy at Northampton, and William was educated at the dissenting academy at Hoxton, a fine institution that in Joseph Priestley's words was 'exceedingly favourable to free enquiry'. William left Hoxton in 1778, a dedicated Christian, and spent five years as a candidate minister serving rural congregations. He was in faith a Calvinist, in politics a Tory.

By the time he had published his major work, *Enquiry Concerning Political Justice* in 1793, and following the French Revolution, he had become an atheist and a passionate rationalist, believing in human perfectibility and rejecting the idea of inherent natural rights upon the argument proposed by Tom Paine that 'every generation is, and must be competent to all the purposes which its occasions require' (*Rights of Man*). An avowed republican, Godwin believed that, by the education of the poor, vice and misery would vanish as natural justice asserted itself.

In the early 1790s he enjoyed a relatively brief, though remarkable, popularity. *Political Justice* earned him the enmity of the Prime Minister, Pitt, and the praise of Coleridge, who addressed Godwin in the sonnet 'O form'd t'illumine a sunless world forlorn' (1794). Crabb Robinson declared that 'no book ever made me feel more generously', while Wordsworth used Godwin as a guide not only to political and social, but also to personal, conduct. During 1794–5 there was frequent personal contact between Wordsworth and Godwin.

By 1796 Godwin's star was waning. Coleridge and Wordsworth rejected his views of reason as a guiding light to perfect justice, though personally remaining his friends. In his letter to 'Caius Gracchus' (*The Watchman*, no. 5, 2 Apr 1796) Coleridge wrote, 'I do consider Mr. Godwin's principles as vicious; and his book as a Pander to Sensuality.' As others' views developed, however, Godwin remained essentially firm to his rationalist principles to the end of his life.

In 1797 Godwin married Mary Wollstonecraft, author of *A Vindication of the Rights of Woman* (1792). The marriage lasted less than a year, Mary dying shortly after the birth of their daughter Mary, later Mary Shelley. Through his family, the aging Godwin became associated with many of the younger generation of Romantic poets. P. B. Shelley admired Godwin even when the philosopher's reputation was at its lowest, first writing to him in January 1812. His visionary and ideological poem *Queen Mab* (1813) is unmistakably Godwinian – man is not born, but made, evil by the pernicious influence of Church and state. Shelley pays tribute to Godwin in the original dedication of *The Revolt of Islam* (1817) to Mary, and *Prometheus Unbound* (1820) continues to show his influence.

Relations between the two men were strained, but never broken, by the elopement and subsequent marriage of Mary and Shelley, following the suicide of his first wife. Mary Shelley's novel *Frankenstein* (1818) owes much to her father's *Political Justice*, with its doctrine that men's characters originate in their circumstances and also to his *Caleb Williams* (1794), with its nightmare world of flight and pursuit.

After the death of his first wife, Godwin married, in 1801, Mary Jane Clairmont, whose daughter by her first marriage, Claire, bore a child, Allegra, to Lord Byron.

Godwin continued to write throughout his life. His novel *Things as They Are: or, The Adventures of Caleb Williams* was warmly praised by Hazlitt. It is an early example of the propagandist novel, a gripping tale of mystery and pursuit and a psychological study showing 'the tyranny and perfidiousness exercised by the powerful members of the community against those who are less privileged than themselves'. Godwin's original preface was omitted after the first edition for political reasons. In the same year as *Caleb Williams* (1794) he published *Cursory Strictures*, defending twelve radicals, including Horne Tooke and Thomas Holcroft, against charges of treason. He also wrote several other novels, including *St Leon* (1799), *Fleetwood* (1805), *Mandeville* (1817), *Cloudsley* (1830) and *Deloraine* (1833); and lives of Mary Wollstonecraft (1798) and Geoffrey Chaucer (1803–4). In 1831 he published a series of essays entitled *Thoughts on Man*, written in the 'full maturity' of his understanding, and showing how the philosophy of *Political Justice* remained fundamentally intact at the end of his life. Essentially utilitarian and necessitarian, a believer in rationalism and human perfectibility, Godwin affirmed that 'human understanding and human virtue will hereafter accomplish such things as the heart of man has never yet been daring enough to conceive'. He remained an atheist to the end.

DAVID JASPER

Goldsmith, Oliver (1728–74)

Goldsmith was the second son of an Irish clergyman. After spending his childhood years at Lissoy, County Westmeath, he attended Trinity College, Dublin, from 1744 till 1749. He was a disorganised and irregular student, and his later medical studies at Edinburgh and Leiden were finally abandoned for a life of wandering about Europe, earning his meals and a bed by playing his flute for his peasant hosts.

In 1756 Goldsmith returned to London to work as an apothecary and a schoolmaster. Finally he settled in London as a general literary hack, translating, reviewing and contributing essays to various periodicals. In 1759 he briefly wrote a periodical paper of his own (modelled on *The Spectator*) entitled *The Bee*.

The publisher Newbery eventually took charge of Goldsmith's chaotic finances, and for his newspaper *The Public Ledger* Goldsmith wrote his first significantly original pieces: the series of essays collected in 1762 as *The Citizen of the World*. These follow a well-known convention by bringing a foreign visitor (in this case Chinese) to London so that he can comment on English manners and values as an outsider.

Goldsmith became a member of Dr Johnson's circle and many of his appearances at meetings of 'The Club' are chronicled by Boswell, who seems, however, to have missed the point of much of Goldsmith's very Irish verbal fooling, and to have cast him (very misleadingly) in his *Life of Johnson* as the jester or court fool of the group. Johnson was instrumental in arranging the publication of Goldsmith's prose tale *The Vicar of Wakefield* (written by 1764), which appeared in 1766. The gentle humour and sentimental charm of this story made it an immediate favourite; and this, together with the success of Goldsmith's topographical, largely autobiographical poem *The Traveller* (1764), made him a celebrity. Goldsmith lived well and helped others generously: consequently his finances were usually in confusion, necessitating endless hack work as a historian, encyclopaedist and editor. But Goldsmith also did more enduring work in the last years of his life. His best-known poem, *The Deserted Village*, appeared in 1770, crystallising nostalgia for a vanishing rural innocence with sharp criticism of those responsible for heartless social change and deprivation of the poor. This latter facet of the poem was to be profoundly influential on both Crabbe and Wordsworth.

Goldsmith's first stage comedy, *The Good-Natured Man* (1768), though too derivative and soft-centred to last, revealed a command of stage situation and lively dialogue which reached maturity in *She Stoops to Conquer* (1773), a play whose solid, entertaining plotting and ability to fuse affectionate irony, warm-hearted sentimentality and wit of every kind, from the farcical to the sparkling (rather in the manner of Restoration comedy), have ensured its survival on the stage to the present day.

Goldsmith's hack work and much of his occasional writing are now forgotten, but at his best, in certain of his essays, he was a unique combiner of light, elegant wittiness with delicate sentimentality. In his plays he shows the same sure grasp of character, and *The Vicar of Wakefield* displays a range of rich character studies which delighted the Romantic generation. Goldsmith learned much from Steele: he was a master of the contemporary vogue for sentimentally involving his readers with the dilemmas of his characters. But he never pushes himself between the story and the reader in the manner of Sterne's narrators.

As a critic Goldsmith was too liable to be swayed by his own emotions – essentially too kindly to be dispassionate. *The Vicar of Wakefield*, *The Traveller* and *The Deserted Village* all raise contemporary issues such as imprisonment for debt, the excessive power of the aristocracy and the effect on the poor of rural enclosures, but Goldsmith's most effective criticism works by involving the reader's sympathy for victims rather than through the power of his argument. At the same time, his best criticism is firmly rooted in social and economic realities: his influence can therefore be felt very widely in the period, from *The Man of Feeling* to *Lyrical Ballads*. Goldsmith's capacity to incorporate felt personal responses within the value structure of formal poetry provides a bridge between Gray's 'Elegy', with its private basis for moralising, and such poems as Wordsworth's 'Michael' and 'The Last of the Flock', which express a strong sense of personal sorrow at a tragedy rooted in public wrongs.

WILLIAM RUDDICK

The Gothic Novel

When applied to fiction the word 'Gothic' is used to designate the eighteenth- and early nineteenth-century novel of terror dealing with mediaeval material, whether Romanesque or Gothic. The Gothic novel developed from the novel of sensibility, carrying that form a stage further by introducing elements of fear and suspense. Edmund Burke's *Philosophical Enquiry into the Origin of our Ideas of the Sublime and Beautiful* (1757) was of decisive importance. In his treatise Burke conferred on terror a major and worthwhile literary role: 'whatever . . . operates in a manner analogous to terror is a source of the sublime'. Thanks to Burke, the excitation of fear became one of the most significant enterprises a novelist could undertake. Thus the Gothic novels employed the agencies of mystery and awe, magic and the supernatural, to arouse terror. Gothic fiction is a literature of nightmare.

It is generally accepted that the Gothic novel began with Horace Walpole's *Castle of Otranto* (1764). Originally the book, published anonymously, consisted of some 150 pages. The subtitle 'A Gothic Story, by Horace Walpole, Earl of Orford' appeared only in the second edition. Walpole set his ghost story in the Middle Ages and brought together the various elements that we now identify as typical: the Gothic machinery, the atmosphere of gloom and terror, and the stock Romantic characters of Gothic tyrant (Manfred) and persecuted heroine (Matilda). But the most marvellous element is the haunted castle, which Walpole makes the main character of the narrative: a very impressive mediaeval castle, with a room containing ancestral portraits and an underground area full of mysterious vaulted passages with trapdoors and secret stairways. Walpole was the father not only of Gothic fiction, but also, it might be argued, of Gothic drama, since in 1768 he wrote *The Mysterious Mother*, the first play in a genre as distinct as that of the Gothic novel. It was followed by a host of other plays, by such distinguished writers as John Home, Hannah More, Richard Cumberland and, in the nineteenth century, Shelley. By means of his Gothic story and drama, and his own Gothic castle at Strawberry Hill, Walpole 'reversed the popular conception of the word "Gothic". He changed it from an adjective of opprobrium into an epithet of praise' (D. P. Varma, *The Gothic Flame*, 1957). But his main achievement was to restore to literature the irrational and the supernatural.

It was not until thirteen years after Walpole's Gothic tale that Clara Reeve (1729–1807) published a literary 'offspring' of *The Castle of Otranto*: *The Champion of Virtue*, better known as *The Old English Baron* (1777). In this 'picture of Gothic times and manners' she combined Richardsonian sensibility with Gothic trappings. The extensive use of dream material helped the author to explore the darker and more disturbing side of human behaviour. When Edmund, the young hero, is asked to spend three nights in the haunted apartment of Lovel Castle to prove his courage, he is assailed by prophetic dreams, but the ghost, 'a man in complete armour', appears in person only when the villains, Sir Walter Lovel's nephews, are on watch. The author's use of the supernatural is so moderate that *The Old English Baron* was dismissed by Walpole as 'a Gothic story, professedly written in imitation of Otranto, but reduced to reason and probability'. Clara Reeve's great contribution to the Gothic novel was to make possible supernatural fiction which does not do violence to human reason.

But the heyday of Gothic fiction was the last decade of the eighteenth century, and an immediate precursor of Ann Radcliffe and M. G. Lewis was Sophia Lee (1750–1824), who wrote *The Recess, or a Tale of Other Times* (3 vols, 1783–5), a story in the manner of Abbé Prévost's *Cleveland*. The title of this semi-historical romance is derived from a subterranean retreat or 'Recess' within the precincts of an old abbey 'of Gothic elegance and magnificence'. It is the story of Matilda and Ellinor, the unfortunate daughters of Mary, Queen of Scots, by a private marriage with the Duke of Norfolk. It is obvious that Sir Walter Scott was greatly influenced by *The Recess* when writing his great historical novel *Kenilworth*. But Lee's direct influence on Gothic fiction is very important too. For instance, the incidents of *The Recess* are introduced by the device of the recovered manuscript, which was to become so common in the Gothic-romance writers. On one of Matilda and Ellinor's first excursions from the Recess their torch flickers out, and they are terrified by the figure of a headless knight on a tomb. The awful aspect of the ruins, which they see on emerging, also affects them deeply. Without *The Recess* the new direction taken by Ann Radcliffe would have been unthinkable.

Mention must also be made of Charlotte Smith (1749–1807), who was first and foremost a poet. Her sonnets had many admirers, including Wordsworth (who commended their feeling for nature, and whom she received in Brighton on his way to France in 1791) and Coleridge (whom she entertained at her London house in February 1800). She was also the first 'Gothic' novelist to combine the attraction of a sentimental love affair with that of landscape. Her characters often find themselves in picturesque surroundings, but also spend part of their time in grottoes, churchyards and subterranean passages. She was fond of Gothic castles and her beautiful description of Rochemorte in *Celestina* (1790) was the obvious model for the castle of Udolpho in Ann Radcliffe's romance. Charlotte Smith was particularly felicitous in her choice of titles, which may have inspired Walter Scott. Examples include *Emmeline, the Orphan of the Castle* (1788), *Ethelinde, or the Recluse of the Lake* (1789), and *The Old Manor House* (1793).

Thus Gothic fiction was well established by the time of Ann Radcliffe and Matthew Gregory Lewis. The novels of Mrs Radcliffe (1764–1823) in particular are in many ways a development from Walpole's *Castle of Otranto*. The very title of her first romance, *The Castles of Athlin and Dunbayne*, published in one volume in 1789, suggests this. The two-volume *Sicilian Romance* (1790) owes a great deal to Lee's *The Recess*, while in the three-volume *Romance of the Forest* (1790) the heroine, Adeline, is haunted by prophetic dreams, like Edmund in Clara Reeve's *The Old English Baron*. With her fourth novel, *The Mysteries of Udolpho, Interspersed with Some Pieces of Poetry* (4 vols, 1794), Mrs Radcliffe established her reputation. It was a tremendous success, to such an extent that the author was called 'the mighty magician of Udolpho' (by Thomas James Mathias) and 'the Shakespeare of Romance writers' (by Dr Nathan Drake). *The Italian, or the Confessional of the Black Penitents* 3 vols, 1797) was the last novel that she published in her lifetime (*Gaston de Blondeville*, written in 1802, appeared posthumously in 1825).

In her first five romances Mrs Radcliffe placed her Gothic castles and abbeys in beautiful romantic landscapes, adding to the imaginary buildings of her predecessors an atmosphere of ruin and desolation. And she made skilful use of the climatic conditions of storm, windblasts and thunder, and other aids to poetical atmosphere such as moonlight. Against this background with its mysterious setting we find the traditional Gothic villain – in most cases of Italian origin (Montoni in *The Mysteries of Udolpho*, Schedoni in *The Italian*); the persecuted maiden (Emily, Ellena); and the young hero (Valancourt, Vivaldi). Supernatural agencies are used to create an atmosphere of terror; and, while in Mrs Radcliffe's fiction most of the mysteries are explained away in the end, the 'explained supernatural' does not play so important a part in *The Italian* as in the earlier books. Her romances, particularly the last three, contain passages of poetic prose, describing 'sublime' scenic impressions or heightening moments of terror, that are more suggestive and evocative than comparable passages in her previous novels. This may be the reason why she was hailed by Walter Scott as 'the first poetess of romantic fiction'.

Matthew Gregory Lewis (1775–1818) was himself very much influenced by Mrs Radcliffe's *Mysteries of Udolpho*. In *The Monk, a Romance*, published anonymously in March 1796, he moved Gothic fiction towards the demonic and the sadistic. The book was extremely successful, and he earned the nickname 'Monk' Lewis. According to Eino Railo (*The Haunted Castle*, 1927), Lewis 'was not a direct, puristic follower of Walpole's school, but the welder into English form of a coarse, German primitive type of romance'. The castle of Lindenbergh is a faithful rendering of a Gothic castle with 'its ponderous walls tinged by the moon . . ., its old and partly ruined towers lifting themselves into the clouds' and the vaults of St Clare's convent have secret passages, narrow and uneven steps and a prison 'with a sepulchral lamp . . . that shed a gloomy light through the dungeon.' Nature plays a far less important part in *The Monk* than in Mrs Radcliffe's work, with the notable exception of the site of Ambrosio's death, which is truly 'sublime'. Instead of beautiful descriptions of nature we find pictures of voluptuous passion, for which Lewis's book was branded immoral. Ambrosio is a lustful and criminal monk who yields to the temptations offered by Matilda, a most beautiful and fatal woman, a kind of *belle dame sans merci* who turns out to be a female demon. Lewis's contribution to the development of what has been called 'ghost romanticism' was to introduce ghosts. The ghost of Antonia's mother, Elvira, come to warn her daughter of approaching death, is already very impressive, but the climax of terror is reached in the scene where the spectre of the Bleeding Nun enters Raymond's bedroom on the stroke of midnight, sits on the edge of his bed and begins to stare at him with hypnotic, fear-inducing eyes. In comparison with this scene, the apparition of Lucifer as, first, a 'youth seemingly scarce eighteen' is not so striking, though Ambrosio is duly terrified when Lucifer reappears 'in all the ugliness, which since his fall from heaven had been his portion'. Like Walpole,

Lewis was the author of a Gothic drama, *The Castle Spectre, a Dramatic Romance* (1797), and of many other Gothic works.

A disciple of Mrs Radcliffe and 'Monk' Lewis, the Revd Charles Robert Maturin (1780–1824) was an Irish clergyman who diverted himself by writing Gothic romances, historical novels and tragedies. He seems to have crowded into his first novel of terror, *The Fatal Revenge, or the Family of Montorio* (1807), nearly every character and incident that had been employed in his predecessors' narratives. Schemoli is a replica of Mrs Radcliffe's Schedoni, and Erminia di Vivaldi is named after the young hero of *The Italian*. The background shifts from the robber's den to the ruined chapel, from the castle vault to the dark passages, the trapdoors and the dungeon of the Inquisition. And at the end the mysteries are explained away. But Maturin's masterpiece is *Melmoth the Wanderer* (1824), which is a collection of tales all dealing with the wanderings and temptations of Melmoth, an Irish Faust figure who has sold his soul to the Devil and is then given a reprieve of 150 years to find someone to change places with him in the afterlife. Like *Montorio* it owes much to the romances of Radcliffe and Lewis. Immalae is another Emily St Aubert, and Melmoth is a mixture of Lewis's Wandering Jew, Ambrosio, Matilda and Marlowe's Dr Faustus. Yet 'it is a wilder, more complicated, and, in many ways, more daring work than most of its predecessors' (Robert Kiely, *The Romantic Novel in England*, 1972). Published some twenty years after *The Italian*, *Melmoth* is a late, if not the last, Gothic novel.

Besides the writers already mentioned, there were hundreds of others who turned out Gothic romances. Most of them are now quite forgotten, including the authors of the 'horrid books' mentioned by Jane Austen in *Northanger Abbey* (1803), a delightful parody of the genre. 'For more than fifty years after the publication of *The Castle of Otranto* the Gothic romance remained a definitely recognised kind of fiction' (E. Birkhead, *The Tale of Terror: A Study of the Gothic Romance*, 1921). Its influence on Romantic writers was incalculable. In Britain no novelist, playwright or poet could escape it, as is evident from Coleridge's 'Rime of the Ancient Mariner' and 'Christabel', Southey's ode 'To Horror', 'The Ruined Cottage', *Thalaba* and *Madoc*, Byron's *Prisoner of Chillon*, and Keats's 'The Eve of St Agnes'. Shelley's youthful enthusiasm for the tale of terror led him to write *Zastrozzi* (1810) and *St Irvyne; or The Rosicrucian* (1811), while his wife Mary produced *Frankenstein, or The Modern Prometheus* (1818). A Gothic atmosphere pervades several of Scott's novels and, later, those of the Brontës and Charles Dickens. On the Continent, the influence of Gothic romance on the early Balzac, on Nodier and Victor Hugo, in France, and on E. T. A. Hoffmann, in Germany, is well known. In the United States Gothic fiction found a most fertile soil, from the novels of Charles Brockden Brown (1771–1810) and the terror tales of Edgar Allan Poe (1804–49) to William Faulkner's *Sanctuary* (1931) and *Absalom, Absalom!* (1936), and the fiction of Truman Capote. Finally, the emergence during the early twentieth century of the new narrative medium of film allowed the effects and aims of late-eighteenth and early-nineteenth-century writers of Gothic fiction to be realised as never before.

PIERRE ARNAUD

Gray, Thomas (1716–71)

Thomas Gray was born in London, in relatively humble circumstances, but because of family connections he was educated at Eton (c.1725–34), where his closest friends were Horace Walpole (the Prime Minister's son), Richard West and Thomas Ashton: they formed themselves into a close-knit artistic group nicknamed 'The Quadruple Alliance'. All but West went on to Cambridge, and, shortly after leaving the university with the intention of studying law, Gray accepted an invitation to accompany Walpole on the Grand Tour to Italy. Their letters home to West describing their crossing of the Alps and the Grande Chartreuse are key documents in the development of a rhetoric of the sublime in eighteenth-century travel literature.

After two years abroad Walpole and Gray quarrelled at Reggio, and Gray hurried back to England alone. His friendship with West grew closer as they criticised each other's verses in Latin and English, but in June 1742 West died, just as Gray was sending him his first mature poem, the 'Ode on the Spring'.

West's death and Gray's sense of his own financial and social difficulties now inspired a succession of poems rooted in a sense of personal melancholy tightly controlled and intensified by his grasp of classical forms. These still constitute the basis for Gray's poetic reputation. The 'Ode on a Distant Prospect of Eton College', the 'Sonnet on the Death of Richard West' and the 'Ode to Adversity' were written in 1742. Gray lived quietly with his family and at Cambridge, residing at his old college, Peterhouse, until 1756, and at Pembroke Hall for the rest of his life. He completed his legal studies but never practised: he read voraciously and studied widely, declaring that study relieved his sense of constitutional melancholy.

In 1745 Gray and Walpole were reconciled. They shared anti-quarian interests, and Walpole was instrumental in arranging for the publication of more than one of Gray's poems. Gray wrote little verse, but the publication of his 'Elegy Written in a Country Church-Yard' in 1751 brought him lasting fame.

In the early 1750s Gray abandoned the reflective mode for a variety of dramatic, violently emotional experimental odes based on his reading of early English, Welsh and Scandinavian history and literature. The *Odes*, published by Walpole at his Strawberry Hill Press in 1757, caught the new mood for 'Gothic' situations and emotions ('The Bard' was to be one of the most frequently illus-trated mid-eighteenth-century poems). Thereafter Gray concen-trated on his historical and botanical studies. He was massively learned, but made no attempt to publish any of his researches. In 1768 he was appointed Regius Professor of Modern History at Cambridge, but lacked the courage to lecture.

Despite his shyness, Gray enjoyed several firm friendships, producing delightful and sometimes sharply satirical letters full of news and comment for his own circle. He also wrote a few satirical verses, inspired by contemporary Cambridge politics, which are among the later masterpieces of the eighteenth-century satirical mode: the lines 'On Lord Holland's Seat' of 1768 are worthy of Pope or Johnson at their sharpest.

In his later years Gray travelled each summer, visiting mountain scenery in Scotland and the Lake District, as was becoming fashion-able in the 1760s. The letters which he wrote to Warton describing his Lakeland visit in 1769 are remarkable for their vivid sensitivity, both to new aesthetic experiences and to the actual physical fea-tures of the landscape. The letter journal (published in 1775 in Mason's *Life*) was rightly praised by Wordsworth.

Gray's remarkable ability to express strong personal feelings (coming close to a sense of desolation and absolute anguish at times) within formal verse structures derived from classical conventions has earned him the praise of modern critics and won him the sympathy of his early readers, though in *The Lives of the Poets* Dr Johnson expressed dislike of both his melancholy and his poetic experimentation (except in the 'Elegy'). The Romantics, however, thrilled to the emotional melodrama, the strong, dark reading of historic forces and the portrayal of the tragic loneness of the poet in Gray's later odes. The lonely poet of the 'Elegy' was provided with appropriately sublime surroundings and given fresh appeal for the Romantic generation by James Beattie's poem *The Minstrel*, a work which Gray read in manuscript and criticised acutely. Gray's letters show him to have been one of the best literary critics of the mid-eighteenth century.

WILLIAM RUDDICK

Greece and the Philhellenes

Greece came under the domination of the Turkish Ottoman Empire by degrees in the fifteenth century. After a long period of acquiescence, during which the country was often a theatre of war between the Turks and the Venetians, the Greeks began gradually to rediscover a distinctive culture and identity. This process accelerated in the eighteenth century, with the founding of new schools and academies, and was given significant impetus by a regular flow of sympathetic travellers from abroad – men such as the politician and educationalist Frederick North, future Earl of Guilford – who, through their books and diaries, encouraged across Europe a belief in the possible regeneration of the modern Greeks as true descendants of the ancient Greeks, heirs to the spirit and virtues of a lost civilisation. The importance of Lord Byron in furthering this upsurge of 'love for Greece', the phenomenon known as philhellenism, cannot be too heavily stressed. His visit to Greece with J. C. Hobhouse in 1809 and 1810 produced the first two cantos of *Childe Harold's Pilgrimage* (published in 1812), and subsequently the series of Grecian–Turkish tales *The Giaour* (1813), *The Bride of Abydos* (1813) and *The Siege of Corinth* (1816).

With the appearance of *Childe Harold*, which made Byron a celebrity overnight, literary philhellenism became a widespread movement, everywhere echoing the poet's romantic theme of the ruin and hoped-for rebirth of a classical splendour:

> Fair Greece! sad relic of departed worth!
> Immortal, though no more; though fallen, great!
> Who now shall lead thy scatter'd children forth,
> And long-accustom'd bondage uncreate?
>
> *(Childe Harold*, II.lxxiii)

Similar sentiments are expressed as late in Byron's career as the famous 'Isles of Greece' stanzas in canto III of *Don Juan*, and also, with a more militant edge, in his translation of the 'Hymn' of the Greek patriot Rhigas (1757–98), 'Sons of the Greeks! let us go / In arms against the foe'.

This interest in Greece, which was shared by historians and art-collectors (the sculptures removed from the Acropolis by Lord Elgin were purchased for the British Museum in 1816), played a crucial part in creating the conditions for the Greek revolt against Turkish rule in 1821. The main responsibility for transforming philhellenic ideals into a political programme belongs, however, to the Greeks themselves, and especially the colonies of expatriate Greeks in Europe, who through organisations such as the Philiké Hetaria, set up in 1815 with centres at Moscow, Bucharest and Trieste, campaigned vigorously and in practical ways for the cause of national independence. The Greek War of Independence which followed the uprising of 1821 divides broadly into three phases. To begin with (1821–4), the Greeks, supported by European volunteers, were largely successful against the forces of the Sultan, but in the second phase (from 1824) the army of Mehemet Ali, Pasha of Egypt, reversed the tide of war in favour of the Turks. Finally, the intervention of the European powers in 1827 led to the establishment of an independent kingdom of Greece in 1833. Among the English Romantics the salient response to the outbreak of hostilities was to be found in Shelley. Whereas Keats's poetry reflected the pure influence of Greek myths, the Elgin marbles, Chapman's Homer, and Lemprière's *Classical Dictionary*, Shelley, who was even more profoundly steeped in Greek art and thought, offered also, in his poetical drama *Hellas* (1821), a direct reaction to immediate events which blended ideas of freedom inherited from the

French Revolution and from millenarian tradition – 'The world's great age begins anew / The golden years return' – with, not least in the Preface, a classic statement of philhellenist concepts and feelings: 'We are all Greeks. Our laws, our literature, our religion, our arts have their root in Greece.' Comparable voices were raised in France, Germany and elsewhere, and, despite the ambivalence of the reactionary governments of Europe towards the liberal exponents of the Greek cause – especially noticeable in the case of the restored Bourbon dynasty in France, which feared revolutionary attitudes though, like others, it saw the drawbacks of standing aside and the danger of potential Russian expansion in the south – expeditions, sponsored by individuals, societies and Greek committees, came out of various countries in the course of the war. Of these, the most extraordinary was the so-called Byron Brigade.

The War of Independence produced an array of fascinating combatants on the Greek side: Baleste, the first commander of the Greek army; the Bonapartist Colonel Fabvier; General Normann of Württemberg; and the British philhellenes Sir Richard Church, General Thomas Gordon and Captain Frank Abney Hastings. It is Byron, however, who stands out in the collective memory, and who most conspicuously bridges the spheres of action and idealism. Lord Guilford built up a Hellenic University in Corfu; Gordon preceded another philhellene, George Finlay, as chronicler of the revolution and of modern Greece; but Byron the poet died Byron the aspiring soldier. Byron went to Greece as one of the two appointed agents of the London Greek Committee, arriving at Missolonghi early in 1824 in scarlet uniform and with full ceremony. His plan seems to have been to unite the several factions and groups of the Greeks and their supporters for an assault upon the strongholds still in Turkish hands. This came to nothing, and the effect of his presence was rather to stimulate a romantic enthusiasm for the struggle which led to an influx of international volunteers to the Byron Brigade itself, whose members ranged from genuine fighting men to dreamers such as Edward John Trelawny who saw himself as a protagonist in a Byron tale. The brigade continued in existence only briefly after Byron's death from illness in April 1824; but in its own way his passing served the Greek cause as forcibly as any military exploit, for it was quickly formed into a popular and inspirational myth of self-sacrifice, giving the poet an enduring place in the pantheon of modern Greek heroes.

The philhellenes remained an active element in the war right through to its final stages, when, in a situation increasingly fraught with uncertain fortunes and civil conflict, the great powers, after years of diplomatic wrangling, agreed by the treaty of London in July 1827 to intervene to stabilise and 'save' Greece. The victory of the allied British, French and Russian fleets over the Turks and Egyptians at Navarino on 20 October 1827 effectively decided the issue. The Turks could not win: Greece was free. In 1834 King Otho conferred the Order of the Saviour on many of the philhellenes, and a monument to those who died was built in 1841 at the church in Nauplia. A register of around 400 names was subsequently compiled by Hilarion Touret and Henri Fornezy, a Swiss, but the numbers of European and American philhellenes in Greece were in truth much greater. Their story has been told in several recent books, including Douglas Dakin's *British and American Philhellenes* (1955), C. M. Woodhouse's *The Philhellenes* (1969), and above all William St Clair's *That Greece Might Still Be Free: The Philhellenes in the War of Independence* (1972), an exhaustive and elegant treatment which extends even to illustrations exemplifying how the war fired the imagination of Delacroix and other contemporary painters.

VINCENT NEWEY

Hazlitt, William (1778–1830)

William Hazlitt was the son of a Unitarian minister, whose strong radical sympathies he adopted in childhood and never gave up. In 1783–4 the Hazlitt family were in America, where Hazlitt senior preached at Philadelphia and Boston. On his return they settled at Wem, in Shropshire, where Hazlitt was first educated. At fifteen he went to the Unitarian college at Hackney to prepare for the ministry. There he developed a strong taste for philosophical studies and began what was eventually to be published in 1806 as his *Essay on the Principles of Human Action*.

Hazlitt gave up all thoughts of the ministry and returned home to study and write in about 1797. In January 1798 he met Coleridge, who had come to preach a series of trial sermons at Shrewsbury; Coleridge's friendship, and those with Wordsworth and Charles Lamb which grew out of it, transformed Hazlitt's life, as he recorded years later in his essay 'My First Acquaintance with Poets'.

For a time Hazlitt studied painting with his elder brother John (a pupil of Reynolds) and he executed several portraits, as well as making copies in the Louvre during the Peace of Amiens in 1802–3. He also read widely and pursued his philosophical studies. In 1808 he married Sarah Stoddart, a friend of the Lambs, and at his wife's cottage at Winterslow, near Salisbury, he began to write steadily. The most notable work of his early years is the *Memoir of Thomas Holcroft* (based on personal acquaintance and the writer's papers), which eventually appeared in 1816.

In 1812 the Hazlitts, with their only son, moved to London, and Hazlitt began to give public lectures, first on philosophy and later on literary topics. These were generally published shortly after being given. Hazlitt wrote parliamentary reports for the *Morning Chronicle* for some time, and then gained some celebrity with his theatrical criticism (beginning with a remarkable series of reports on Edmund Kean's first London appearances) in the same newspaper. Contacts with other newspapers followed, together with the important connection with John and Leigh Hunt, in whose periodical *The Examiner* he first began to appear regularly as an essayist.

Between 1815 and his death in 1830, Hazlitt wrote enough essays to fill several volumes, besides the published lectures, book reviews, art criticism and other work, covering an extraordinarily wide range: all were distinguished by a constant flow of ideas, sharply focused, original judgements, spontaneity and freshness of style, and a real concern to identify and fix the imaginative character of the artists, authors, politicians and other public figures of whom he wrote.

In the late 1810s and early 1820s, Hazlitt (often directly inspired by the ideas of Lamb or Coleridge) produced his finest treatments of the drama, Elizabethan and seventeenth-century literature, and the early novelists, whose work he loved. The published lectures succeeded one another in rapid succession. The series *Characters of Shakespeare's Plays* (1817) was followed by *Lectures on the English Poets* (1818), *Lectures on the English Comic Writers* (1819) and *Lectures on the Dramatic Literature of the Age of Elizabeth* (1821). At this time Hazlitt also began to collect his occasional essays together into volumes.

An unhappy love for a much younger woman was recorded in *Liber Amoris, or the New Pygmalion* (1823), one of the most striking English confessional works of the Romantic period, but one which damaged Hazlitt's contemporary reputation. He married again soon afterwards and drew on his wife's money to travel on the Continent, sending home material which was collected in *Notes of a Journey through France and Italy* (1826). At this period he also embarked on a vast compilation, his *Life of Napoleon Buonaparte*, intended to counteract the Tory viewpoint of Scott's and other biographies, and this work was only completed as he died.

Hazlitt was a member of the brilliant group of writers recruited by John Scott for the *London Magazine* at the beginning of the 1820s. Like his friend Charles Lamb he became one of the great Romantic exponents of the art of memory, vividly recreated within the brief compass of the familiar essay. *Table Talk* (1821–2) shows him at the height of his powers as a communicator of the individual viewpoint. *The Spirit of the Age* (1825) contains some of the most brilliant criticism (and polemic) of the Romantic age.

The Spirit of the Age

Hazlitt's series of essays on his contemporaries first appeared in magazines (especially the *New Monthly*) during 1823–4 and was collected in a single volume towards the end of 1824, though the title page gives the date as 1825. The Galignani edition, which

Hazlitt supervised in 1825, rearranged the material slightly and added a fine essay on George Canning.

Hazlitt's subjects range from such arch-Tory legal dignitaries as the Lord Chancellor, Lord Eldon, to notable philosophical figures and reformers such as William Godwin and Jeremy Bentham. They, together with several other political figures, including Horne Tooke, Canning, Sir James Mackintosh and Malthus, represent the generation (generally closer in age to Hazlitt's father than to himself) who, from their diverse ideological standpoints, fought the great political battles of Hazlitt's childhood in the opening years of the 1790s, when enthusiasm for the French Revolution gave way to widespread approval for Pitt's policy of repression and his attempts to silence any sources of radical opinion. Hazlitt brilliantly portrays the intellects of several who were prominent under the shadow of Burke's fame and influence in the early 1790s. He also provides memorable portrayals of leading intellectual figures from the years that followed.

At no point in *The Spirit of the Age* is Hazlitt's understanding of the intellectual, political and cultural forces which he groups under the title phrase stated directly and comprehensively. But it is clear from repeated comments that Hazlitt has in mind a strong 'force for change' hostile to the pre-1789 political order and apparent both in the words of those who desire to further it and in the deeds of those who wish to suppress it. The new force is democratic and progressive; it appeals to the imagination and human experience and not to systems. It rejects privilege of all kinds; it has hopes and visions for the human race.

Hazlitt's subjects are measured against the new forces. Wordsworth's poetry illustrates them, though his own life does not. Walter Scott's novels teem with examples of the *Zeitgeist*, and his Tory politics are powerless to modify their effect. Coleridge, the source of inspiration for other searching minds, has lost his way amid his own researches. Jeremy Bentham fails because, though he sees the need for democracy, his system-building lacks any understanding of real human life.

The essays in *The Spirit of the Age* show Hazlitt's mind at its most brilliantly interpretative and his prose style at the highest point of its development. A characterisation emerges from the recording of a handful of childhood traits ('The Right Hon. George Canning') and is developed through a flow of assertions and impressions which derive from them. Paragraphs build up through a succession of assertive phrases or short sentences to establish a characterisation of remarkable solidity and power. Or, at times, the prose will move into a sustained flow of impressions as remarkable as any of the great prose-poetic structures of De Quincey: the essay on Coleridge shows this technique at its most brilliantly sustained and original. Hazlitt's early training in philosophical logic also stands him in good stead in several of the more closely argued essays. That on Malthus affords a classic demonstration of hard reasoning made immediately comprehensible and convincing for a popular audience.

WILLIAM RUDDICK

Hogg, James (1770–1835)

James Hogg was born in Ettrick, an area of the Scottish Borders, and became a shepherd there. As a young man, he collected folk poems and ballads, which he sent to Walter Scott for *The Minstrelsy of the Scottish Border*. Scott, who mentions Hogg in the notes to 'Auld Maitland' in the *Minstrelsy*, encouraged Hogg as a writer; Hogg's first book, a collection of his ballad poems, was published in 1807 under the title of *The Mountain Bard*. He became known as 'the Ettrick Shepherd', and subsequently moved to Edinburgh, where he became a contributor to *Blackwood's Magazine* and a noted man of letters. He retired to a farm at Yarrow, in the Scottish Border country, in 1816. His best-known works are *The Queen's Wake* (1813), a series of poems supposedly recited by various poets to Mary, Queen of Scots, at Holyrood Palace; and, above all, *The Private Memoirs and Confessions of a Justified Sinner* (1824).

The Private Memoirs and Confessions of a Justified Sinner is a brilliant exercise in morality and the macabre. It deals with an antinomian, one who believes that the 'elect' of God can do no wrong, and that because he is 'justified' he can commit no crime. The story is of two brothers, Robert and George Colwan: Robert is brought up by a narrow-minded minister, and 'taught to pray twice every day, and seven times on Sabbath days; but he was only to pray for the elect'. George is brought up by his father, the Laird of Dalcastle, and becomes a good sportsman and a brave and friendly young man. He is followed and persecuted by Robert, and finally murdered by him in mysterious circumstances.

The book narrates these events, and then retells the story from another point of view in Robert's 'Confessions': it is thus an original and experimental work which has complex narrative perspectives and different points of view. Robert meets an acquaintance who is very attractive to him because he speaks the language of the elect, and who comes to resemble him in appearance. He is the Devil, or Robert's second self, his *Doppelgänger*; he causes Robert to murder a good minister, Mr Blanchard, and subsequently his brother. After the murder Robert takes possession of the estates of Dalcastle, but he is pursued and possessed by the Devil.

In a short final section, the story is concluded, one hundred years later, by an account of Robert's suicide (as remembered in local tradition), of the exhumation of his body, and of the discovery of the manuscript of his confessions. His skull is found to be of an 'almost perfect rotundity, with only a little protuberance above the vent of the ear' (i.e. horns).

The story of such devilish temptation suggests the influence of the Gothic novel. However, the novelty of the work lies in the importance of the 'second self', which opens vistas on the subject of double personality and subconsciousness. It is also a devastating portrayal of an extreme Calvinism which led to intolerance and self-righteousness, of the kind which had prompted Burns to write of the 'unco guid'.

JACQUES BLONDEL

Hunt, James Henry Leigh (1784–1859)

Leigh Hunt was educated at Christ's Hospital (1791–9). His early poems, *Juvenilia*, were published in 1801. A selection of his pieces as drama critic for the *News* was published in book form in 1808. In the same year, with his printer brother John, he began to edit *The Examiner*, which became the voice of liberal and reforming opinion, as well as a platform for literary talent. He also briefly edited *The Reflector* from 1810. Both brothers were imprisoned (1813–15) for libelling the Prince Regent, but Leigh was permitted to have his family with him and to continue to edit *The Examiner*. He was visited by friends, including Byron, Moore and the Lambs.

While living at the Vale of Health, Hampstead, in 1816 he ripened his friendship with Shelley and Keats, introducing them to one another. Hunt published Keats's sonnet 'O Solitude' in *The Examiner* in May, and in December 'On First Looking into Chapman's Homer'. His praise in an article, 'Young Poets', initiated his lasting campaign to gain recognition for them. His own most important poem, *The Story of Rimini* (about Paolo and Francesca), was published in that year, and was followed by the volumes *Foliage* (1818), *Hero and Leander* and *Bacchus and Ariadne* (1819). At first, Hunt's encouragement and delight in 'the realms of gold' were invaluable to Keats. Later Keats felt that the failures of taste and the self-indulgent 'lusciousness' in Hunt's poetry had been a damaging influence on his own; but, after some coolness, Keats found refuge, friendship and nursing with the Hunts in the summer of 1820 shortly before leaving for Rome. The literary periodical *The Indicator*, edited by Leigh Hunt (1819–21), published Keats' 'La Belle Dame sans Merci' in 1821.

Formerly frequently persecuted for their political protests against oppression, the Hunt brothers now had to contend with attacks on their literary probity. In 1817 in *Blackwood's Magazine*, J. G. Lockhart, hiding behind the pseudonym 'Z', launched a campaign of quite extraordinary virulence, misrepresentation and sheer snobbish spite on what he called 'the Cockney School of Poetry'. His targets were particularly Leigh Hunt, Hazlitt and Keats. In the *Quarterly Review* Gifford waged a similar war. In the Preface to *Adonais*, Shelley – mistakenly – attributed Keats's death to this damning criticism.

In 1821 Shelley wrote from Italy that Byron 'proposes that you should come and go shares with him and me in a periodical Work' and pressed Hunt to come. He arrived with his wife and seven children early in July 1822; on 8 July Shelley was drowned. Byron and Hunt produced *The Liberal*, with *The Vision of Judgment* in the first number and contributions by Byron, Shelley, Hazlitt, Hunt and Hogg (among others) in the three following numbers; but their relationship was far from happy and in 1825 the Hunts returned to England. During the separation Leigh and his brother John had also broken off their collaboration in serious disagreement, mainly about money.

Leigh Hunt seems to have been always impecunious and improvident but it was not for want of hard work. After his return from Italy he edited, and wrote much of, *The Companion* (1828), a new *Tatler* (1830–2), and *Leigh Hunt's London Journal* (1834–5). In *The Tatler* he praised at length the young poets Alfred and Charles Tennyson and in his last instalment judged Alfred the greater of the two. Hunt's best-known poem (with 'Jenny kissed me'), 'Abou Ben Adhem', was published in an anthology, the *Book of Gems*, in 1838. His play *A Legend of Florence* was put on at Covent Garden in 1840 and was well received. Hunt's other works include the anti-war poem *Captain Sword and Captain Pen* (1835), his *Poetical Works*, and, in prose, *Imagination and Fancy* (1844), *Wit and Humour* and *Stories from the Italian Poets* (1846), *Men, Women and Books* (1847), *A Jar of Honey from Mount Hybla* and *The Town* (1848), *An Autobiography* (1850), *Table Talk* (1851), *The Old Court Suburb* and a version of Beaumont and Fletcher (1855). In 1847 he received a civil-list pension.

In his essays his object was 'to reap pleasure from every object in creation' and particularly from art, music and literature. His enjoyment is infectious, but he lacks the depth of the greater contemporary essayists. Though suffering from the piecemeal way it was constructed, originating in *Lord Byron and Some of his Contemporaries* (1828), Hunt's *Autobiography* of 1850 (revised 1859) is still of great interest, and the idealism and charm of his personality come over in it. Great harm was done to Hunt when his friend Dickens used 'the light externals' of his character for Skimpole in *Bleak House*. At Thornton Hunt's request, after his father's death, Dickens made a public denial that it had even occurred to him that 'the admired original would ever be charged with the imaginary vices of the fictitious creature'. Hunt's greatest importance lies in his editing, at first as fearless champion of freedom, later in his remarkable ability to recognise and nurture literary talent in others and often to discern correctly the comparative stature of his contemporaries.

MARY WEDD

Imagination

All Romantic artists were fascinated by the function and workings of the imagination, and throughout the eighteenth century emphasis on its creative powers grew, in line with the increased attention paid to the creative subject. Three essential assertions recur in most writings of the period: the centrality of the subject; the communicative power of imagination; and man's ability to turn imagination into an autonomous order of existence, thus making utopia the prime motive for human progress, and 'symbol' a specific mode of reading the world. But such praise for the imaginative power can also be viewed as the concealment of a sense of loss in a universe felt as overwhelmingly complex.

When the subject is no longer seen as the mirror of a stable outer world, imagination is defined in terms of interaction rather than reflection; both perceiver and perceived are in the making, and partake of the same vital dynamism. Blake even claims that 'Nature is Imagination itself', denying the poet's dependence on external objects. But the main impulse towards a new systematic inquiry into the workings of imagination came from German philosophers. Kant's analysis of perception as an active synthesis in space and time, and Schelling's insertion of the imaginative act within the infinite flux of life strongly influenced Coleridge and hastened his rejection of empiricism. In his *Biographia Literaria*, Schelling's views are interwoven with his own insights as a poet and thinker. Coleridge's definitions of imagination have remained landmarks in the history of criticism: he claims that, even at the lowest perceptual level, a 'primary' imagination is at work, as 'the living Power and prime Agent of all human Perception', and a 'repetition in the finite mind of the eternal act of creation in the infinite I AM'. As for the artist's 'secondary' imagination, it is a heightened process, 'an echo of the former, co-existing with the conscious will', and differing from the 'primary' stage 'only in degree, and in the mode of its operation. It dissolves, diffuses, dissipates, in order to recreate. . . . It is essentially vital' (ch. 13). Such definitions are obviously not meant for practical criticism: their aim is to situate the artist's creative power within the vital impulse pervading both common perception and its willed, highly elaborate 'echo' – the artist's vision. This focus on will and internal growth makes it impossible to reduce artistic creation to a mere process of copying existing objects. Hence Coleridge's distinction between 'imagination' and 'fancy', the latter being only 'a mode of memory' working with 'ready made materials', whilst imagination necessarily implies a profound disturbance, a 'dissolving' and radical questioning of experience, as in the repeated symbolic occurrences of 'The Rime of the Ancient Mariner' or 'Kubla Khan', with their successive contradictory visions of the same referents. Similar questionings are also to be found in Wordsworth's poetic narratives of crises ('blank desertion') when our familiar surroundings lose their everyday naturalness to become unutterably alien, and a new imaginative recognition emerges; many poignant episodes of *The Prelude* are devoted to such subliminal states where the relation between 'inner' and 'outer' is not to be taken for granted, but is felt as an imaginative leap. All Romantic poets tried to render the unstable, flickering quality of such moments of 'self-exposition' (Coleridge), when the mind is like a 'fading coal' (Shelley's metaphor). Whatever the metaphor, its tenor is always a momentary revelation, akin to religious experience.

The lurking danger of solipsism arising from the emphasis on individual imagination is counterbalanced by a constant concern for the communication of the artist's imaginative experience. Both Wordsworth and Coleridge, positing a universal, 'primary' imagination, refer to a common ground between poet and reader. Wordsworth's definition of the poet as 'a man speaking to men', in the Preface to *Lyrical Ballads*, is echoed by Shelley's pages on the artist as 'prophet', in *A Defence of Poetry*; Keats, by likening imagination to a process of assimilation, 'distillation' or vital nourishment, also asserts its capacity to refine the consciousness of the whole human community. Thus imagination is never equated with an individual mode of escapism. Wordsworth and Shelley even believe that it can also handle the data of contemporary science. Moreover, such experiences as Wordsworth's 'spots of time' or Keats's 'moments of intensity' involve the reader in subconscious modes of communication, in experiences of imaginative synthesis only utterable in 'symbolic' terms, requiring a specific 'education' of the self in order to be understood.

These views point to a potential imaginative evolution or revolution, leading to new modes of free 'uncircumscribed' life, as envisioned in Shelley's *Prometheus Unbound*, for example. Whether political or poetic, Romantic 'utopia' involves imagination as an agent of integration, capable of organically binding all levels of consciousness, of attuning man to the 'life of the universe' (Coleridge). Keats's depiction of death in terms of sensuous enjoyment and Shelley's vision of a future humanity 'exempt from awe' exemplify the utopian striving to overcome anguish or guilt.

Yet such triumphant assertions of the power of imagination may also be seen as a way of exorcising their opposite, i.e. a deep awareness of the potential dangers of an increasingly utilitarian society, stifling the imaginative impulse, fragmenting and dissociating the human psyche. It may be significant that the 'abysmal' versions of Romantic imagination should exert a kind of fascination on twentieth-century readers: De Quincey's 'Piranesian' spaces, Wordsworth's sense of loss and Coleridge's compulsive repetitions of his mariner's awesome visions bear witness to an undercurrent of anxiety and tend to undermine belief in the unifying power and 'sublime' energy of 'imagination'. Hence the present tendency to detect 'irony' in most Romantic statements on imagination. Byron paved the way for this when bringing down to earth the monumental illusions or seriousness of the previous generation. Yet even Byron's ironic distance in *Don Juan*, for example, may be viewed as a triumph of imagination – of the kind that 'dissipates . . . in order to recreate'. From Blake to Byron, the word 'imagination' implies a will to struggle against any tendency to reduce the world to 'fixities and definites'. Whether seriously utopian or ironically lucid, 'imagination' in Romantic literature no longer labels a 'faculty', but signifies 'transitory flashes' of experience, when man attempts to overcome the otherness of the world or to contemplate his own limits.

DENISE DEGROIS

The Industrial Revolution

The term 'Industrial Revolution' is widely and rather vaguely used to describe a complex period of social, ergonomic and economic change from around 1750 to around 1830. Its principal characteristic was the invention and application of means to produce goods more efficiently and cheaply. This process was facilitated by periods when capital was available for the building of machines and factories. It also coincided with a steep rise in population (caused principally by a reduction in the rate of mortality), which provided more markets, more mouths to feed, more employees for a workforce, and attendant problems of housing and economic welfare.

The first signs of change from an agricultural economy and individual production were found in early attempts to improve textile machinery, and in Thomas Newcomen's beam engine of 1708, used to pump water out of coal pits. At almost the same time, Abraham Darby began to use coke smelting to produce pig-iron, which led to improved metal-working in iron and steel. Ironworks on a large scale were concentrated in areas such as the west Midlands, south Wales, and the Clyde valley, where the great Carron ironworks began operation in 1760.

In textiles, James Hargreaves of Blackburn during the 1760s invented a simple spinning-machine, or 'jenny'. Richard Arkwright's frame of 1768 enabled the cotton to be woven, and in 1771 Arkwright set up a factory at Cromford, near Derby, driven by water power. Samuel Crompton's 'mule' followed, and the production of cotton textiles became regularised and mechanised. James Watt's development of the steam engine (1765–9) released the factory-owners from a dependence on water power, and allowed them to set up factories in places where there were no rivers, or with slow-flowing water.

With the coming of steam power to the manufactories, many workpeople could be brought under one roof; and the transport of the goods which they had made became easier with the improvement of roads and the construction of canals. The Grand Junction Canal, for example, opened in 1777, allowed Josiah Wedgwood's Etruria works near Stoke-on-Trent a convenient means of transport for its raw materials and for its products. Other canals were dug, mainly in the north and the Midlands, and London was reached in 1790, enabling large loads to be moved cheaply, and relatively quickly. Road improvements, chiefly by Thomas Telford and John Loudon Macadam, also facilitated communication and the movement of goods. The application of steam power to moving engines or steam carriages was first attempted by Richard Trevithick in 1803; but it was not until the construction of the Stockton and Darlington railway in 1821 and of George Stephenson's *Locomotion* that a practical and useful railway engine entered service (the line had been constructed for horse-drawn wagons). Later, Stephenson's *Rocket* won the competition for an engine to run on the Liverpool and Manchester railway, and the 'railway age' began.

The increase in manufacturing on a large scale led to a steep rise in the demand for labour, and to the employment of women and children as well as men in the factories. Some factory-owners provided humane conditions of employment and good housing, but in other places the drift to the towns and cities to find work caused the development of poor housing, with inadequate sanitation. Thus physical conditions were often poor, although there is some evidence that (contrary to popular belief) the factory workers were better off in the cities than they would have been if they had remained in the country: in the textile industries, for example, workers were certainly more prosperous than the agricultural poor. However, there was a widespread sense of deracination, and of mechanisation, which had to be engaged with: workers had to come to terms with the new experience of living in cities, and with the repetitive processes of industrial production.

Francis Klingender's *Art and the Industrial Revolution* (1947, rev. 1968) describes the effect of these changes on art and literature. Klingender quotes from John Dyer's *The Fleece* (1757) describing

> Th'increasing walls of busy Manchester,
> Sheffield, and Birmingham, whose redd'ning fields
> Rise and enlarge their suburbs.

The factories and their processes were used as subjects by certain painters, most notably Philip James de Loutherbourg (1740–1812) and Joseph Wright of Derby (1734–97), both of whom were fascinated by the dramatic effects of furnaces and other heavy-engineering effects, especially when seen at night. Burns wrote at the door of the Carron ironworks,

> We cam na here to view your warks,
> In hopes to be mair wise,
> But only, lest we gang to Hell,
> It may be no surprise. . . .

Other Romantic writers referred to the Industrial Revolution, either in passing, as Shelley does in *Queen Mab*, referring to the early industry of nail-making, or as Keats does in 'Isabella' (stanza

XIV, later used by F. Scott Fitzgerald in *Tender is the Night*). Blake's 'satanic mills' in 'And did those feet in ancient time' are sometimes taken to refer to factories (although Blake also used the symbol of the mill as a grinder of the soul). The most sustained and serious treatment of the Industrial Revolution (until the protest of Elizabeth Barrett Browning in 'The Cry of the Children') was that of Wordsworth in book VIII of *The Excursion*. This passage (ll. 87–230) acknowledges the progress made by industry and applauds the power of science, but also describes the 'darker side' of the Industrial Revolution, picturing the building of factories, the change-over from day shift to night shift, the employment of women and children as well as men, and the subjection of human need to the profit motive. It is interesting to compare this section of *The Excursion* with James Thomson's description, written before the Industrial Revolution, of 'Industry, rough power!' in *The Seasons* ('Autumn', ll. 43–150).

J. R. WATSON

The Jacobin Novel

The English 'Jacobin novel' was a product of the 1790s, inspired by the events in France and fuelled by the controversies which followed Burke's *Reflections on the Revolution in France*. During these arguments, the supporters of the Revolution were branded as 'Jacobins' or extremists by their opponents; they accepted the name with pride. They were members of a loose group of sympathisers which included William Frend (Coleridge's tutor at Cambridge, who was sacked for his radical views and for his Unitarianism), John Thelwall, Richard Price, Thomas Holcroft, William Godwin, Mary Wollstonecraft and others; Wordsworth was an occasional member of the group at some point, certainly in 1795. They often met at Frend's house, or Godwin's, or at the house of Joseph Johnson, the radical publisher. The principal novelists of the group were William Godwin and Thomas Holcroft, and to them must be added the names of Elizabeth Inchbald, an actress and writer, and Robert Bage, who lived outside London. The most significant novels of the group were

Elizabeth Inchbald, *A Simple Story* (1791)
Thomas Holcroft, *Anna St Ives* (1792)
Robert Bage, *Man as He Is* (1792)
William Godwin, *Things as They Are; or, The Adventures of Caleb Williams* (1794)
Thomas Holcroft, *The Adventures of Hugh Trevor* (1794)
Robert Bage, *Hermsprong; or, Man as He Is Not* (1796)

The recurrence of the 'as he is' or 'as he is not' element in these titles is some indication of their preoccupation with human character and its relation to society. In dialogues and monologues, or other kinds of discussion, political and social questions are raised; and the characters are often examples of vice or virtue according to Godwinian principles. The novels often contain privileged tyrants, or parents who are preoccupied with class distinction; these are opposed by those, such as Hermsprong, who stand for natural goodness, courage and decency. Hermsprong has been brought up by the American Indians, which is an indication of the way in which the novels were sometimes 'philosophical' in their belief that character was formed by incidents and external circumstance (Gary Kelly, *The English Jacobin Novel, 1780–1805*, 1976). The novelists imitated Richardson and Fielding in their methods, but used the novel form to express their opinions on political and social questions.

Thus Elizabeth Inchbald's *A Simple Story* is an account of a young girl, Miss Milner, and of her upbringing and education. Its exploration of a young woman's love is remarkable, and it has been suggested that *A Simple Story* was an influence on *Jane Eyre* (S. R. Littlewood, *Elizabeth Inchbald and her Circle*, 1921). The beautiful and intelligent Miss Milner falls in love with her guardian, Dorriforth, later Lord Elmwood; the second part of the novel deals with her daughter, Matilda. In both parts there is a strong interest in the place and role of women, and in the needs of an emotional and intelligent person. As a novel it has been described as 'pre-Jacobin' (by Kelly), but in addition to its interest for a portrayal of a feminine intelligence *A Simple Story* showed the Jacobin novelists how to treat issues such as education and behaviour in society with psychological realism. Elizabeth Inchbald is best known for the play *Lovers' Vows*, which was the play chosen for the theatricals in Jane Austen's *Mansfield Park*; but she deserves to be more widely read.

Thomas Holcroft had been a shoemaker and stocking-weaver, among other trades. *Anna St Ives* is an epistolary novel after the manner of Richardson, exploring the conflict between virtue and wickedness with psychological realism. It is Holcroft's Godwinian thesis that passions can be overcome and made to give way to reason: the main plot describes Anna's rejection of Clifton Coke, and her preference for Frank Henley (the name 'Frank' is significant), but the novel also includes satirical attacks on a number of traditional evils. Holcroft's *The Adventures of Hugh Trevor* is a picaresque novel, again concerned to recommend the qualities of reason and reform. Its figures of Turl, the rationalist engraver, and Evelyn, the philanthropist, are contrasted with the corrupt and dissolute Lord Idford; while Hugh Trevor, the hero, has views that are similar to those in Godwin's *Political Justice*.

Godwin's own novel, *Things as They Are; or, The Adventures of Caleb Williams,* was also related to his political philosophy. In it, the virtuous and attractive Falkland kills the tyrannical and loutish squire Tyrrel, who has ruined his tenant Hawkins. When Tyrrel's body is found, Hawkins and his son are tried and executed. Caleb Williams, the self-educated and intelligent secretary to Falkland, suspects that the latter is Tyrrel's murderer: Falkland then begins to persecute Williams, and ensures that he is imprisoned and discredited. Eventually, however, Falkland is so influenced by Williams' goodness that he confesses his guilt (in an earlier, more remorseless version, Falkland continued to assert his innocence, and Williams continued in gaol). The Preface, not published in 1794, describes the book as 'a general review of the modes of domestic and unrecorded despotism, by which man becomes the destroyer of man'.

Godwin also describes the book as 'a study and delineation of things passing in the moral world'. This could serve as a description of many of the Jacobin novels, and very clearly of Robert Bage's *Hermsprong*. This novel is conspicuous for its good humour and amiable satire, although it has a serious purpose, and sets the noble, courageous and virtuous Hermsprong in opposition to the established society of eighteenth-century England. The chief representatives of this, fossilised in their establishment roles, are the tyrannical and vicious Lord Grondale, and the sycophantic parson, Dr Blick; and, as with Elizabeth Inchbald, the principal women are portrayed as sympathetic, clever and intelligent. Caroline Campinet is gentle and sweet-tempered, and Maria Fluart resourceful and witty. Miss Fluart is a triumph of the novelist's art, a figure who anticipates the splendid Marian Halcombe in Wilkie Collins' *The Woman in White*; while Hermsprong himself is a conspicuous example of 'man as he is not', at least not in the England of the 1790s which the radicals were trying to change.

J. R. WATSON

Keats, John (1795–1821)

The youngest of the major Romantic poets, John Keats was born on 31 October 1795 in London, into a fairly well-to-do lower-middle-class family. The accidental death of his father in 1804, and the long illness of his mother, who died of tuberculosis in 1810, impressed him with the acute sense of the precariousness of life which haunts his poetry. For most of his boyhood and adolescence, Keats's real home was the liberal and progressive Enfield Academy where his passion for reading was encouraged and directed by the headmaster's son, Charles Cowden Clarke, who did more than any other person to influence the future course of the poet's life. In 1811 Keats left school to be apprenticed to a surgeon of Edmonton, Thomas Hammond; in October 1815 he registered as a student at Guy's Hospital, and in July 1816 was licensed to practise as an apothecary and surgeon. In confirming the empirical bent of his mind, Keats's medical training helped to determine the particular complexion of his poetic genius: a yearning for the ideal was balanced by a strong sense of the actual.

While still at school, Keats had become a reader of Leigh Hunt's weekly, *The Examiner*, and espoused its editor's liberal politics and literary tastes. Since 1814 he had been attempting to write verse; the first of his poems to be published, the sonnet 'O Solitude', appeared in *The Examiner* for 5 May 1816. Cowden Clarke, a personal acquaintance of Hunt's, showed him some of Keats's manuscripts, and arranged a meeting between the two men (Oct 1816). Hunt's encouragement, and his public praise of Keats in *The Examiner* for 1 December 1816, probably determined him to give up medicine and to devote himself to poetry. His brief creative life was a heroic effort to achieve the goal he had set himself, to be 'among the English Poets' after his death (*Letters*, ed. H. E. Rollins, 1958, I, 394) – a phrase which expresses not only his ambition but also his sense of the continuity of English culture, his passionate love of its language, and his desire to revitalise it by returning to its early triumphs in Chaucer, Spenser and Shakespeare. The result of

this ardent dialogue with the 'great of yore' was a poetic development of unparalleled extent and rapidity.

Keats's first published collection (*Poems 1817*), dedicated to
Leigh Hunt, is largely derivative. His second volume, the 'poetic
romance' *Endymion* (1818), for all its profusion of memorable images, partly deserves its author's severe censure: 'a feverish attempt, rather than a deed accomplished'. The third and last
volume, *Lamia, Isabella, The Eve of St Agnes, and Other Poems* (1820),
contains some of the greatest poems in the language.

Keats's primary gift was an intense sensitivity to natural beauty
and the ability, which conscious effort helped him to perfect, to
create suggestive verbal analogues of sensuous experience. But he
was also endowed with a searching intellect whose restless questionings soon led him to perceive the shortcomings of his first
mentor, and spurred him on to aspire to the high seriousness of
epic poetry. To prepare himself for the task, he read Milton and
Dante (in Cary's translation, 1814) and, with his friend Charles
Brown, went on a strenuous walking-tour of the Lake District and
Scotland, with the express purpose of loading his memory with
sublime images of natural grandeur (June–Aug 1818).

The two decisive years in his brief creative career (1818–19) were
saddened by his brother George's emigration to America (June
1818), and by the illness and death of his brother Tom (1 Dec 1818),
while in the autumn of 1818 Keats's dedication to poetry found a
rival in his passion for Fanny Brawne, to whom he became engaged in December 1819. Their marriage plans were thwarted by
money problems and by the onset of Keats's last illness. The
tuberculosis he had probably caught while nursing Tom began to
move into its active stages in the autumn of 1819; severe haemor-
rhages occurred in February and June 1820. Attended by his friend
the painter Joseph Severn, Keats left for Italy on his physicians'
advice in September 1820 and died in Rome on 23 Feb 1821.

The whole of Keats's *oeuvre* may be approached as a spiritual
autobiography, the record of the passionate quest of a sceptical
mind for the certainties it could not find in any received doctrine,
philosophical or religious. Its very unassertiveness, its freedom
from any 'palpable design' on the reader, is one of the qualities
which have kept it from ageing. Keats's response to the beauty of
nature or art was analogous, in intensity and effect, to religious
contemplation; and his celebration of beauty, once he had left
behind his early Huntian escapism, was conceived as a cure for
'despondence' (one of the key words of *The Excursion*, whose

author, Wordsworth, Keats both admired and criticised), the *mal du siècle* generated by the decline of religion and the collapse of revolutionary ideals. The fervour of *Endymion* proceeds from the feeling that beauty must be the sensuous token of ultimate truth, the temporal manifestation of the unknown reality that gives meaning to the world of experience. Keats's growing awareness of natural evil soon undermined this early aesthetic idealism, which the great odes of 1819 look back upon with a mixture of scepticism and nostalgia. But even this critical aestheticism was relinquished in *The Fall of Hyperion*, in which the relationship between beauty and truth is finally severed. The spiritual vacuum created by this dissociation was filled by an anguished yearning for 'knowledge', the metaphysical knowledge that could alleviate the 'agony' of ignorance.

Most critics now accept the reality of this evolution, but some view Keats's poetic progress as a series of vacillations between polar opposites: 'negative capability' versus the desire for 'knowledge'; the amoral detachment of the 'camelion Poet' versus the ambition to achieve a definite poetic and ethical identity. Some would even argue that the symbolic drama of *The Fall of Hyperion* (1819), where these polarities are debated, is sketchily foreshadowed by 'Sleep and Poetry' (1816). Others feel that undue emphasis on Keats's interest in ideas, so conspicuous in the remarkable *Letters* (ed. H. E. Rollins, 1958), obscures the true character of his poetic gift, his uniquely rich sensuousness. The most fruitful approach, perhaps, is that which does full justice to the two sides of Keats's genius, and recognises the importance of the creative tension between 'Sensations' and 'Thoughts' (*Letters*, I, 185) which in effect helped him to secure a permanent place 'among the English Poets'.

Endymion: A Poetic Romance

Keats's first long narrative poem, begun in mid-April 1817 (*Letters*, I, 134) was completed by the end of November 1817; revised in January–March 1818; and first published towards the end of April 1818. Conceived by the poet as 'a test, a trial of [his] Powers of Imagination' (*Letters*, I, 169), *Endymion* is markedly uneven in poetic quality but invaluable as a record of self-analysis. The two versions of the Preface (a self-castigating original draft was cancelled at the request of Keats's publishers) reveal the part played by exacting self-criticism in the astonishingly quick maturing of Keats's poetic powers. A first allusion to the myth of Endymion and Phoebe appears in 'I stood tip-toe upon a little hill' (ll. 181–204); its attractions for Keats were that it could serve to express his sensuous love of nature and his yearning for transcendence. The coupling of 'Greece and England' (II.254) reflects the larger ambition of revitalising English culture by an appeal to the imaginative vigour of Greek mythology.

Endymion, a young 'shepherd prince', confides to his sister Peona that he has fallen in love with a mysterious maiden whom he has seen in a dream. Pining for this lost vision of supernatural loveliness, Endymion is now bowed down with melancholy and unable to attend to his kingly duties. In a long speech which has attracted more critical attention than the rest of the poem, Endymion justifies his decision to devote himself to a quest for his visionary love. The hero's pilgrimage takes him along subterranean regions (book II), beneath the sea (book III) and through 'the giddy air' (book IV). The narrative potential of the classical legend is exhausted by the end of the first book; and the hero's various encounters – with Adonis asleep in his bower (II.387–587), with Arethusa and Alpheus (II.936–1017), with Glaucus mourning for dead Scylla (III.193–923) – are partly called in by the necessity of postponing the poem's climax; but they may also be viewed as variations on the central fable, bound together as they are by a common theme, the triumph of love. Keats's most important addition to the traditional myth, however, is the introduction in book IV of the Indian Maid, a lovelorn mortal whose charms Endymion finds irresistible. The hero's heart is now torn between his heavenly and his earthly loves. His dilemma is miraculously solved for him by the metamorphosis of the Indian Maid, who turns out to have been none other than Phoebe in disguise. The transfiguration occurs, significantly perhaps, only after Endymion has resolved to desert his dream for the sake of his human love.

None of Keats's known statements suggests that he had a definite allegorical intention in mind, but the poem's fable lends itself readily to neo-Platonic interpretations, Endymion's quest for Phoebe symbolising the soul's yearning for ideal beauty and immortality. Late-nineteenth-century readers found support for this view in Keats's use of such Platonic-sounding phrases as 'fellowship with essence' (I.779). Later critics have rebutted this reading with the argument that the language of the poem conveys its heavenly aspirations far less convincingly than it does its un-

abashed eroticism. The prevailing opinion now is that to ove
emphasise either aspect is to blind oneself to the poem's meaning
which lies precisely in an attempt to formulate a religion of beauty
which, instead of divorcing the spiritual from the sensual, presents
them as connatural degrees of a continuous scale of values. With a
characteristic mixture of enthusiasm and uncertainty, *Endymion*
celebrates the 'Principle of Beauty' (*Letters*, I, 266), not in a spirit of
self-sufficient aestheticism but as a cure for 'despondence'. The
creed stated in the proem derives its urgency from an acute
awareness of the sufferings it is meant to alleviate. Broadly similar
in intention to *The Excursion* (1814), *Endymion* transposes Wordsworth's idealism into a mythological mode, at once serious and
non-assertive. Its labyrinthine windings confusedly foreshadow
the debate about the 'truth of Imagination' (*Letters*, I, 184) which
was to take definite shape in Keats's mature poetry.

'The Eve of St Agnes'

Keats's third romance was composed between 18 January and
2 February 1819 at Chichester, Sussex, and Bedhampton, Hampshire (*Letters*, II, 58–9). The poem's central thematic contrast, between death and the ardour of young love, is widely held to echo the
death of Keats's brother Tom (1 Dec 1818) and the poet's love for
Fanny Brawne, with whom he had come to an 'understanding' on
Christmas Day 1818 (*Letters*, I, 411). Its 'mediaeval' atmosphere
may owe some visual details to Chichester Cathedral, but its main
sources are to be found in Keats's reading: the more obvious
influences are *The Faerie Queene*, *Romeo and Juliet*, Mrs Radcliffe's
Gothic romances, Scott's 'The Lay of the Last Minstrel' (1805) and
Coleridge's 'Christabel' (1816). The distinctive character of 'The
Eve' (in Leigh Hunt's phrase, 'rather a picture than a story') is
largely determined by the Spenserian stanza, which, without im-
peding the movement of the narrative, is ample enough to allow
for self-contained *tableaux vivants*. Keats' original draft, most of
which has been preserved, provides an illuminating insight into
his methods of composition (see M. R. Ridley, *Keats's Craftsman-
ship*, 1933).

Keats may have found details of the 'popular superstition' (*Letters*, II, 139) on which the story is based in John Brand's *Observations
on Popular Antiquities* (1777, 1813) or in Robert Burton's *Anatomy of
Melancholy* (1621). According to the legend, virgins who conform '

the proper observances may see their future husbands in their dreams during the night of St Agnes' Eve (20 Jan). In Keats's narrative, Porphyro, in love with Madeline, the daughter of his family's worst enemy, steals into her father's castle amidst the boisterous revelry of a feast held on St Agnes' Eve. He learns from Angela, Madeline's aged nurse, that the girl intends to perform the superstitious rites. An unpremeditated stratagem occurs to Porphyro's passionate mind: let him be concealed in Madeline's bedchamber, and he will justify her confidence in the legend by appearing to her in the flesh during the night. Angela is prevailed upon to secrete him in a closet, well-stored with luscious dainties, from which Porphyro watches Madeline prepare herself for bed and fall into a deep sleep. After carefully setting the stage for the enactment of the legend, Porphyro plays on Madeline's lute, arousing her to a condition between sleeping and waking in which she confesses her love for him, which emboldens Porphyro to 'melt into her dream'. As the consummation takes place, the moon sets and a storm arises. Fully awakened, and undeceived by Porphyro, Madeline consents to elope with him. Undetected by the drunken guards, the lovers steal out of the castle 'into the storm'. This bald summary is enough to suggest the pitfalls that the poem triumphantly avoids. What could all too easily have become a risqué anecdote of voyeurism is sublimated into a passionate celebration of young love.

In revising the poem at Winchester in September 1819, Keats clearly felt that its idealisation of sexual passion laid it open to ridicule; that, as he wrote about 'Isabella' (Letters, II, 174), there was 'too much inexperience of life, and simplicity of knowledge in it'. He added a stanza (after no. vi in the textus receptus of 1820) making the legend more explicit, and explaining in particular why Porphyro prepares a rich feast of delicacies (xxx–xxxi) before he rouses Madeline from her enchanted sleep. This addition, like the crucial rephrasing of ll. 314–22, leaving the reader in no doubt as to the consummation of the lovers' union, was also designed to add masculine bitters to the sweets of the tale. Finally, the revised reading of the last three lines of the poem gave further emphasis to their reminder of the physical reality of death, in order 'to leave on the reader a sense of pettish disgust', as Richard Woodhouse recorded (Letters, II, 162–3). Conscious of the harm that this outspokenness might do to Keats's reputation, his publishers persuaded him to restore the original version (see Letters, II, 182–3).

While some present-day readers regard this as an act of censorship, others feel that Woodhouse and Taylor saved Keats from destroying the delicate balance of his first draft. For 'The Eve' is a sophisticated romance which manages to weave its pleasurable spells without disguising the fact that they are illusions. The deliberate puncturing of the world of make-believe in the concluding stanza widens a psychic distance which is subtly maintained throughout the poem by the narrator's tone. It is this balance between scepticism and the will to believe which, together with the poem's more obvious contrasts, secures its hold on the reader's imagination.

The most remarkable feature of 'The Eve' is its poetic idiom, which creates elaborate pictorial and musical effects with memorable synaesthetic richness. And indeed the poem was praised in Victorian days as a triumph of art for art's sake. In reaction to such excesses, some twentieth-century readers have burdened 'The Eve' with a weight of metaphysical meaning. Porphyro's progress to Madeline's chamber and his entrance into her dream have been viewed as a symbolic ascent of the imagination to a paradoxical reconciliation of time and eternity. Reversing this perspective, more sceptical critics have presented the poem as an anti-romance, and brought into relief its latent ironies to contend that Madeline is meant to be seen as the self-deluded victim of her imagination. Others still have argued that the poem is a subtle exploration of the psychology of imaginative experience. No one would deny that the very structure of the plot was such as to bring into play some of Keats's deep-seated concerns about the relation between dream and reality. The central sequence (xxvi–xxxvii) is at once a dramatisation of 'Adam's dream' – Keats's metaphor for the visionary imagination (*Letters*, I, 185) – and its ironic subversion: while Madeline's dream comes true, what substantiates it is a trick. In the same manner, Keats's fantasy of 'an immortality of passion' (*Endymion*, II.808) is first allowed to take a plausible shape, and then dismissed as an illusion. In these respects, 'The Eve' points forward to the great odes of 1819.

Hyperion: A Fragment

Keats's unfinished epic was begun in the autumn of 1818 (*Letters*, I.387) and abandoned in April 1819; it was published in 1820. A second version, recast in the form of a dream-vision related by a first-person narrator, was begun in July 1819 (*Letters*, II, 132) and left incomplete in September (*Letters*, II, 167); *The Fall of Hyperion: A Dream* was first published by Richard Monkton Milnes in 1857. Allusions in *Endymion* (Preface and IV, 774) and in the letters (I, 207) show that the subject had been in Keats's mind since late 1817. In a letter of December 1818 (II, 12), he refers to the earlier version as 'the fall of Hyperion', which confirms that his attention was even then focused on the idea of the Fall. And indeed the mythological framework of the two poems – the Titanomachia, or dethronement of the older race of Titans by the new Olympian gods – provided the poet with little more than that basic myth.

The first version, Keats's most ambitious attempt at myth-making, boldly invites comparison with *Paradise Lost* in presenting a new, epic treatment of the Fall in solemn, Miltonic blank verse. The second version, influenced by Dante's *Purgatorio*, superimposes biblical on mythological elements in a cultural palimpsest. Its opening Canto, a different poem in many respects, relinquishes epic objectivity for an anguished lyricism that rises to a poetic intensity seldom surpassed in Keats's verse. In both versions, the poet's imagination – obliquely or directly – faces the enigmatic figure of Mnemosyne (Moneta is her sterner counterpart in *The Fall*), the depositary of the 'knowledge of good and evil', a Keatsian symbol of the tragic 'mystery' of man's fate.

Hyperion opens *in medias res* with a depiction of the already fallen Saturn, grief-stricken and bewildered. He rouses himself to re-newed resistance in a dialogue with Thea, Hyperion's wife, who leads him away to the den where the 'bruised Titans' are assem-bled. The second part of Book I shows the yet unvanquished Hyperion, obsessed with ominous images of 'death and darkness' (I.242), preparing to fly to Saturn's help. Deliberately echoing Mil-ton's description of the fallen angels in Hell, Book II presents the council of the defeated Titans, pictured in sculptural poses, vainly trying to come to terms with the woe of mortality. The only positive response to mutability is voiced by Oceanus who urges his dejected brethren to accept inevitable change as the necessary condition of progress. Timid Clymene's lament raises the objection that philosophical arguments are powerless to dispel grief, while confirming Oceanus's wisdom in hailing the advent of Apollo's superior beauty. Book III is devoted to the deification of Apollo who receives from Mnemosyne the 'knowledge enormous' (III, 113) of the 'giant agony of the world' (*The Fall*, I.157).

Abruptly curtailed in *Hyperion*, the assumption of the burden of tragic consciousness is fully dramatised in the opening Canto of *The Fall*; the epic fable of *Hyperion* III is converted into a personal myth. After 'pledging all the Mortals of the world' (I.44) in a gesture of universal sympathy, the dreamer drinks a 'full draught' of a cool nectar which drives him out of the sensuous paradise of an earlier vision and gives him access to the temple of knowledge. After this fortunate fall, another ordeal is imposed on the narrator who is finally granted permission to gaze at the 'high tragedy' of Moneta's face, where he will read the events related in *Hyperion*. Conceived as a prelude to the uncompleted epic, Canto I drew on concerns so vital to the poet that it grew into a spiritual autobiography. The debate between the narrator and Moneta questions the value of Keats's poetic achievement and expresses his final ideal of the poet as a disinterested sage, whose all-embracing consciousness can offer the consoling wisdom of a justification of evil.

Despite their stylistic differences, the two fragments present complementary versions of the fall of Keats's imagination into the 'Purgatory blind' (*Letters*, I.262) of metaphysical anguish. In the earlier poem, Hyperion symbolises Keats's agonised awareness of evil, Apollo the poet he wished to become. While the epic was probably designed to embody the parable of a necessary fall from innocence leading through suffering to the achievement of a new equilibrium, its triadic pattern fails to appear distinctly because the superiority of Apollo is asserted rather than shown. Similarly, while it dramatises the 'Agony . . . of ignorance' (*Letters*, II.81) with convincing intensity, *The Fall* fails to transmute the contradiction of beauty and evil into the desired harmony.

Several critics have felt that it was this ambition to formulate a secular 'system of salvation' (*Letters*, II.102), rather than the difficulty of writing in an artificial, Miltonic idiom (Keats's own reason, see *Letters*, II.167), which prevented him from completing *Hyperion*. The death of his brother Tom (1 December 1818) is widely admitted to have been the cause of the abrupt change of tone which marks off the third Book from the first two. There is a return to the style and imagery of *Endymion*, and transcendental longings replace the mundane outlook of the opening books. It may be that in his 'lonely grief' Keats lost faith in the optimistic finalism expressed by Oceanus (II.173–243). A parallel explanation is that he found himself unable to establish the prevalence of beauty over evil on rational foundations.

Recent criticism treats these two fragments with respect; it is his own letters that provide the best gloss to the two poems. The parable of human life as a 'Mansion of Many Apartments' (*Letters*, I.280–1) admits the inevitability of a fall from innocent delight in beauty to a perplexed consciousness of evil. In the same letter (3 May 1818 to J. H. Reynolds) Keats expresses a tentative belief in 'a grand march of intellect' (I.282). But, in a journal-letter to the George Keatses (II, 101–4) written at or about the time when *Hyperion* was abandoned, Godwinian perfectibility is viewed as a shallow evasion of metaphysical issues; there follows another parable of the world as 'the vale of Soul-making' (II.102), justifying evil as a Providential obstacle that the soul must learn to overcome. In an earlier passage of the same letter (II.79–81) appears Keats's ideal of 'disinterestedness', or selfless sympathy, and one of the many expressions of his desire for the 'knowledge', or 'lore of good and ill' (*Letters*, I, 262), that would enable him to live up to his new, exacting ideal of the poet as 'a sage, / A humanist, physician to all men' (*The Fall*, I.189–90).

'Lamia'

'Lamia' was the last narrative poem to have been completed by Keats. Part I was finished by 11 July 1819 (*Letters*, II, 128). Part II was written between mid-August and 5 September 1819 (*Letters*, II, 157). The interval was taken up by the composition of *Otho the Great*, in collaboration with Charles Brown, and by the remodelling of *Hyperion*. 'Lamia' was further revised in March 1820 (*Letters*, II, 276). The poem was designed to gratify the public's taste for the sensational ('What they want is a sensation of some sort' – *Letters*, II, 189); it was given pride of place in the 1820 volume, possibly at the suggestion of Keats's publishers, a preference which does not go undisputed at the present time. Keats himself had a high opinion of the poem, partly because he felt that its brisk, sophisticated detachment was invulnerable to critical sarcasm (*Letters*, II, 174). His choice of Dryden's heroic couplets as a metrical model reflects the same desire to avoid sentimentality of any kind.

The fable was borrowed from a passage of Robert Burton's *Anatomy of Melancholy* (1621) which was printed at the end of the poem in 1820. Keats's most significant modification of his source is the addition of an introductory episode (I.1–145) providing an ironic counterpoint to the main story. Hermes, in amorous pursuit of a nymph, chances upon Lamia, at this point a grotesque snake, 'a gordian shape of dazzling hue' (I.47), whose serpent head boasts 'a woman's mouth with all its pearls complete' (I.60). The Lamia's ambiguity goes beyond her strange physique, since she is both helplessly enclosed 'in the serpent prison-house' (I.203) and gifted with powers of clairvoyance and magic. In return for Hermes' promise to restore her to her former human shape, Lamia undoes the spell by which she had made the nymph invisible, thereby promoting the god's speedy conquest. Metamorphosed into a 'lady bright' (I.171), but retaining her magic powers, Lamia is translated from Crete to the neighbourhood of Corinth, where she seduces Lycius, a young philosopher whom she had dreamed of, and fallen in love with, in her former state (I.200–19). Lamia snatches Lycius from the world of reality to a fairy palace where the pair are united in an 'immortality of passion' (*Endymion*, II.808). Part I thus concludes on the promise of a 'long immortal dream' (I.128), like the opening episode of Hermes and the nymph. But in part II 'a thrill of trumpets' pierces the magic shell in which the lovers are enclosed, tempting Lycius to substantiate his dream by making it public. Despite Lamia's fears and admonitions, Lycius insists on inviting his friends to a wedding-feast which the re- signed bride's spells adorn with 'fit magnificence' (II.116). An unbidden guest, Lycius's former tutor, the philosopher Apollo- nius, forces his way into the banquet room. Lamia, knowing his ability to distinguish illusion from reality, had begged her lover to keep him out. Under the philosopher's piercing gaze, Lamia with- ers and fades away at the moment when her ambivalent nature is disclosed. Unable to survive the dissolution of his dream, Lycius dies, his marriage robe turned into a shroud.

Presenting as it does a series of contrasts – between dream and reality, imagination and reason, poetry and philosophy, the sense of wonder and scientific rationalism – the poem invites allegorical interpretation; but its inner contradictions defeat all attempts to reduce it to a neat diagram. Its peculiar elusiveness is partly due to Keats's desire to shield himself from mockery by an assumed detachment which occasionally verges on cynicism. And, indeed, in publishing the source of his story, he may have wished to distance himself further from it. But the borrowed plot was in fact thoroughly naturalised by Keats's imagination, which clothed its bare outline with his own private obsessions. Lamia's dream comes true, like Endymion's, and her magic palace provides a shelter where the intensity of human passion can be reconciled with permanence. The dream of an 'immortality of passion' is no sooner restated, however, than it is exposed as an impossibility – a dismissal which, as Keats himself knew, was an act of self-division (see *Letters*, II, 234). In 'Lamia' his determination to resist the attractions of the myth is matched by his attachment to it, an imperfectly mastered tension which jeopardises the poem's unity and results at times in self-parody.

To read 'Lamia' as a repudiation of romantic dreaming is to disregard the narrator's outburst against 'cold philosophy' (II.229–38); to read it as a plea for the poetic imagination is to overlook the fact that the 'tender-personed' Lamia (II.238) is first described as a grotesque monster. Neither Lycius nor Apollonius is such as to invite the reader to endorse his limited view of the central character, whose ambivalence seems to colour the whole poem – an ambiguity which, intentionally or not, precludes any serious response to the myth.

Biographical interpretations point out that 'Lamia' reflects the 'gordian complication of feelings' (*Letters*, I, 342) that sexual love inspired in Keats, and that it mirrors the conflicting attitudes generated by his passion for Fanny Brawne. (See for instance *Letters*, II, 123.) As in 'La Belle Dame sans Merci' – of which *Lamia* has been said to be a baroque elaboration – the sexual and poetic themes are inseparable, although love and poetry appear as mutually exclusive choices in some of the letters to Fanny Brawne.

The most illuminating parallels are perhaps those which can be drawn with canto I of *The Fall of Hyperion*. In Lycius there is something of the weak 'dreamer' (*The Fall*, I.162) that Keats at the time feared he himself had become; and 'Lamia' as a whole (once described as '*Endymion* in reverse') shows him trying to reject the romantic 'luxuries' of his early poems, just as the narrator of *The Fall* has to leave behind the delights of his enchanted garden in order to gain access to the temple of knowledge. But the unresolved tensions of 'Lamia' reveal that the reflective part of Keats's genius was doing violence to his deepest creative impulses.

The Odes

His most original achievements, Keats's major odes include 'To Psyche', 'To a Nightingale', 'On a Grecian Urn', 'On Melancholy' and 'On Indolence'. 'To Autumn', though not so designated by Keats, is commonly held to belong to the group. The order of their composition is conjectural; in the above list, which follows conventional opinion, only the first and the last poems can be dated with any precision. 'Psyche' was written towards the end of April 1819 (*Letters*, II, 105); 'To Autumn' around 19 September 1819 (*Letters*, II, 167). In transcripts, 'Nightingale' is dated 'May 1819', 'Grecian Urn' and 'Melancholy' '1819'; 'Indolence' was composed in the spring of 1819 (*Letters*, II, 78–9, 116). 'Nightingale' and 'Grecian Urn' were first published in *Annals of the Fine Arts* in July 1819 and January 1820 respectively. 'Nightingale', 'Grecian Urn', 'Psyche', 'Autumn' and 'Melancholy' were published in the 1820 volume, in that order. 'Indolence' first appeared in *Life, Letters and Literary Remains of John Keats*, edited by Richard Monkton Milnes (1848). Their stanzaic structure derives from Keats's experiments with the sonnet; apart from 'Psyche', which has a loose Pindaric form, the spring odes use a ten-line stanza generally combining a Shakespearean quatrain with a Petrarchan sestet, a pattern lengthened to eleven lines in 'To Autumn'.

With the odes, Keats invented not only a new and influential mode of symbolic poetry but also discovered the form most appropriate to his agnostic, questing genius. For these odal hymns, despite their individual differences and the variety of their overt subjects, are lyric debates in which the heart's desires are submitted to the scrutiny of the sceptical mind. In the most original of the group, 'Nightingale' and 'Grecian Urn', this intensely personal dialogue tends towards dramatic objectivity in that its course and outcome seem to be determined by the central symbol's incapacity to sustain the meanings that the poet would wish it to bear.

For all its vibrant affirmations, 'Psyche' is not the straightforward restatement of the religion of beauty of *Endymion* that it may appear to be. The introduction of the earlier poem asserts that the despondent mind can find refuge in 'a thing of beauty'; in the ode, the relationship is significantly reversed: the poet vows to provide a shelter for his private symbol of beauty 'in some untrodden region' of his mind. While it remains latent in 'Psyche', Keats's scepticism is openly voiced in 'Nightingale' and 'Grecian Urn' – two parallel, and eventually unsuccessful, attempts to conceive a convincing reconciliation of beauty and permanence. A meditation on the poetic experience, *Nightingale* critically reviews the moments of aesthetic transcendence in which beauty had seemed to be the sensuous promise of supersensuous truth. A reflection on the meaning of art – that human activity which is based on the assumption that beauty is truth – 'Grecian Urn' both acknowledges the illusory character of such an assumption and mitigates the pain of that admission by affirming the consolatory value of this necessary illusion. In 'Melancholy' no attempt is made to grant beauty the status of a symbol of transcendent truth; it is now viewed as one of the contradictory aspects of a reality which is inescapably Janus-faced; to confront 'Beauty that must die' is to face the essence of man's fate. What appears in 'Melancholy' as a strained, tense imperative becomes in 'Autumn' a serene contemplation, or acceptance, of transience.

This inevitably reductive summary is meant to suggest that the odes may be read as a sequence expressing in brief compass the movement of Keats's thought, from his early aesthetic idealism, through growing scepticism, to a final reconciliation with the limitations of life. Arguments for and against this global approach have made up a large part of the continuing critical debate about Keats's masterpieces. The indeterminacy of their symbolic mode allows for wide divergences of interpretation. Each of them may be viewed as the record of a particular mood; their obvious interrelations tempt the reader to treat them as the successive stages of a long quest for a surrogate religion.

The major interpretative crux remains the punctuation and meaning of the last two lines of 'Grecian Urn', which are now generally held to be spoken by the urn to 'man'. 'Indolence' has been criticised for its uncertainty of tone, but also praised for its exploration of a mood of creative reverie akin to Wordsworth's 'wise passiveness'. While no general agreement is conceivable, nor indeed desirable, about the meaning of the odes, their pre-eminence in the Keats canon is now undisputed.

JEAN-CLAUDE SALLÉ

Lamb, Charles (1775–1834)

Charles Lamb was born in London at Crown Office Row, where his father was clerk to Samuel Salt, a Bencher of the Inner Temple. Charles was educated from 1782 to 1789 at Christ's Hospital in Newgate Street, where he began a lifelong friendship with Coleridge; then he worked briefly for Joseph Paice and in the South Sea House before, in 1792, entering the East India House, where he remained until he retired in March 1825.

On 22 September 1796 his elder sister, Mary (1764–1847), killed their mother in a fit of insanity. Charles, despite opposition from his brother John (1763–1821), saved her from permanent incarceration by undertaking responsibility for her. Her illness recurred more frequently and for longer periods as she grew older, but all their friends testified to her intelligence and wisdom when in her normal state.

Before the 'day of horrors' Lamb had been writing love sonnets to 'Anna' – probably Ann Simmons, who lived near Blakesware ('Blakesmoor') in Hertfordshire where his grandmother was housekeeper. In his first extant letter to Coleridge, Lamb attributes to this infatuation his own brief spell of six weeks 'in a madhouse', though 'my head ran on you' almost as much. Once he had decided to devote his life to Mary, he gave up for many years any idea of marriage; but in 1819 he proposed to the actress, Fanny Kelly, who refused his offer though she remained a lasting friend.

Partly because of Mary's illness – 'We are in a manner marked' – the Lambs frequently changed lodgings, but their evening parties, at which might be met many of the key figures of the age, have been made famous by contemporary accounts, notably in Hazlitt's 'On the Conversation of Authors' and 'Of Persons One Would Wish to Have Seen'. Lamb's retirement disappointed his expectations, his move to Enfield and Edmonton having exiled him from friends and the London he loved, and Mary's increasing illness leaving him lonely. Wordsworth said that Coleridge's death in July 1834 hastened Lamb's. He died of erysipelas after a fall on 27 December of that year. Mary lived on until 1847.

Coleridge's *Poems* published in 1796 (2nd edn 1797) contained also verses by Lamb. In 1798 *Blank Verse* by Charles Lloyd and Charles Lamb included Lamb's best-known poem, 'The Old Familiar Faces'. In the same year he published his prose *Tale of Rosamund Gray*. In 1802 came *John Woodvil, a Tragedy*, written in verse in the Elizabethan style. In 1806 he had a farce, *Mr H–*, put on at Drury Lane, but, when it proved a failure, he helped to hiss it off the stage.

With his sister he wrote *Tales from Shakespear* (first published 1807 and never out of print since), *Mrs Leicester's School* and *Poetry for Children* (1809). These and several individual fairytales, retold by Charles, as well as his *Adventures of Ulysses*, were written for Mrs Godwin's Juvenile Library. Of *Specimens of English Dramatic Poets who Lived about the Time of Shakespeare* (1808), 'done out of old plays at the Museum', Lamb in his 'Autobiographical Sketch' (1827) says, 'He also was the first to draw the Public attention to the old English

Dramatists . . .'. Though the revival of interest in Elizabethan and Jacobean drama, then generally unknown, was not solely due to him, his influence predated and inspired the renaissance furthered by Coleridge and Hazlitt.

Lamb contributed to many papers and periodicals, including Leigh Hunt's *Reflector, Examiner,* and *Indicator.* In the first appeared such well-known essays as 'On the Tragedies of Skakespeare' and 'On the Genius and Character of Hogarth'. Because of his full-time employment, Lamb's criticism has to be culled not only from his essays but from his letters and the reports of his friends (it can be found in *Lamb as Critic,* ed. Roy Park, 1980). Contemporaries regarded him as one of the finest critics of his age and both Coleridge and Wordsworth followed his advice on their poetry. Lamb's review of Wordsworth's *Excursion* (1814), though mangled by the editor of the *Quarterly Review,* remains illuminating. His scattered pieces of verse and prose to date were published in 1818 as *The Works of Charles Lamb.* In 1820 he was asked to contribute to the new *London Magazine* edited by John Scott and he continued under successive editors. For these familiar essays Lamb used the pseudonym 'Elia', and collections of them were published as *Essays of Elia* (1823) and *Last Essays of Elia* (1833).

In 1831 Lamb wrote of 'our first boyish heats kindled by the French Revolution'. In youth he, with many then radicals who were and remained his friends, was pilloried in verses and a Gillray cartoon in *The Anti-Jacobin* (1798), and he contributed to the radical paper *The Albion* (1801). Though never seduced by political systems, in his essays Lamb expressed his continuing concern for and indignation on behalf of the poor and underprivileged: sometimes overtly, as in the Popular Fallacy 'That Home is Home though it is never so Homely'; more often by the use of 'that dangerous figure – irony' – 'dangerous' because easily missed by careless readers.

After a hundred years of popularity, Lamb's reputation was almost eclipsed by the attacks of Leavisite critics from the 1930s to the end of the 1950s. However, since the early 1960s detailed study and analysis of individual essays have revealed Lamb as exercising a characteristically Romantic imagination in the creation of a subtle and unsuspectedly tough prose equivalent of the art of contemporary poets. In examination of his criticism Roy Park finds 'a strong independent mind' and 'a penetrating insight into what is of permanent and lasting value in literature'. Probably the easiest introduction for a new reader is through the humour and humanity of Lamb's *Letters*, in the process of being edited by Edwin W. Marrs for Cornell University Press, which have been called 'the most fascinating correspondence in our language'.

Essays of Elia **and** The Last Essays of Elia

Essays of Elia, containing twenty-seven of Lamb's contributions to the *London Magazine* since 1820 and one essay ('Valentine's Day') previously published in Leigh Hunt's *Examiner* and *Indicator*, was published by Taylor and Hessey in January 1823. Revisions for this book were minor except that three articles titled 'The Old Actors' were cut and reshaped into 'On Some of the Old Actors', 'On the Artificial Comedy of the Last Century' and 'On the Acting of Munden'. Many of the essays show his keen interest in the theatre.

The Last Essays of Elia, published by Edward Moxon in April 1833, collected essays under the 'Elia' pseudonym contributed to the *London Magazine* till Lamb left it in August 1825 and then to the *New Monthly Magazine*, the *Englishman's Magazine* and *The Athenaeum*. There were minor omissions and revisions, but the only major change was to the second edition of 1835, after Lamb's death, when 'A Death-Bed', concerning the death of Randal Norris, was omitted at the request of his family and 'Confessions of a Drunkard' substituted. This had already been published in *The Philanthropist* in 1813, then in *Some Enquiries into the Effects of Fermented Liquors* in 1814 and 1818. It was resurrected under the 'Elia' signature in the *London Magazine* of August 1822, when Lamb was in France and had no new contribution ready. A note was added which countered suggestions that Lamb himself was an habitual drunkard. This was important to him because of his work and because his life was insured. Nevertheless, Lamb's recourse to alcohol for consolation and to loosen his stammering tongue was a weakness against which he fought without much conviction.

In the 'Elia' essays Lamb, like other Romantic writers, used certain facts of his own life as a means of expressing universal human experience. Though writing in the first person, he manipulated the fictional persona of 'Elia' both to distance himself and thus avoid egotism, and also to incorporate the experience of others – as, for example, when in 'Christ's Hospital Five and Thirty Years Ago' Elia relates as his own the schoolboy feelings of Coleridge, which were very different from Lamb's. The Elia 'character' is also an implement in expressing his ambivalent attitude to the world. Members of his family appear disguised: his father as Lovel in 'The Old Benchers of the Inner Temple'; his brother as James Elia in 'My Relations' and 'Dream Children'; and Mary as 'my cousin Bridget' in 'Mrs Battle's Opinions on Whist', 'Mackery End in Hertfordshire' and 'Old China'. 'Old China' illustrates Lamb's capacity to make an everyday object an 'objective correlative' and his tea-cup has been compared to Keats's Grecian urn.

Lamb shared with Wordsworth a sense of the importance of childhood experiences and of memory, as illuminating both the transience of human life and the immortality conferred by the creative imagination (see particularly 'Blakesmoor in H—shire'). Despite an underlying seriousness, often signalled by devastating irony, which can be turned also against himself, the humour, wit and sheer fun of the essays should not be undervalued. Both aspects of Lamb's art can be found in, for example, 'A Dissertation upon Roast Pig', as can his sophisticated use of learned allusion. He is also capable of a touching pathos without sentimentality, as in 'Dream Children'.

Hazlitt commented, 'The style of the Essays of Elia is liable to the charge of a certain mannerism' ('Elia', in *The Spirit of the Age*), and Lamb's use of archaisms is still a stumbling-block to some readers. But Hazlitt excuses it in 'On Familiar Style': 'Mr Lamb is the only imitator of old English style I can read with pleasure; and he is so thoroughly imbued with the spirit of his authors, that the idea of imitation is almost done away.' Wordsworth said of the Elia essays, 'his works are our delight, as is evidenced better than by words – by April weather of smiles and tears whenever we read them' (letter to Crabb Robinson, 14 Nov 1833).

Recent criticism has emphasised the rhetorical skill with which Lamb manipulates the reader–author relationship, the careful structure of the best essays aided by the build-up of appropriate imagery and allusion and by sensitive word choice, as well as the ambiguity of his ironic stance which refuses didacticism and punctures pretension. Stress has been laid on Lamb's influence on the nineteenth-century novel and in particular on Dickens. The World's Classics *Elia and Last Essays of Elia* edited by Jonathan Bate contains a useful introduction and helpful notes.

MARY WEDD

Landor, Walter Savage (1775–1864)

Walter Savage Landor was born at Warwick in 1775. As a writer he was highly regarded by a few, but was known to most of his contemporaries as a 'character': an impetuous and headstrong man (caricatured as Boythorn by Dickens in *Bleak House*) holding in his youth extreme radical views. He left Oxford without a degree after being involved in a fracas. Later, he became bankrupt as a result of his over-bold management of the estate he had inherited in Monmouthshire, and went into exile in Italy in 1813. He resided in Florence, then Fiesole, returning to England in 1835 after breaking with his wife, and had to leave Bath again in 1858 under the threat of a trial for defamation of character in a private feud. He spent his last days in Florence, in a solitude cheered by the friendly protection of Robert Browning, and died there in 1864.

His first important poem was *Gebir* (also written in Latin as *Gebirus*) in 1798, a romantic 'oriental' tale, treated so as to convey revolutionary and humanitarian ideas, and written in a compact style contrasting with the conventional poetic diction of his earlier attempts. Too difficult to be popular, though it won the admiration of Southey, De Quincey and later Shelley, it is a little-known but interesting landmark in the change of taste that led to Romanticism. There is great vigour in some parts at least of the dramatic poetry he wrote next, especially his 'closet tragedy' *Count Julian* (1812). His main prose work consists in the successive series of *Imaginary Conversations of Literary Men and Statesmen* (1824, 1828, 1829): either brief dramatic scenes from history, or longer discussions between historical or contemporary characters, ranging freely over a variety of topics, from politics and morals to literature and philology. The *Pentameron and Pentalogia*, in 1837, including Boccacio and Petrarch as characters, is of the latter type. Some more conversations appeared in the *Collected Works* of 1846, and Landor went on using the form as a vehicle for essays, mostly political. But in later life he returned to poetry with his *Hellenics*, published in the *Works* of 1846, and separately in 1847, together with additional poems ('Heroic Idyls') translated from earlier poems written in Latin. They treat pastoral scenes or minor episodes from the classical legends through an imaginative reconstruction of ancient times, and are infused (as Swinburne remarked) with a truly pagan spirit of naturalism. Throughout his life Landor wrote brief lyrics and epigrams ('Rose Aylmer', 'Dirce') which have found their way into anthologies.

PIERRE VITOUX

Landscape

In eighteenth-century England, three quarters of the population lived in the country and, except in London, no town-dweller was more than an hour's walk from the wildest places. Natural scenery was part of everybody's daily experience, but it was taken for granted and little attention was paid to it. Mentioning Bagshot Heath, Defoe found it 'horrid and frightful to look on, not only good for little, but good for nothing'. That was a widespread feeling, especially at the beginning of the century, when it was considered that 'the proper study of Mankind is Man' (Pope), and when a hierarchy existed among painting-genres, with portraits and historical scenes at the top of the scale, landscapes and still-lifes at the bottom. When nature was referred to – or enjoyed – in literature or in life, it was either in the tradition of poetic diction, or reminiscent of the soft scenery of south-eastern England, with meadows strewn with flowers, or verdant hills and dales (images influenced by Milton's description of the Garden of Eden in book IV of *Paradise Lost*).

Particularly unattractive to contemporary taste was a mountainous landscape, a repulsion partly resting on a metaphysical basis. For Thomas Burnet (*Sacred Theory of the Earth*, 1684–90) and many thinkers of his time, the universe before the Fall was a flat paradise, with a smooth surface. Mountains and oceans appeared later, together with human corruption and sin. 'Uncouth', 'inhospitable', 'Earth's tumours', 'blisters', 'warts', were terms commonly applied to mountainous scenes.

The development of taste came with the growing practice of the Grand Tour after the 'Glorious Revolution' of 1688. Crossing the Alps became a vivid experience for Englishmen on their way to Italy; fear was mixed with fascination: Burnet himself, Lord Shaftesbury, Joseph Addison and John Dennis were among the first to express that mixture of pleasure and awe, a new aesthetic approach to landscape. In Italy, the English visitors discovered Claude Lorrain, with his peaceful, harmonious compositions, but they also enjoyed Salvator Rosa, the painter of grand scenery and strong emotions. Under these combined influences, a new taste for wild landscapes developed in England, promoted by the improvement of communications. Travellers visited Wales, the Lake District, the Highlands of Scotland. Most poets – Gray, Walpole – took similar journeys, writing diaries and letters for the benefit of their more sedentary contemporaries, who gave the warmest welcome to that travel literature; the English were moving 'from mountain gloom to mountain glory' (Marjorie Hope Nicolson, *Mountain Gloom and Mountain Glory*, 1959).

Landscape gardening was going through similar changes at the same time. The great landowners modelled and remodelled their estates; fortunes were spent in the process and scores of volumes were devoted to the subject. Sir William Temple's *Upon the Garden of Epicurus* (1685), John James's *The Theory and Practice of Gardening* (1712), Stephen Switzer's *Iconographia Rustica* (1718, 1742), Horace Walpole's *On Modern Gardening* (1770) and Sir William Chambers' *Dissertation on Oriental Gardening* (1772) were among the best-known publications on the subject. In the first decades, under Dutch, French and Italian influences, the 'formal garden' had prevailed: a rectangular plan, straight alleys crossing at right angles; long avenues, terraces, costly parterres, waterworks and topiary art; everywhere a desire for order and symmetry – a style cultivated, and still surviving, at Blenheim and Hampton Court, as at Versailles.

But the monotony, the artificiality and stiffness of such landscapes could not please the English long. Charles Bridgeman, William Kent and Lancelot ('Capability') Brown were the artisans of change, the fathers of the 'English' or 'natural' garden: wavy lines, trees and grass growing at liberty – or so it seemed – and running water. Working 'with a painter's eye and a poet's feeling', they imitated scenes from Claude Lorrain, as at Stourhead, Wiltshire, evoking moods of pleasure or melancholy in those taking a walk through such landscapes. Then, when contemporary taste was surfeited by 'an excess of naturalness', as practised by Brown's heir and successor, Humphry Repton, attempts were made to introduce the picturesque into gardening; but neither Sir William Chambers, nor Uvedale Price on his own estate at Foxley, Herefordshire, was really successful in such ventures.

Landscape-painting showed a similar evolution, emerging from a position of inferiority accepted and enforced by Sir Joshua Reynolds, President of the Royal Academy (Discourse V), to the triumphant heights attained by Turner. The topographers represented the early stage: portraits of places and houses in the tradition of Wenceslaus Hollar. Bird's-eye views were executed by Leonard Knyff and Johannes Kip (*Britannia Illustrata: or, Views of Several of the Queen's Palaces, as also of the Principal Seats of the Nobility and Gentry of Great Britain*, 1707). John Wootton's *Stoke Park, Gloucestershire*, George Lambert's *Wolton Park* (1739), Richard Wilson's *Croome Court* (1758), Paul Sandby's views of Windsor Castle (1760–71), Canaletto's views of the Thames and Paul Scott's London scenes are all examples of the topographical style.

The classical or ideal landscape was to a large extent an imaginary one developed under the influence of Claude Lorrain, with many Italianate features. 'This sort of painting is like Pastoral in Poetry' (Jonathan Richardson, *An Account of the Statues, Bas-Reliefs, Drawings, and Pictures in Italy, France, etc.*, 1722). Wootton and Lambert were at their best in this style; Richard Wilson (*Rome and the Ponte Molle*, 1754) and John Robert Cozens (*Lake Albano and Castel Gandolfo*, 1783–8) also tried their hands at it; and Turner himself worked in the same vein (*Crossing the Brook*, 1815; *The Bay of Baiae with Apollo and the Sybil*, 1823), as well as introducing the sublime.

A special mention must be made of 'visionary' landscape, the ultimate stage in that development. Contact with reality was blurred; imagination took the upper hand, after observation and emotion. Something fantastic characterises James Ward's *Gordale Scar* (1811–15) and John Martin's pictures inspired by the Bible (*The Deluge*, 1834; *Sodom and Gomorrah*, 1832; *The Great Day of His Wrath* (1852). A similar fascination with the irrational, the supernatural, fired Fuseli (Füssli) and Blake, but their nightmarish pictures centred on human beings rather than landscapes. (see also entry on the PICTURESQUE.)

SUZY HALIMI

Macpherson, James (1736–96)

James Macpherson had the ability to create in words an outline of heroic action, and a vaguely Celtic atmosphere. His most admired writings – ostensibly translations from ancient Gaelic epics by Ossian, the son of Fingal – appealed strongly to Goethe, Schiller and Herder; and Napoleon became a devotee of Ossian. A line by Keats which has no direct connection with Macpherson – 'huge cloudy symbols of a high romance' – could perhaps be used to describe what is most characteristic about Macpherson's work and style. His indistinctness, which seemed so exciting to many readers in the late eighteenth and early nineteenth centuries, was no accident. He is most famous today as a literary forger.

After publishing a tedious long poem, *The Highlander* (1758), Macpherson persuaded his fellow Scottish authors the dramatist John Home and critic Hugh Blair that he had access to Gaelic epic poetry of the distant past. His brief work *Fragments of Ancient Poetry Collected in the Highlands of Scotland and Translated from the Gaelic* appeared in 1760. Two years later he brought out *Fingal; an Ancient Epic Poem in Six Books*, and this was followed in 1763 by *Temora, an Ancient Epic Poem in Eight Books*. Thereafter, controversy as to the genuineness of his literary claims raged both within and beyond Scotland. Thomas Gray wrote to Thomas Warton, 'this man is the very Demon of Poetry, or he has lighted on a treasure hid for ages'. For his part, Dr Johnson agreed with David Hume's comment that 'he would not believe the authenticity of *Fingal*, though fifty barearsed highlanders should swear it', and Macpherson threatened to beat Johnson with a stick if he met him.

The arguments continued for the rest of his days. Meanwhile, ever the opportunist, Macpherson had left Scotland for public life in the south, becoming a member of Parliament, writing in defence of Lord North's ministry, and serving at different periods as Surveyor-General of the Floridas and agent to the Nabob of Arcot. His *History of Great Britain from the Restoration* was published in 1775. He was buried in Westminster Abbey at his own expense.

In 1805 a committee chaired by Henry Mackenzie concluded that Macpherson had liberally edited Gaelic poems, inserting many passages of his own. After carefully examining the evidence afresh in our own time, Professor Derick Thomson has reached a broadly similar verdict. Macpherson, it would appear, did have originals in

Gaelic ballads for much of his work, but he also freely adapted and expanded, and was 'neither as honest as he claimed nor as inventive as his opponents implied' (*A Companion to Gaelic Scotland*, 1983).

DONALD A. LOW

The Marvellous and Occult

The origins of the interest in the marvellous in the Romantic period were similar to those which led to the Gothic novel. The principal one was the desire to subvert prevailing hierarchies of established values and systems of belief: to this end, Thomas Gray described the strange figure of the bard, standing on a mountaintop and defying the armies of the conquering Edward I. The bard was a figure from an earlier age (like the minstrel, whose departure was lamented by Blake and Scott), unusual, prophetic, isolated, strange. In the same way ghosts, magicians and witches all acted as figures who helped to challenge the accepted norms of social behaviour and intellectual belief. They acknowledged the irrational as opposed to the rational, and allowed into fiction and poetry (and occasionally drama) areas of feeling and action which were distinctly out of the ordinary.

In recognising this, critics of the eighteenth century linked the idea of magic with that of the poetic imagination. Richard Hurd, in *Letters on Chivalry and Romance* (1788), noted that the poet had 'a supernatural world to range in . . . Gods, and Fairies, and Witches at his command'. And Scott, writing of Horace Walpole's *The Castle of Otranto*, described the author's intentions as 'not merely to excite surprise and terror, by the introduction of supernatural agency, but to wind up the feelings of the reader till they become for a moment identified with those of a ruder age, which "Held each strange tale devoutly true"' (Ioan Williams, *Sir Walter Scott on Novelists and Fiction*, 1968). Scott saw one function of the marvellous as that of introducing an element of excitement that would 'wind up the feelings of the reader'; his remarks about a 'ruder age' are connected with an interest in the Middle Ages as being not only an age of excitement and chivalry, but also an age in which poetry was given a different kind of attention, a credulity which (in Scott's eyes) had disappeared during the rationalism of the eighteenth century. For all of these reasons, Spenser was a favourite poet with the Romantics, not only for his high moral teaching, but also because of his knights and ladies, his magicians and dwarves, and the atmosphere of enchantment which he created and to which the reader was expected to surrender.

To read Spenser was to suspend disbelief, and the suspension of disbelief was identified by Coleridge as that which 'constitutes poetic faith' (*Biographia Literaria*, ch. 14). Coleridge's share of the enterprise of *Lyrical Ballads*, he made clear, was the supernatural: 'and the excellence aimed at was to consist in the interesting of the affections by the dramatic truth of such emotions, as would naturally accompany such situations, supposing them real' (*Biographia Literaria*, ch. 14). The result was 'The Ancient Mariner', that poem of marvellous, strange and supernatural action, which exercised its own mysterious spell on many contemporary readers; Lamb said that it dragged him along 'like Tom Piper's magic whistle', and the association of magic and the poetic imagination is one that Coleridge would have recognised and welcomed. Indeed, as Anya Taylor has pointed out, the metaphor of magic was often employed to describe imaginative poetry in this period (*Magic and English Romanticism*, 1979). In other poems, 'Kubla Khan' and 'Christabel', Coleridge continued to use the marvellous: in the former he describes the vision of the poet who sees the damsel with a dulcimer, and in the second the figure of Geraldine is mysterious and frightening, though perhaps not malignant. In all these poems the strange happenings or visions engage the emotions, or (in Scott's phrase) 'wind up the feelings': they were part of the curiosity in the unusual features of human behaviour shown by the Romantics – the unexpected and unpredictable combined with the human to make a heady mixture of the exciting and the credible. In the same way, there was a strong interest in madness, seen in the poems of Wordsworth (such as 'The Mad Mother'), and the paintings of Géricault (*La folle, Le fou*).

The interest in magic is found in Keats's 'Lamia' where the serpent is metamorphosed into a beautiful woman, only to be destroyed by the cold philosophy of Apollonius; indeed, Keats often yearns for a world of magic, dream and fancy. Scott also portrays the magician Michael Scott at the end of *The Lay of the Last*

Minstrel. In many cases such magic was associated with the supernatural, and the possibility of ghosts was also one which fascinated the Romantics. 'We talk of Ghosts', wrote Mary Shelley (Journal, 18 Aug 1816), referring to Byron, Polidori, Shelley, herself, and M. G. ('Monk') Lewis, who was staying with Byron near Geneva. *Frankenstein*, which was the one remarkable work to come out of these conversations, was not a ghost story; but in its particular treatment of science fiction it has strong affinities with the interest in the marvellous.

Stories in which the victims of the crime reappear as ghosts (as in the case of Banquo) link the ghost story with the Romantic poets' interest in crime and punishment: a notable example is Crabbe's 'Peter Grimes'. Similarly, the power of the Devil to appear in different shapes and disguises provides another dimension to the occult, that of the religious, as it does in James Hogg's *Private Memoirs and Confessions of a Justified Sinner*; Burns treated the subject comically in 'Tam o' Shanter'. In yet another manifestation, the occult and marvellous are associated with various forms of dangerous experience, such as vampires, or fatal and doomed figures, or Shelley's interest in the Medusa (see Mario Praz, *The Romantic Agony*, 1930, ch. 2).

Wordsworth, as Anya Taylor has pointed out, did not like the association of poetry with magic, preferring 'the simple produce of a common day'; and Scott disliked the extreme forms of the occult: 'I think the marvellous in poetry is ill-timed & disgusting when not managed with moderation & ingrafted upon some circumstance of popular tradition or belief which sometimes can give even to the improbable an air of something like probability' (*Letters*, ed. H. J. C. Grierson *et al.*, 1932–7, I, 121). The occult and the marvellous were also connected with the Romantic taste for the exotic and the oriental. The influence of *The Arabian Nights' Tales* (see under ORIENTALISM), with the stories of magicians, genies, magic rings, and other marvellous properties, was pervasive; and in some of the major oriental poems of the period, such as Southey's *Thalaba*, magic and its power play a substantial part.

J. R. WATSON

Maturin, Charles Robert (1780–1824)

Charles Robert Maturin was born in Dublin, into a well-to-do family of French Huguenot origin; he was educated at Trinity College. In 1803 he was ordained in the Anglican Church of Ireland. He married Henrietta Kingsbury and served as curate of St Peter's, Dublin, from 1805 until the end of his life. His first novel, *The Fatal Revenge; or, The Family of Montorio* (1807), was praised by Walter Scott in the *Quarterly Review* (1810), which marked the beginning of a long correspondence and a real literary friendship between the two men. *The Wild Irish Boy* (1808), written in imitation of Lady Morgan's *The Wild Irish Girl* (1806), was, like *The Fatal Revenge*, published at the author's own expense. Financial difficulties began in 1809 when his father lost his position in the Irish Post Office. Being 'a high calvinist in [his] religious opinion' (letter to Scott, Jan 1813), a man of eccentric behaviour and an author of fiction, Maturin was denied preferment by his Arminian superiors. He opened a school to prepare students for Trinity College, and needed to write for money. *The Milesian Chief* (1812) brought him £80, and his first play, *Bertram* (1816), was a success at Drury Lane, where Kean played the title role. Unfortunately most of the money went to the creditor of a person for whom he had acted as guarantor. His next two plays, *Manuel* (1817) and *Fredolfo* (1819), were failures. He returned to fiction with *Women; or, Pour et Contre* (1818), *Melmoth the Wanderer* (1820) and *The Albigenses* (1824), a venture into the genre of the historical novel. He also published two volumes of sermons: *Sermons* (1819) and *Five Sermons on the Errors of the Catholic Church* (1824). His natural improvidence, material setbacks and the critics' indifference or overt sarcasms combined to weaken an already strained constitution, and he died at the age of forty-four.

To say that Maturin is a Gothic novelist does not do justice to his versatility: he tried Gothic fiction and drama, sentimental fiction and Irish tales with 'scenes of actual life' (Preface to *The Milesian Chief*), and historical fiction. He imitated all his predecessors in different genres: 'Monk' Lewis and Ann Radcliffe, Lady Morgan, Maria Edgeworth and Walter Scott. Among his novels of Irish life, *The Wild Irish Boy*, with its incoherent narrative technique and gross characterisation, is a pot-boiler which has sunk into well-deserved oblivion. Madame de Staël's *Corinne* (1807) undoubtedly served as a model for *Women; or, Pour et Contre*, but what carries conviction in this book is the subtle analysis of the gentle Eva's feelings, the depiction of De Courcy's 'feeble and oscillating mind' (Scott, *Edinburgh Review*, June 1818) and the satire on the excesses of Calvinistic Methodism. *The Milesian Chief* is the story of the passion that unites Connal O'Morven, the only descendant of a Milesian chieftain, and Armida, the daughter of an English usurper. Placed against the background of the 1798 uprising, exploring the themes of honour, pride and rebellion in a now-bombastic, now-forceful language, 'full of sound and fury', *The Milesian Chief* is primarily a tale of Ireland, a country 'where the most wild and incredible situations are hourly passing before modern eyes' (Preface). If *The Albigenses* owes much to Scott's novels, as a literary achievement it is well below *Ivanhoe* (1820), and the conflict between the Catharists and the Crusaders is a pretext for a display or blind anti-Catholicism. Admittedly, *The Albigenses* is a historical novel, but the amount of supernatural material introduced into the story shows that to the end Maturin remained in essence a Gothic writer.

In his Gothic works Maturin shows that he is a master of the fantastic, and in the dedication in *The Milesian Chief* he writes pointedly, 'If I possess any talent, it is that of darkening the gloomy, and of deepening the sad; of painting life in extremes, and representing those struggles of passion when the soul trembles on the verge of the unlawful and the unhallowed.' The plot of *The Fatal Revenge* is so complicated that the readers 'wander, bewildered, baffled and distracted through labyrinthine mazes' (E. Birkhead, *The Tale of Terror: A Study of the Gothic Romance*, 1921), as the author explores the fear arising 'from objects of invisible terror' (Preface). The book is in the tradition of Ann Radcliffe: all the seemingly supernatural events are explained at the end. Yet it is a powerful story of mental persecution, and the haunting presence of Orazio/Schemoli, an avatar of the necromancer of the German *Schauerroman* (horror novel), is felt throughout the book. For him 'time is ever beginning, suffering is ever to be'. Similarly, Bertram is 'the champion of despair' whose resolution to wreak vengeance on his enemies is strengthened by supernatural agency. Both foreshadow Melmoth, whom Baudelaire called 'the great satanic creation of the Reverend Maturin' ('De l'essence du rire', 1855). No book is more labyrinthine than *Melmoth the Wanderer*. A fusion of Goethe's Faust and Mephistopheles, of Ann Radcliffe's Schedoni and Godwin's St Leon, of Lewis's Ambrosio and Wandering Jew, Melmoth is an embodiment of 'the spirit that denies' (Goethe, *Faust*), doomed to wander over the earth, a reprobate who regrets his original innocence. Ultimately, Maturin's rebels become pathetic figures, strangers to the world, to their fellow creatures and to themselves. All of them, unlike Prometheus, fail. Maturin made of Melmoth the very prototype of modern derision, and the Wanderer's final fall points to the intellectual and metaphysical bankruptcy of contemporary fiction.

CLAUDE FIEROBE

Mediaevalism

Mediaevalism was an attitude that resulted from improved knowledge and understanding of the Middle Ages, which in the early eighteenth century was generally viewed as the 'Dark Ages' and so as uncouth, barbaric and 'Gothic'. Ignorance, and the political and religious changes that had occurred since the Reformation, had contributed to this picture. The development of mediaevalism was part of a larger antiquarian interest in all facets of the nation's past: literature, architecture, warfare and costume, for example.

Literary mediaevalism was bound up with the change in aesthetics signalled by Burke's *Philosophical Inquiry into . . . the Sublime and Beautiful* (1757). The word 'Gothic' came to indicate anything that was related to the Middle Ages, the feudal social order, and the deep faith and sense of mystery to which the mediaeval cathedrals, for example, bore witness (see J. Addison, *The Spectator*, nos 110 and 419). If in the early years of the century Roman Catholicism had been considered as an enemy, at least on political grounds, the new emphasis on the imagination and the development of notions such as the sublime made way for a new perception, and led to a re-evaluation, of the mediaeval faith. Thus K. L. Morris, in *The Image of the Middle Ages in Romantic Victorian Literature* (1984), can speak of a 'religious mediaevalism'. This new attitude towards the old faith was accompanied by a willing suspension of disbelief in the supernatural. When later on the Goths came to be regarded as the direct ancestors of the English people, what was 'Gothic', or believed to be so, became worthy of interest.

The changing tastes that accompanied the rise of mediaevalism were manifested in the collection of old texts and the creation of new ones purporting to be old or having a 'Gothic' flavour. The Middle Ages (then taken as extending as far as the Elizabethan period) inspired works ranging from Sir Richard Blackmore's Arthurian epics and Pope's 'Eloisa to Abelard' (1717), to Horace Walpole's *The Castle of Otranto*, the first 'Gothic' novel (1764). As late as 1789, Richard Hole composed *Arthur, or the Northern Enchantment*. Whatever involved the legends of the Round Table, or knighthood and tournaments generally, was welcome: William Warburton, the future Bishop of Gloucester, wrote a *Dissertation on the Origin of Chivalry and Romance*, published together with Jarvis's translation of *Don Quixote* into English (1749). Inspiration could be Celtic, Germanic or Scandinavian: Gray's odes rang of Odin and his warriors; James Macpherson sang the supposed feats of Ossian. Literary criticism demonstrated the beauties of *The Faerie Queene* (in Thomas Warton's *Observations*, 1754), and celebrated the golden days of Queen Elizabeth (Bishop Hurd's *Letters on Chivalry and Romance*, 1762). Thomas Tyrwhitt, in his edition of *The Canterbury Tales* (1775), was the first to make Chaucer fully intelligible to the eighteenth century.

Following the example of Macpherson, Thomas Percy (1729–1811) a country curate, first translated *Five Pieces of Runic Poetry* (1763), and then collected old English and Border ballads, with the assistance of the poet William Shenstone. These he published as *Reliques of Ancient English Poetry* (3 vols, 1765). They are a mixture of genuine old ballads, and of poems adapted or arranged to suit the taste of the time, avoiding the 'grossness' of former ages. Among the old ballads included is 'Chevy Chase', which Addison had praised in *The Spectator* (1711) and which Burke had instanced as capable of rousing very strong passions. Percy then translated Mallet's *Northern Antiquities* (1769), full of details of the age of knighthood. Percy's dedication of the *Reliques* to the Countess of Northumberland is a perfect illustration of the spirit presiding over the enterprise: although he describes these ballads as 'the barbarous productions of unpolished ages', he also claims that 'these poems are . . . the first efforts of ancient genius . . . exhibiting the customs and opinions of remote ages . . . No active or comprehensive mind can forbear some attention to the reliques of antiquity.'

The fact that the ballad had never died out facilitated collection. Newspapers and novels proved that people still carried on the tradition of ballad-singing about the mediaeval heroes of history or legend – Guy of Warwick, Robin Hood, the Seven Champions of Christendom, and so on – and Percy demonstrated to his contemporaries the continuity between their ancestors and themselves: ballad-singing was a time-honoured activity. Percy manipulated his authorities, but his doing so shows that he also researched them. Some of his texts tell of the age-old feud between Scots and English in the Border country, while others are traceable to such exotic sources as a Spanish romance. His 'Essay on the Ancient Minstrels', at the end of volume I, set out to prove that the mediaeval minstrels were direct heirs and successors of the old Druidic bards. The essay is full of anecdotes and supplemented by abundant learned notes. A second appendix traces 'The Origin of the English Stage', reviewing theatrical activities from the first moralities down to the Restoration and gives Shakespeare's age its due importance.

The fame that Percy's researches earned him also brought him preferment (he became Bishop of Dromore) and spurred others into collecting ancient ballads. Evan Evans edited the *Specimens of the Antient Welsh Bards* (1764), translated from the Welsh. Charlotte Brooks collected and published *Reliques of Irish Poetry* (1769), and Scotland produced a particularly rich crop of old songs and ballads, such as the collections by David Herd and John Pinkerton. Robert Burns was himself an avid collector (see under BURNS).

Percy was not the only one to take an interest in the history of the stage. Thomas Warton's *History of English Poetry* (1774), a monument of scholarship in its time, deals with all forms of poetry, including that recited on the stage. Early in the century amateurs, including Lady Mary Wortley Montagu, started collecting old plays, and in 1744 Robert Dodsley published a *Select Collection of Old Plays*. The most important playwright to be studied, collected and edited was Shakespeare: during the century twelve different editions of his work were produced, from that of Nicholas Rowe to Edmond Malone's scholarly text. 1769 was the year of the Shakespeare Jubilee in Stratford-upon-Avon, and the Revd Joseph Greene carried out important research into the bard's biography. All this was part of the interest in the past that paved the way for Romanticism.

GEORGES LAMOINE

Metre and Form

The poetry of the Romantic period exhibits wide variations in the handling of metres and forms. The major poetic forms which were inherited from the seventeenth century and earlier in the eighteenth included blank verse, the heroic couplet, the sonnet, the Spenserian stanza, and a number of lyric forms; there was an instinctive awareness, also, of the importance of decorum, of the importance of metre and the appropriate diction and register. The Romantic poets inherited these conventions, but adapted them to their own purposes, and modified them where necessary.

Blank verse, used by Milton in *Paradise Lost*, was subsequently employed by James Thomson in *The Seasons* for the purposes of nature poetry; Cowper's *The Task* used it more freely and informally for reflection, in ways which anticipate Coleridge's 'conversation poems'. Wordsworth used blank verse in *The Prelude*, probably in deliberate imitation of Milton, and Keats wrote *Hyperion* and *The Fall of Hyperion* in blank verse. It was used by Shelley for *Alastor*, and for his verse dramas, *Prometheus Unbound* and *The Cenci*, and Southey used it for *Madoc*.

The heroic couplet was chosen by Wordsworth for his first published poems, *An Evening Walk* and *Descriptive Sketches* (1793). It was used by Keats in a number of early poems ('I stood tip-toe', 'Sleep and Poetry'), in *Endymion* and in 'Lamia'. Clare used the heroic couplet in his savage satire *The Parish*, and Crabbe used it in most of his verse tales. It was metre of such poems as *The Pleasures of Memory*, by Samuel Rogers, and *The Pleasures of Hope*, by Thomas Campbell, and it was employed by Thomas Moore for substantial parts of *Lalla Rookh*.

Byron used blank verse in his early satires, *English Bards and Scotch Reviewers* and *Hints from Horace*, but switched to the Spenserian stanza for *Childe Harold's Pilgrimage*: he said that he found the nine-line stanza form easier to write. The Spenserian stanza was reminiscent not only of Spenser and *The Faerie Queene*, but also of eighteenth-century poems employing the same form, such as James Thomson's *The Castle of Indolence* and James Beattie's *The Minstrel*, which led to John Clare's *The Village Minstrel*. The Spenserian stanza was also used by Wordsworth in the 'Salisbury Plain' poems, and by Shelley in *Laon and Cythna* (later *The Revolt of Islam*) and in *Adonais*. Keats used it in 'The Eve of St Agnes', and Campbell in *Gertrude of Wyoming*.

The sonnet, in both its English and Italian forms, was widely used in the period, most notably by Wordsworth (who greatly admired Milton's sonnets), Keats and Clare. Clare experimented restlessly with variations on the traditional sonnet form. Another kind of experiment, J. H. Frere's *Whistlecraft*, showed Byron the possibilities of using the Italian form of *ottava rima* (ABABABCC) for comic and satiric verse in English.

The influence of other forms, such as the ode and the hymn, is evidence of the way in which the Romantic poets took a received form and used it for their own purposes, often making it irregular or even fragmentary. The ballad, too, was very popular, both for its narrative speed and economy and for its use of simple and dramatic diction. Popularised by Percy's *Reliques of Ancient English Poetry* (1765), the ballad was used by Wordsworth for many of his contributions to *Lyrical Ballads* and by Coleridge in 'The Ancient Mariner'. Lyric metres were more conventionally employed, although Blake produced his own inspired variations on lyric and hymn forms in *Songs of Innocence and of Experience*.

In *Jerusalem*, Blake explicitly rejected blank verse: 'in the mouth of a true orator, such monotony was not only awkward, but as much a bondage as rhyme itself. I therefore have produced a variety in every line, both in cadence and number of syllables.' He thought that 'Poetry fettered fetters the human race', and it is clear that many of the Romantic poets sought forms and metres that would give them more freedom of expression. Southey experimented (in a way that anticipates Hopkins' better-known 'sprung rhythm') with short syllables taking the time of one in *Thalaba*, and Coleridge similarly (in 'Christabel') counted by accents and not by the number of syllables. His experiments were admired by Scott, who developed them extensively in his narrative poems (his debt to Coleridge is acknowledged in the Preface to *The Lay of the Last Minstrel*).

What Scott, in this preface, called 'the attempt to return to a more simple and natural style of poetry' was responsible for a loosening of traditional metres and forms, and the creation of hybrid or original kinds of poetry. The term 'Lyrical Ballads' is an example; another is the transformation of the ode into what M. H. Abrams has called 'the greater Romantic lyric'. This was a derivative of the irregular Pindaric ode, common in English literature since the time of Cowley. But perhaps the most remarkable departure from tradition in the period was the use of the fragment: the publication of apparently unfinished poems, such as Coleridge's

'Kubla Khan' or Keats's *Hyperion*, and the increasing provision of texts ending with a question-mark, emphasise the growing tendency of the Romantic poets to produce texts that are open-ended and which require a creative response from the reader. Others, such as Byron's *Don Juan*, offer endless digressions and entertainments, as well as a loosely picaresque narrative.

J. R. WATSON

Moore, Thomas (1779–1852)

Unjustly neglected since the Victorian Age, Thomas Moore now appears as one of the best of the lesser Romantics. A versatile and highly gifted writer, he enjoyed in his lifetime a popularity that rivalled that of Byron and Scott.

The son of a Dublin grocer, educated at Trinity College, he went to London to study law at the Middle Temple, but soon devoted all his efforts to the development of his talents for music and poetry. His translation in verse of the *Odes of Anacreon*, dedicated to the Prince Regent, was favourably received (1800). Then his sentimental and amorous *Poems of the Late Thomas Little* (1801), a juvenile production meeting the taste of the day, won him popular acclaim. Appointed in 1803 Admiralty Registrar at Bermuda, he spent some months visiting the island and travelling in America. Then, committing his official duties to a deputy, he returned to England, bringing back his *Epistles, Odes and Other Poems* (1806) inspired by this experience.

In 1808 he found his true poetic vocation in the poetico-musical enterprise launched by W. Power in Dublin. To traditional Irish tunes (harmonised by Sir John Stevenson) Moore adapted English lyrics to replace the old popular Gaelic texts. The composition of his *Irish Melodies* – a real feat of prosodic virtuosity – was a lifelong pursuit, extending over ten numbers published between 1808 and 1834. The musician–poet himself sang his melodies with great charm and feeling in the drawing-rooms of high society. By his nostalgic vision of Erin suffering under the English yoke, his nationalistic appeals and patriotic message, he served the cause of

Ireland in England, and the melodies, enthusiastically received on all sides, soon became part of the common heritage in English-speaking countries.

A liberal and a friend of the Whigs led by Lord Holland, a brilliant conversationalist and wit endowed with great social powers, Moore was fêted and lionised, became the intimate of all the writers and men of note of his time, and introduced the young Lord Byron – whose close friend he was to become – into the literary circles of the Regency.

The remarkable lyricist was also a master of satire. His pungent political squibs regularly published in the *Morning Chronicle* (1812–42) delighted the general public, as did his *Twopenny Post-Bag* (1813) and later, in the same vein, his *Fudge Family in Paris* (1818) and *Fables for the Holy Alliance* (1823).

Moore acquired a European reputation with *Lalla Rookh*, a long oriental poem in four tales, which appeared with great applause in 1817 and for which he received the fabulous sum of £3000. The gorgeous descriptions, dramatic episodes and erudite evocations of Eastern scenes and manners, as well as the sensual grace, flowing cadences and luxuriant imagery, were attuned to the taste of a generation fascinated by the East.

In 1818, through the dishonesty of his deputy in Bermuda, Moore was involved in a loss of £6000 and had to go abroad to escape imprisonment for debt. He travelled in Italy with Lord John Russell, visiting Byron at La Mira, near Venice (who made the gift of his memoirs to him), then settled in Paris for nearly two years. During this exile he wrote *The Loves of the Angels*, published in London in 1823, on a subject much in vogue among English and French Romantics.

In 1822, after his debts had been cleared, Moore returned to England. Now at a turning-point in his literary career, he confined himself almost entirely to prose. *The Memoirs of Captain Rock* (1824), denouncing the distress of Irish peasants, had a strong impact.

The news of Byron's death at Missolonghi (Apr 1824) came as a shock and a personal blow. The poet, too remiss in his negotiations with John Murray, Byron's publisher, to recover possession of Byron's memoirs, on which money had been advanced to him, was caught up in an impossible situation. Byron's wife, half-sister and closest friends, all fearing that the documents might contain some scandalous revelations detrimental to their reputations, agreed to have them destroyed. Moore strongly opposed the move and desperately fought to prevent the burning, which took place at Murray's. Most unfairly he was later held responsible for this destruction. He then laboured incessantly for several years to serve the memory of his friend, collecting manuscripts, letters and papers, and gathering information. All this went into a penetrating analysis of Byron's personality, a fair presentation of his life, and a remarkable appraisal of his works and genius: *Letters, Journals of Lord Byron, with Notices of his Life* (1830).

Two other biographies by Moore are worth mentioning: *Memoirs of the Life of the Right Hon. R. B. Sheridan* (1825), and *Life and Death of Lord Edward Fitzgerald* (1831). Moore's only novel, *The Epicurean* (1827) was not a success; nor was his *History of Ireland* (1835–46). His extensive *Memoirs, Journal and Correspondence*, edited by his friend Lord John Russell (1852–6), is a mine of information on the character and society of the times.

An upright and a lucid man, Moore never allowed himself to be dazzled by fame and popularity, and sensed with admirable foresight that his most original and enduring contribution to art and literature would be his *Irish Melodies*. A born lyricist, he remains unrivalled in the art of fusing music and poetry.

THÉRÈSE TESSIER

Nature

The word 'nature has many meanings and none of them need to be taken as characteristic of Romanticism. Yet the importance of the theme cannot be denied. Where does the originality of Romantic nature lie? One approach may consist in opposing two meanings of the word and following the shift which led to their occasional confusion. First, nature is the sum of what we can perceive, the world, natural objects, 'the forms of nature'; but nature also means the principle or power that animates or even creates the objects of nature, and we speak of the laws of nature, sometimes spelt Nature. In other words we may oppose the product and the producer, the creation and the creator, *natura naturata* and *natura naturans*.

The interest in natural objects grew in the eighteenth century, perhaps because of the developing Industrial Revolution, and was helped in the second half of the century by the improvement of roads, which made it easier to visit places of natural beauty. It was the century of the English garden, of the first guide books, of the picturesque and the sublime. Nature ceased to be a mere background to men's doings and became more closely related to them. It became nature as seen, as heard and as felt by man. The intercourse between man and nature began. Poets saw natural objects through their own eyes and no longer through the writings of Theocritus and Virgil. Conventions gave way to a genuine longing to perceive and describe what Wordsworth was to call 'the mighty world / Of eye, and ear', and 'the beauteous forms' of nature ('Tintern Abbey'). Men in general became more sensitive to the effects of nature on their minds and their hearts. Shaftesbury, as early as 1711, speaks of fields and woods as 'my refuge from the toilsome world of business'. In the same passage of *The Moralists* he sees nature as 'supremely fair and sovereignly good', and thus expresses one of the beliefs of the century, the benevolence of nature, later taught by Rousseau and the Romantics, with occasional reservations (Keats loved nature but in the 'Epistle to John Hamilton Reynolds' was sorry to see the gentle robin ravening on the worm). Shaftesbury goes on with words about 'all loving [nature] and all lovely, all divine . . . a wise substitute for Providence'; though the exact meaning of the word 'substitute' is debatable, his conception points to the further evolution of the idea of nature, the tendency to find more in nature than natural objects. But at first the goodness of nature is thought of as a proof of the goodness of its creator. Young found it 'Christian' (*Night Thoughts*, IV.704) and it was a 'glass reflecting God' (IX.1267). Cowper, in *The Task* (I.749), introduced a clear divide: 'God made the country, and man made the town'. God, external to his creation, still transcended it.

But when men sought refuge in nature they came so close to it, and felt its beauty so intensely, that they loved it, and longed to become united with it: *Childe Harold's Pilgrimage* contains the lines 'I live not in myself, but I become / Portion of that around me' (III.lxxii). Poets began to inquire into the nature of nature, and about the sources of its bounteous gifts. Coleridge in 'Dejection' privileged the mind's power: we find love and joy in nature because we love and enjoy, 'we receive but what we give'. Wordsworth in *The Prelude* spoke of the mind as 'Lord and Master', but he also, after Locke and other empiricists, saw what nature gave us and celebrated a 'wise passiveness' in which 'impulses' from nature taught us more than books. How much nature and self bring each other is best summed up in ll. 106–7 of 'Tintern Abbey', describing the half-perceptive, half-creative process. Nature and man cooperate in their noble interchange. Their close intimacy with nature induced poets to seek the essence of what they variously felt as the presences or powers of nature.

They discovered active powers in nature, pervading and animating it. Philosophy and science offered concepts and fairly convenient terms to account for what they felt. In Coleridge and the later Shelley, nature is often interpreted in Platonic or neo-Platonic terms. Coleridge borrowed from the Cambridge Platonists the notions of 'power' and 'soul' and the adjective 'plastic' (i.e. shaping, forming). In 'Religious Musings' he imagines a 'plastic' power that 'rolls' through matter 'In organising surge', but this power remains at God's disposal. More pantheistic is the 'presence' of Wordsworth's 'Tintern Abbey', which also 'rolls through all things', but obviously does so without God's help. The word 'roll', as used in these two poems recalls Newton's gravitation. Though they rejected the mechanical aspects of Newton's physics, some Romantics felt attracted by the ideas of force or pervading ether. Shelley relied on scientific concepts to convey his views in *Prometheus Unbound*. The word 'motion' is often used by Wordsworth to express his feelings about nature's 'life'. Coleridge also felt that we and the world were 'One Life'. 'Soul' and 'spirit' are often used to convey a sense of 'the life of things'. 'Spirits' in the plural, as used in the earlier versions of *The Prelude*, betray a form of animism. 'Spirit' in the singular refers to a more general active power such as that felt in Wordsworth's 'The Old Cumberland Beggar', 'a spirit and pulse of good', A life and soul, to every mode of being / Inseparably linked' (ll. 77–9). Thus the forms of nature (*natura naturata*) came to be quickened with life and at times with divine power (*natura naturans*): this awareness of the presence of an active principle in nature, of the infinite in the finite, of the eternal in the transient, best characterises the Romantic conception of nature.

MARCEL ISNARD

Negative Capability

The phrase was coined by Keats in a letter to his brothers in December 1817 (*The Letters of John Keats*, ed. H. E. Rollins, 1958, I, 193), shortly after he had completed the first draft of *Endymion*. The text has been preserved only in a transcript by John Jeffrey, and may well be corrupt, for all its convincing Keatsian ring:

> several things dovetailed in my mind, & at once it struck me, what quality went to form a Man of Achievement especially in Literature & which Shakespeare possessed so enormously – I mean *Negative Capability*, that is when man is capable of being in uncertainties., Mysteries, doubts, without any irritable reaching after fact & reason – Coleridge, for instance, would let go by a fine isolated verisimilitude caught from the Penetralium of mystery, from being incapable of remaining content with half knowledge. This pursued through Volumes would perhaps take us no further than this, that with a great poet the sense of Beauty overcomes every other consideration, or rather obliterates all consideration.

This enthusiastic conjecture attempts to define what Keats then regarded as the quality of mind essential to the creative writer, the capacity to resist the urge to systematise, a freedom from meta-physical prepossessions allowing the poet to contemplate reality without trying to reconcile its contradictory aspects, a willingness to receive the isolated insights of the poetic experience as tentative 'speculations' which do not require demonstration, and which lose nothing from their inability to be fitted into a closed, rational system.

The concluding sentence suggests that what inclined Keats to regard this ability to do without the comforts of positive certainty as an invigorating openness of mind was the intensity of his ecstatic 'Sensations, of the Beautiful, the poetical in all things' (*Letters*, I, 179). What silences the desire to know the unknowable is the momentary exaltation of the poetic experience which, while it lasts, puts the unquiet mind at rest. Unlike the exclamations of the celebrated letter to Bailey of 22 November 1817 (*Letters*, I, 184–5), the definition of Negative Capability does not include the claim that the poet's imagination provides him with transcendental insights; but it dates from a period when Keats was willing to believe that beauty is the sensuous token of ultimate reality.

The passage provides so apt a formulation of Keats's early poetic creed that readers have often been tempted to regard it as the definitive summation of his poetics. It is no more, in fact, than a tentative answer to the problem which engrossed Keats's restless spirit throughout his brief career: the relation between beauty and truth, the respective validity of intuition and discursive reasoning, of poetry and philosophy. In this very passage, the final qualifi-cation ('or rather obliterates all consideration') already implies what a letter of March 1819 was to make explicit – that beauty and truth belong to two distinct orders of experience, the emotional and the cognitive:

> Though a quarrel in the streets is a thing to be hated, the energies displayed in it are fine; the commonest Man shows a grace in his quarrel – By a superior being our reasoning[s] may take the same tone – though erroneous they may be fine – This is the very thing in which consists poetry; and if so it is not so fine a thing as philosophy – For the same reason that an eagle is not so fine a thing as a truth. (*Letters* II, 80–1)

It will be noted that the implicit hierarchy of December 1817 finds itself reversed here, and that 'beauty' and 'truth' are now clearly dissociated. Rather than a permanent credo, then, Negative Capability should be viewed as a stage in the evolution of Keats's thinking, as the definition of an aesthetic quietism which his growing scepticism eventually led him to qualify and relinquish.

Nor should Negative Capability be confused with the idea that the 'camelion Poet' (*Letters*, I, 387) has no definite personal identity, although the two notions are complementary aspects of a coherent conception of creative genius, based on the assumption that the essence of poetry lies in disinterested, amoral, selfless contemplation. Some of their interrelations appear in scattered passages of the letters, notably in Keats's criticism of poetry 'that has a palpable design upon us' (*Letters* I, 224), in his analysis of dogmatism as a form of self-assertion, and in his belief that settled convictions stunt the mind's growth (*Letters*, II, 213).

Although the two notions reflect Keats's familiarity with Hazlitt's ideas, they are firmly grounded in introspection. The roots of Negative Capability are to be found in Keats's agnosticism, in his 'capability of submission' (*Letters*, I.184); those of the 'camelion Poet' in his capacity for in-feeling or imaginative self-projection. The genesis of both concepts may have owed something to a young man's trust in the creative potentialities of the future. The poet of *Endymion* regarded the 'mystery' as a stimulating challenge prompting the mind to pursue 'solitary thinkings' that 'dodge / Conception to the very bourne of heaven' (I.294–5). The feverish narrator of *The Fall of Hyperion* yearns for the comforts of certain knowledge, tortured as he is by the fear that his 'speculations' may be useless dreams. In the latter poem, Keats's spiritual testament, the exhilarating freedom of doubt has been converted into the agony of ignorance.

JEAN-CLAUDE SALLÉ

Nelson, Horatio (1758–1805)

Horatio Nelson was born on 29 September 1758 at Burnham Thorpe, a village in Norfolk of which his father was rector. He went to sea in 1770, aged twelve, and served his early apprenticeship in ships of the line, merchantmen and exploration vessels. He became a lieutenant in 1777, and commanded a frigate at the age of twenty in 1779. He spent some years as a post-captain in the West Indies, but was recalled to Europe in 1793, on the outbreak of war with France. He served in the Mediterranean fleet, losing the sight of one eye as the result of an action off Corsica in 1794, and playing a hero's part in the battle off Cape St Vincent (14 Feb 1797); he lost an arm at Santa Cruz de Tenerife in July 1797. On his recovery from this wound, he returned to the Mediterranean, and won his first major naval victory at the battle of the Nile (1 Aug 1798), which effectively put an end to Napoleon's plans for expansion eastward.

Nelson became a vice-admiral in 1801, serving under Admiral Sir Hyde Parker at the battle of Copenhagen; it was during a daring attack on the city that Parker signalled a command to retreat, but Nelson is supposed to have put the telescope to his blind eye. During the Peace of Amiens (1802–3), Nelson lived on his estates in Surrey with his friends Sir William and Lady Hamilton, whom he had met in Naples during his Mediterranean station. His relationship with Emma, Lady Hamilton, had led to his separation from his wife in 1801. On the resumption of hostilities, Nelson left for Spanish waters, and blockaded the French fleet under Villeneuve. Anxious to gain control of the English Channel, the French fleet sailed on a diversionary voyage to the West Indies, but on their return the ships were once again blockaded in Cádiz. Their attempt to escape was Nelson's opportunity, and he destroyed the French fleet at the battle of Trafalgar (21 Oct 1805), during which he met his own death.

His death in battle set the seal on a career that was charismatic, brilliant and heroic: the triumph at Trafalgar was tempered by national mourning at the admiral's death. His famous signal 'England expects that every man will do his duty', and his widely reported last words, 'Thank God I have done my duty', became part of the legend of the heroic and noble warrior, and linked the concept of duty with patriotism in a way that was to prove very influential. He was buried with great ceremony in St Paul's Cathedral.

Nelson was the subject of a number of poems, most notably Wordsworth's 'Character of the Happy Warrior'; he also appears, less significantly, in Wordsworth's delightful 'Benjamin the Waggoner'. Scott was asked to write a patriotic song about Nelson and Trafalgar, and later asked to write a heroic poem; he tried to write the first, but did not succeed, and felt that the second was beyond his powers because he had had no experience of seafaring or of naval warfare, and also that 'the fate of the hero of the Nile, of Copenhagen, and alas! of Trafalgar is almost too grand in its native simplicity to be heightened by poetical imagery' (letter, 29 Dec 1805). He compensated for these refusals by paying a warm tribute to Nelson (together with Pitt and Fox) in the introduction to canto I of *Marmion* (1808). He was mentioned briefly, but non-committally, by Byron in *Don Juan*, I.

Southey wrote an article on Nelson for the *Quarterly Review* in 1810, which was later expanded into a book (published in 1813); and Thomas Campbell described the action off Copenhagen in his 'Battle of the Baltic' (1804–5), beginning 'Of Nelson and the North'. Campbell's best-known poem, 'Ye Mariners of England', first published in 1801, originally had Grenville as one of the heroes: after Trafalgar, Nelson's name was substituted. This poem was a good example of the idea of naval heroism that Nelson's career and death did so much to foster. A more subtle tribute to the navy, and to its national importance, is found in Jane Austen's *Persuasion* (1818).

J. R. WATSON

Orientalism

An interest in the Orient as an exotic and wonderful place had been common in Europe since the Middle Ages and the astonishing voyages of Marco Polo; Milton described one of the delightful curiosities of the Far East when he wrote of

> the barren plains
> Of Sericana, where Chineses drive
> With sails and wind their cany wagons light. . . .
> (*Paradise Lost*, III.435–9)

But Milton, like other seventeenth-century Puritans, also associated the East with luxury and indulgence, so that he depicted Satan

> High on a throne of royal state, which far
> Outshone the wealth of Ormus and of Ind,
> Or where the gorgeous East with richest hand
> Show'rs on her kings barbaric pearl and gold. . . .
> (*Paradise Lost*, II.1–4)

Milton's phrase 'the gorgeous East' was picked up by Wordsworth, in his sonnet 'On the Extinction of the Venetian Republic', written probably in 1802, which begins, 'Once did She hold the gorgeous east in fee'. Into that word 'gorgeous' are subsumed many of the qualities which made the East attractive to everyone except the Puritans: its beauty, its colour, its richness, its magic and its mystery. Eighteenth-century writers, in particular, eagerly explored the imaginative possibilities of Eastern settings, Eastern characters, and Eastern plots. In 1710–12 Ambrose Philips translated his *Persian Tales* from a French compilation, *Mille et un jours*, and this influenced Samuel Johnson in the choice of Abyssinia as a setting for *Rasselas* (1759), and in the theme of the search for happiness. At the same time, the architect and landscape gardener Sir William Chambers (1726–96), who had visited China as a young man, laid out the gardens at Kew with Chinese constructions, including the notable pagoda, which still stands. His *Designs for Chinese Buildings* (1757) was influential in the formation of taste, but his *Dissertation on Oriental Gardening* (1772) was less successful.

Other writers explored oriental themes. In 1742 William Collins (1721–59) printed his *Oriental Eclogues*, which he is said to have written at school after reading the relevant part of Thomas Salmon's *Modern History: or, the Present State of All Nations*, a monumental work in thirty-one volumes published between 1725 and 1738. Collins' description of oriental writers in his Preface gives an idea of the richness and fertility which the East was thought to possess: 'There is an Elegance and Wildness of Thought which recommends all their Compositions; and our Genius's are as much too cold for the Entertainment of such Sentiments, as our Climate is for their Fruits and Spices.'

The fascination with the East continued in prose and poetry (see M. P. Conant, *The Oriental Tale in England in the Eighteenth Century*, 1908), giving the authors scope for fine description, excitement and magic. It reached its most extraordinary form in William Beckford's *Vathek*, written in French in 1782, when Beckford was twenty-one. It was translated into English (1786), and found many imitators, including Keats's friend John Hamilton Reynolds (*Sofie, an Eastern Tale*, 1814). Byron acknowledged a debt to *Vathek* in *The Giaour* (1813). *Vathek* tells a sensational story about the Caliph Vathek, who becomes the servant of Eblis, the Devil; after numerous astonishing incidents, he gains entrance to the halls of Eblis, only to discover that his heart bursts into flame within his body.

Coleridge's 'Kubla Khan' is probably the single most distinguished example of orientalism in the Romantic period: its brilliant imagery of Kubla's 'pleasure-dome' and fertile gardens, its description of the savage chasm outside through which the sacred river bursts, and its portrayal of the damsel with a dulcimer 'singing of Mount Abora', are all evidence of the strong imaginative pull of Eastern settings and actions. Almost contemporaneous with it was Landor's *Gebir* (1798, rev. 1803); this is set in Egypt, and concerns the fate of two brothers: Tamar the shepherd, who wrestles with a sailor nymph and who is carried away by her beyond the ordinary world; and Gebir the prince, who conquers Egypt, falls in love with the queen, Charoba, and is killed during the marriage feast through the treachery of her nurse Dalica.

Byron's *The Giaour* (1813) is one of several 'Turkish tales' which occupied him from 1813 to 1816. The word 'Giaour' was a term of contempt used by Turks to describe non-Muslims, and the hero of Byron's story is an outsider in the society of the tale. He is loved by Leila, a slave; when her master, Hassan, finds out, he has her bound in a sack and thrown into the sea. The Giaour kills Hassan in revenge, and then enters a monastery. This was followed by *The Bride of Abydos* (1813), *The Corsair* (1814), *Lara* (1814) and *The Siege of Corinth* (1816). Byron had visited the Near East during his tour of 1809–11 (described in the first two cantos of *Childe Harold's Pilgrimage*, 1812), and these tales gain much of their appeal from his enthusiastic memory of the land and seascapes, the people, and specific events (Byron rescued a girl who was about to suffer the same fate as Leila, for example, and was himself nearly captured by pirates – see Leslie Marchand, *Byron, a Portrait*, 1971). The tales are those of heroism and outlawry, of passionate love and gloomy introspection, of pirates and violent death, of murder and revenge. They were influenced by the verse tales of Scott, which were extremely popular, but with the action and scenery transferred to another clime and atmosphere, and given even more excitement. The result was a series of tales that were immensely successful.

'Stick to the East', Byron said to Thomas Moore on 28 August 1813: 'the North, South, and West, have all been exhausted' (*Letters and Journals* ed. L. Marchand, III, 101). Apart from minor figures such as Henry Gally Knight (1786–1846), who wrote *Ilderim, a Syrian Tale* (1816) and *Phrosyne, a Grecian Tale* (1817), the only other poet to have attempted to use the East, Byron noted, was Southey. Byron had little respect for him, and called his oriental poems 'Southey's unsaleables': he was referring to *Thalaba* (1801) and *The Curse of Kehama* (1810). 'His personages don't interest us', Byron told Moore, 'and yours will.'

In fact *Thalaba* is one of Southey's better poems, with a skilful handling of a flexible and supple metre, and fine imagery: like Tolkien's *Lord of the Rings*, it describes Thalaba's destruction of the kingdoms of the evil magicians by the power of a magic ring. Thalaba is killed in the process, but is rewarded in paradise. The poem had a second edition in 1809, which suggests that it was not quite so unsaleable as Byron supposed. *The Curse of Kehama* has an Indian setting, and describes the downfall of the cruel Raja Kehama.

When he received Byron's 'Stick to the East' letter, Thomas Moore was already at work on *Lalla Rookh* (1817). This is a loose collection of stories told during the journey of the Princess, Lalla Rookh, to be married to the Prince of Cashmere. Feramorz, the young poet, beguiles the evenings with imaginative and touching stories, and the Princess falls in love with him. The prose narrative between the stories describes the progress of the royal party, and contains some enjoyable comedy at the expense of the Great Chamberlain, Fadladeen, who is suspicious of anything that interferes with his concept of order and decorum, including poetry and young love. At the end, the Princess arrives at Cashmere to be married, and discovers that Feramorz the poet is, in fact, the Prince whom she is destined to marry.

The charm and good nature of Moore's work shows that the oriental tale can be used in the Romantic period for many different purposes. In Byron's case, the heroes and heroines are little more than Gothic villains, or outlaw heroes translated to another environment, and given more spectacular opportunities for action. Landor and Southey are concerned with the progress of good in the face of evil; and Moore is able to use the oriental setting as an exotic and attractive background to a story of affection and love. In every case, however, the East is able to reflect the longing of the self for something beyond itself, for a world which was known to exist but which, for most people, was inaccessible, except in the world of poetry, imagination or dream; in Coleridge's case, it was found in a travel book, Purchas's *Pilgrimages*, which entered his dream world under the influence of opium. His cry about the 'damsel with a dulcimer' is

> Could I revive within me
> Her symphony and song
> To such a deep delight 'twould win me . . .

and this delight is characteristic of the effect of oriental imagery, settings and characters on the Romantic imagination.

J. R. WATSON

Paine, Thomas (1737–1809)

Paine's best-known works are *Common Sense* (1776) and *The Rights of Man* (parts I and II, 1791 and 1792) – the first being part of his written contribution to the American Revolution, and the second his defence of the French Revolution in face of Edmund Burke's stringent attack. But Paine did more than write: shortly after publishing *Common Sense* he was appointed aide-de-camp to General Green and took a position with Congress after the Declaration of Independence; and while awaiting trial for sedition in 1792 (brought by the British government in respect of the second part of *The Rights of Man*) Paine left England in order to take up a seat in the revolutionary Convention in France. As he wrote to George Washington, 'A share in two revolutions is living to some purpose' (16 Oct 1789).

The striking contrast between Paine's humble origins and the political eminence that he achieved seems to exemplify the possibilities open to all in the kind of egalitarian meritocracy with which he sought to replace existing aristocratic societies. The son of a Quaker staymaker from Norfolk, and first employed in his father's trade, Paine left England for America in 1774 aged thirty-seven, having 'failed in business, in marriage, and in vocation' (Isaac Kramnick, Introduction to *Common Sense*, 1976). America in the first throes of revolution was truly Paine's land of opportunity. For, whereas his early political activity in England had cost him his position as an excise man, he quickly achieved political and literary success in the New World. Over the following ten years he played an active role in revolutionary America, publishing the *Crisis* pamphlets (1776–83), visiting France on diplomatic missions, helping to re-establish what afterwards became the Bank of North America, and, in the meantime, working at the scientific interests which seemed almost obligatory for radical thinkers of the period.

Paine left America for France in 1787 and in September of that year he crossed the Channel to the country that he had left thirteen years previously. He arrived in England no longer an obscure excise man but able to make contact and become friendly with such political figures as Charles James Fox and Edmund Burke. During the first two years of the French Revolution, Paine alternated between France and England, but never returned to the country of his birth after being outlawed in December 1792. His position in

France became increasingly dangerous as the more moderate Girondins were displaced by the Jacobins. In 1793 he opposed the execution of Louis XVI and, with the outbreak of war with England, the figure the British government had sentenced to death for revolutionary writings was imprisoned as a potential enemy of the Revolution. Paine was released in 1794 after the intervention of James Munroe, American Ambassador to France, and lived in Munroe's house for the next few years, writing and publishing part II of *The Age of Reason* (1795; part I had been written in prison) and *Agrarian Justice* (1797). Paine returned, in 1802, to an America quite different from the one he had known; his attacks on Christianity in *The Age of Reason* alienated an American people that had already forgotten its revolutionary past. Paine died in America in 1809 in relative obscurity: 'The staymaker from Thetford who had shaped the world as few have ever done . . . was laid to rest in a quiet pasture with no ceremony, no fanfare, no appreciation' (Kramnick).

Common Sense was published at a crucial stage in the build-up to the American Revolution. The American colonies were at war with Britain but, as Kramnick puts it, 'the colonists were far from clear about what they were fighting for'. The more militant among them saw the war as a struggle for independence, others as an attempt to resecure the rights and freedoms supposedly due to all British subjects. *Common Sense* appeared in January 1776, argued passionately and persuasively for independence, quickly sold 120,000 copies, and arguably played a major part in influencing the Continental Congress's adoption of the Declaration of Independence on 4 July – the language of which bears 'the indelible imprint of *Common Sense'* (Henry Collins, Introduction to *The Rights of Man*, 1969). Although Paine claimed that his pamphlet offered 'nothing more than simple facts, plain arguments, and common sense' (*Common Sense*, p. 81), for the most part its effects were achieved through an inspiring rhetoric which convinced Americans (and is still used to convince them) that they have the power and opportunity 'to begin the world over again' (p. 120).

The rhetoric and argument of *Common Sense* – especially the discussion of the origin of government and reflections on the British constitution (pp. 65–71) and the attack on monarchy (pp. 71–81) – form a kind of apprentice work for the two parts of *The Rights of Man*. Resident in England from May to September 1789, Paine witnessed the early enthusiasm there for the French

Revolution. Paine travelled to Paris in September and associated with those who had drafted the Declaration of the Rights of Man (which was greatly influenced by America's Declaration of Independence). When he returned to England, Paine was faced with the beginnings of an about-turn in the English perception of the Revolution, culminating in the publication in November 1790 of Edmund Burke's *Reflections on the Revolution in France*. The two parts of Paine's *Rights of Man* form the most famous of the many 'replies to Burke' which appeared over the next two years. Henry Collins argues that 'In the persons of Burke and Paine, eighteenth-century England produced two superb prototypes of conservative and radical thought and started a debate which is still in progress.' The sales of *The Rights of Man* were even more startling than those of *Common Sense*: E. P. Thompson is 'inclined to accept the claim of 200,000 sales in England, Wales and Scotland . . . in 1791 to 1793' (*The Making of the English Working Class*, 1963). Such huge sales – and the fact that it was made available to the poorer classes through cheap editions – seems to have influenced the government's response as much as the book's blatantly revolutionary contents. As 'the first modern political writer to express himself in the language of ordinary men' (Collins), Paine was perceived as endangering Britain's traditional order by encouraging 'the people' fully to enter political discussion and activity.

Part I of *The Rights of Man* is primarily a refutation of Burke, a critique of Britain's much-vaunted parliamentary system, an attack on monarchy and aristocracy, and a defence of the French Revolution which seeks to 'correct' Burke's 'misinterpretations' of crucial events such as those of 5 and 6 October 1789. Burke's attempt to rouse sympathy for the French clergy or for Marie-Antoinette is actually a heartless deception given the immense amount of suffering the poor endured under the *ancien régime*: Burke 'pities the plumage, but forgets the dying bird' (p. 73). Against Burke's argument that the traditional hereditary system in Britain follows the pattern of nature, Paine stresses the unnaturalness of primogeniture and of the attempt to make the documents of 1688–9 bind succeeding generations to a certain form of government for all time. On the contrary, Paine contends 'for the rights of the *living*, and against their being willed away, and controlled and contracted for, by the manuscript assumed authority of the dead' (p. 64). Thus the rights of men are presented as *natural* and in contradistinction to the artificially maintained differences in rank which sustain traditional societies. Part II of *The Rights of Man* presents an even greater radical critique of the English constitution and opens by challenging the government to bring criminal charges against it, punning that 'it must be *criminal* justice indeed that should condemn a work as a substitute for not being able to refute it' (p. 177). Paine's early position as an excise man comes to the fore in this part in that he proposes to revolutionise the very notion of taxation. Paine's ideal society would use taxation not to rob the poor to make the rich richer but to redistribute the nation's wealth in order to offer what many have seen as a prototype model of the welfare state.

Paine thus seems modern in many ways: he wanted to replace aristocratic society with democracy; he defended the rights of nation states to political autonomy; and he wanted to remove venerable traditions and institutions which impeded the rise of men of ability, inhibited capitalist enterprise, and infringed the rights of the individual. It is possible, therefore, to see potential contradictions in Paine's thought which only became apparent when the Industrial Revolution produced its own unique repression and deprivation of the working classes or when the United States became the immense world power he predicted it would be. Paine is thus both a hero of the working-class movement and a figure who can be quoted when Americans seem to be losing their vision of themselves, as being at their greatest in 'the times that try men's souls' (*Crisis* papers).

TOM FURNISS

Painting

The greatest English painting of the Romantic period is that of the landscape-painters Constable and Turner, who are discussed separately, as is Palmer. But during his lifetime Constable's exclusive attention to English landscape was thought to be a disadvantage, and when he was elected to the Royal Academy he was told by the President, Sir Thomas Lawrence that he should consider himself fortunate.

Lawrence (1769–1830) was a successful and popular portrait-painter, who was an expert with the aristocracy. Portrait-painting had several distinguished practitioners in eighteenth-century England, including Hogarth, Reynolds and Gainsborough, and Lawrence carried on the tradition with great technical skill, like Sir Henry Raeburn (1756–1823) in Scotland. Lawrence's portraits of actors such as John Philip Kemble (in the roles of Hamlet and Coriolanus) were dramatic, in more than one sense; but he was chiefly known for his ability to portray the Prince Regent, later George IV, without making him look too fat. In postures of substantial grandeur, often imitated from the Apollo Belvedere or other classical models, the Prince appeared noble and heroic. The age admired heroism, naturally, and both Nelson and Wellington were frequently painted; in France, the wonderfully talented Jacques-Louis David painted his heroes and martyrs of the Revolution (Lepellétier, Marat), and later produced vivid and glamorous portraits of Napoleon Bonaparte.

Heroic painting was usually associated with portraits of individuals, which is a significant fact in itself; but there were other painters who attempted to revive the art of heroic painting in England. In the late eighteenth century, the most notable examples were James Barry (1741–1806), who decorated the Society of Arts with immense pictures (1777–83), and who criticised British art for being concerned with 'little inconsequential things – portraits of dogs, landscapes &c'; and the American-born Benjamin West (1738–1820), who became President of the Royal Academy and who painted many classical, biblical and historical subjects. During the Romantic period, Barry and West were succeeded by Benjamin Robert Haydon (1786–1846), a painter with noble ideas about high art who was overwhelmed with excitement at the arrival of the Elgin marbles in London (1808) and who aimed to raise historical paintings to a new height in Britain. Haydon was an aspiring and enthusiastic man, who won the friendship of Wordsworth and Keats; but his paintings were on the whole unsuccessful, and his grandiose schemes got him into debt. He finally committed suicide in front of one of his unfinished canvases. He is chiefly remembered for his fine portrait of Wordsworth (now in the National Portrait Gallery), and for his astonishingly lively and frank *Life* from his autobiography and journals (edited by Tom Taylor and published in 1853). This includes a wonderful description of an 'immortal dinner' given by Haydon in December 1817 at which Wordsworth, Keats and Lamb, among others, were present.

Haydon saw the Elgin marbles at the same time as his friend and teacher Henry Fuseli (Johann Heinrich Füssli, 1741–1825), a Swiss-born artist who came to London and who became a Professor at the Royal Academy. Fuseli painted many illustrations to the works of Shakespeare and Milton: he was particularly interested in *Macbeth*, with its opportunities to depict crime, remorse and witchcraft. His paintings were preoccupied with such subjects: his interest for the student of the Romantic period is in the treatment of subjects such as *The Nightmare* (1782), in which the night-mare pushes its face through the bed curtains, and an incubus sits beside the twisted figure of a tormented sleeping girl (see under DREAMS), and *Young Woman Imprisoned with a Skeleton*. Fuseli's pictures, which are often fantastic, Gothic or erotic, are concerned with feelings that exist beyond the normal parameters of established and respectable society.

Another strange and individual painter was John Martin (1789–1854), a painter with an acute eye for the dramatic event and the spectacular subject. His first major work, *Sadak in Search of the Waters of Oblivion* (1812), shows a solitary figure in a vast and surreal landscape. Martin also painted *The Bard* (1817), the solitary figure from Gray's poem; he loved painting catastrophes, including *The Fall of Babylon* (1819), *Belshazzar's Feast* (1821), *The Destruction of Herculaneum* (1822) and *The Seventh Plague* (1823). A picture of 1839, *The Last Man*, takes up a theme which was also explored by Mary Shelley, Byron and Thomas Campbell.

Martin wanted his pictures to be grand, arresting, stupendous: his last works – *The Last Judgement*, *The Great Day of his Wrath*, and *The Plains of Heaven* – are vast canvases, attempting to achieve the sublime by spectacle and size. At the other extreme was the narrative and genre art of David Wilkie (1785–1841). Wilkie made his early reputation with *Pitlessie Fair* (1804), which encouraged him to move from his native Scotland to London, where he continued to be very successful with *The Village Politicians* (1806) and *The Blind Fiddler* (1807). His observation of village life was primarily influenced by the Dutch School, though it also has affinities with Wordsworth's interest in humble and rustic life; but it lacks Wordsworth's sense of the tragic and sublime, and looks forward more directly to Victorian narrative art. Later Wilkie's subjects, while remaining those of narrative and genre, became associated with heroic or historical events, as in *Chelsea Pensioners Reading the Gazette of the Battle of Waterloo* (1817–21) and *The Duke of Wellington Writing a Despatch* (1836). Wilkie was present in Edinburgh at the celebrated visit of George IV in 1822, and painted *The King's Entrance to Holyrood* (1830); his other principal Scottish picture of these years was a subject from the Reformation, *The Preaching of John Knox before the Lords of the Congregation, 10 June 1559.*

Martin's energies were directed towards the sublime and the apocalyptic; Wilkie's towards the provincial and historical. Other painters attempted to engage more directly with the contemporary world of science and industry, painting the new industrial landscape with its unusual machinery and its lurid light. Philip James de Loutherbourg (1740–1812), a scene-painter at Drury Lane Theatre, became famous for his spectacular lighting-effects in his Eidophusikon, an illuminated series of scenes. He travelled to Shropshire to see Coalbrookedale and Madeley, and painted the industrial scenes there (see Francis Klingender, *Art and the Industrial Revolution*, rev. ed, 1968). The principal painter of the new science was Joseph Wright of Derby (1734–97), who had a taste for subjects strongly lit by candlelight or by forge fires, as in *An Iron Forge* (1774). In *Arkwright's Cotton Mill* (1783) the mill is seen by moonlight, with all its windows illuminated. He is best known, however, for his paintings of scientific experiments, such as *A Philosopher Giving that Lecture on the Orrery, in which a Lamp is Put in the Place of the Sun* (1766), or *An Experiment on a Bird in the Air Pump* (1768), showing a vacuum pump at work. In these paintings Wright used chiaroscuro effects (strong contrasts of light and shade, as in Caravaggio) to celebrate the beginnings of industrial science.

J. R. WATSON

Palmer, Samuel (1805–81)

Samuel Palmer was born in London, the son of a bookseller; he showed early promise in painting and drawing, and was given lessons; but his true education in art dated from his meeting with Blake's friend John Linnell in 1822. Through Linnell, Palmer met Blake a number of times before the latter's death in 1827, and was greatly influenced by him. He said that to walk with Blake in the country 'was to perceive the soul of beauty through the forms of matter' (Robert Melville, *Samuel Palmer*, 1956). Palmer became one of the youthful admirers and followers of Blake known as 'the Ancients' (because they thought the ancients superior to the moderns), a group including such figures as Linnell, Edward Calvert and Frederick Tatham. In 1826 Palmer and Tatham went to live at Shoreham, in Kent. For Palmer this became his 'Valley of Vision' (see Geoffrey Grigson, *Samuel Palmer: The Visionary Years*, 1947): he saw it through the eyes of a follower of Blake as 'Inchanted ground' and 'the Countrey of Beulah', taking his imagery and sensibility from the Bible, *Paradise Lost* and Bunyan's *Pilgrim's Progress*. His paintings of these years are some of his finest: they contain landscapes seen with an eye that is visionary and profoundly religious. In 1832 he began the process of moving back to London, which he completed in 1834; thereafter his work, though often very fine, 'never reattained the intensity which Shoreham had wrought in him' (Raymond Lister, *Samuel Palmer*, 1974). He married Linnell's daughter Hannah, and made painting-tours in Wales and Italy; in the years that followed he worked hard to maintain himself as a painter and illustrator, and provided some of the illustrations for Dickens' *Pictures in Italy*. But these years were troubled by arguments with the assertive Linnell, and it was not until his late drawings, illustrations for Milton's 'L'Allegro' and 'Il Penseroso', and for Virgil's Eclogues, that he recaptured the earlier greatness.

Palmer's religious sensibility is found in many of his pictures. Some of them, such as *The Repose of the Holy Family* (c.1824-5) and *Ruth Returned from the Gleaning* (1828), have biblical subjects; others, such as *The Valley Thick with Corn* (1825), have biblical references attached to them (in this case to Psalm 65). More often, the paintings of this period are not specifically religious, but full of imagery of abundance and richness, with trees in full blossom (*In a Shoreham Garden*, 1829) or plump and cared-for sheep (*The Magic Apple Tree*, 1830) or harvest fields full of sheaves (*Cornfield by Moonlight, with the Evening Star*, c.1830). The moon appears in many of Palmer's paintings, often casting a rich and unusual light over the landscape (cf. Coleridge's remarks about moonlight and landscape in *Biographia Literaria*, 14). Another frequent feature is a church with a spire, found in *The Magic Apple Tree, Hilly Scene with Church and Moon* (c.1826), and the well-known *Coming from Evening Church* (1830). In all these paintings the features of the landscape combine with the figures of shepherds, or families (in *Coming from Evening Church*) to give an overwhelming impression of peace and fertility, abundance and benevolence. 'The earth is full of thy richness', he wrote in one of his sketch books, and his work is constantly infused with the sense of a Divine Creator whose hand is visible in his works. In Robert Melville's words, Palmer 'was able to convert all his impressions into emblems of God's grace'.

J. R. WATSON

Pantheism

The word 'pantheism' (signifying 'all [Greek *pan*] is God [*theos*]'), was coined by John Toland in 1705. Pantheism is a kind of monism: all that exists is one single substance and this substance is God. The concept rejects all distinction between the creator and its creation; therefore God is no longer distinct from the world, no longer transcendent, but merely immanent in the world. As a consequence of the confusion of God and nature, pantheism has sometimes been defined as 'the worship of nature as divine'. Frequent slack usage of the word suggests a need for caution.

Suggestions of pantheism are often found in Romantic poems. John Toland himself heralded the later practice of associating pantheism with nature-worship when he wrote in his *Pantheisticon* that the pantheists might be called 'the Mystæ and Hieophants of Nature'. The worship is not necessary; the presence of an immanent active power is. Shelley in *Queen Mab* saw throughout the world 'wide diffused / A spirit of activity and life / That knows no term, cessation or decay' (VI.147–9). When Emerson writes in his *Journals* (26 May 1837), 'I see my being imbedded in [the One Universal Mind]; as a plant in the earth so I grow in God. I am only a form of him. He is the soul of me', he may well be a pantheist, but his meaning must be determined by reference to his work. When Coleridge writes his famous words about the One Life – 'every Thing has a Life of it's own & . . . we are all *one Life*' (letter to Sotheby, 10 Sep 1802) – he may only be giving voice to an idea of the times. In the same year the French doctor Cabanis regarded as chimerical Buffon's distinction 'between living and dead matter'. But Coleridge and Wordsworth had discussed Spinoza's doctrine that God is the world – *deus sive natura*: an all-pervading presence not unlike the 'something far more deeply interfused . . . that rolls through all things' in 'Tintern Abbey'. The words 'presence' and 'life' soften the feeling in this poem, but the beginning of the ninth book of *The Excursion* claims the existence of an '*active* Principle' which 'subsists / In *all* things, in *all* natures . . . in *every* pebbly stone . . . from link to link / It circulates, the Soul of *all* the worlds' (emphasis added). The omnipresence of this active principle is strongly similar to pantheistic conceptions, though Wordsworth, in a letter of 1815, refused to admit that anywhere in the poem he failed to distinguish 'between nature as the work of God and God himself'. But there is a discarded passage of *The Prelude* where the identity of the creation and the creator – the confusion of the One and the many – and its implicit rejection of a transcendent deity, together with the assertion of the unity of all things, define pantheism: Wordsworth speaks of the one interior life

> In which all beings live with God, themselves
> Are God, existing in the mighty whole
> As indistinguishable as the cloudless east
> At noon is from the cloudless west when all
> The hemisphere is one cerulean blue.
> (*The Prelude*, ed. E. de Selincourt, 1926, p. 525)

Such passages and even milder expressions of the same ideas, irritated Coleridge, who responded to them as orthodox Christians do: he wrote that the trait he disliked most in Wordsworth was 'the vague, misty, rather than mystic confusion of God with the world, and the accompanying nature-worship' (Thomas Allsop, *Letters, Conservations and Recollections of S. T. Coleridge*, 1864, pp. 57–8). He tried to fight those tendencies in his friend and partly succeeded. Lines 253–68 of book XIII of the 1805 *Prelude* show a transition from a somewhat ambiguous assertion of the unity of the world to a clear one of the separation of God and his creation. Coleridge's

attitude illustrates the attitude of theologians. They strongly condemn pantheism. In *Aids to Reflection* (1825) he voices, as it were,
their repugnance: his hand, he says, trembles as he writes of those
who reduce the creator to a mere *anima mundi* and who would
substitute some sort of universal principle for the personal God of
Christianity.

K. C. F. Krause (1781–1832), in order to solve those difficulties,
coined the term 'panentheism'. The introduction of the Greek
particle *en* was to emphasise that all is not God, but that all is *in*
God. Some theologians and several scholars readily accepted the
notion. Others showed that it might as well be used of Spinoza and
argued that it made no real difference. From the point of view of
the student of literature, what really matters is the correct appreciation of the ideas of the poet more than the giving of a name. And
from the point of view of the theologian the nature-worshipper is
no more a Christian than the sun-worshipper is. Yet the word
'panentheism' has also been thought less morally fatal than 'pantheism', because its deterministic consequences do not imply a
denial of evil.

MARCEL ISNARD

Peacock, Thomas Love (1785–1866)

Thomas Love Peacock, like his own life-span, fits into no literary
category. He began as the friend of Shelley, and ended as the
father-in-law of Meredith; he was a man of business who, as Chief
Examiner at the East India Office, was at the top of the vast
concern which Charles Lamb, as a clerk there, characterised as
'South Sea House'. He kept his private and public life reticently
apart from his literary self, using in his writing only the figures,
fads and idiosyncrasies of the major Romantics as the material for
his witty, humorous, unmalicious and compact body of prose
fiction.

Normally classified as a novelist, in that his work comprises
dialogue, characters and a modicum of plot, he can be grouped
only with figures as mutually discrepant as Rabelais, Bunyan, the

Landor of *Imaginary Conversations*, the dramatists Goldsmith and Sheridan, and the librettist W. S. Gilbert, with whom he shares, variously, a relish for good living, a reliance on direct speech, delight in humour and nonsense, and a deft way with verbal dexterity and unmalicious satire.

He began his writing-career as a poet, trying various modes, but the only poems of his that are still read today are the various elegant lyrics and songs that intersperse his novels; some of these, such as 'The War-Song of Dynas Vawr', have become anthology pieces. His friendship with Shelley destroyed his poetic pretensions, in that Peacock recognised the genius of Shelley (producing, in 1858, the perceptive *Memoirs of Percy Bysshe Shelley*) and in that Shelley immediately offered perpetual exercise for Peacock's sympathetic sense of the ridiculous, and material for it to work on.

Shelley, in some guise and under some name, appears in more than half Peacock's novels, along with most of the major Romantics – Wordsworth, Coleridge, Byron, Southey and Leigh Hunt most conspicuously – and with them their theories, ideals and poses, sharpened though not distorted to the ends of delighted ridicule. Diverted, like Jane Austen's Elizabeth Bennet, by follies and inconsistencies, in his work he consistently laughs at them, and at all the oddities of the Romantics, from Shelley's vegetarianism, Byron's youthful misanthropy and Hunt's pantisocracy to the incongruity – as he sees it – of Southey's and Wordsworth's idealistic theories and their subsistence on governmental appointments. But he never jibes at the private man, only at the public published image. The friend of Shelley and both his wives divulges nothing of the private scandals and distresses, and little even of Byron's far more public ones. The impulses that formed the English Romantic movement as a whole are the main impulse for this unromantic and humane writer.

Despite his long life, his writing-career was a compact one of sixteen years: beginning with *Headlong Hall* in 1816, then continuing with *Melincourt* (1817), *Nightmare Abbey* (1818), *Maid Marian* (1822), *The Misfortunes of Elphin* (1829) and *Crotchet Castle* (1831). His last novel, *Gryll Grange*, published twenty-nine years later, in 1860, is an old man's *recherche du temps perdu*, still relishing his beginning-of-the-century youth, and showing no sign that the Victorian novel exists.

Standing somewhat apart from his usual mode are the two mediaevalising and antiquarian works *Maid Marian* and *The Misfortunes of*

Elphin, the one preceding and wholly unlike Scott's *Ivanhoe* in handling the Robin Hood story, the other using Welsh and Arthurian legend long before Charlotte Guest's *Mabinogion*.

Peacock's other works are in general expanded comic symposia, set in a country house and relying above all upon entertaining dialogue and debate. The pattern requires the presence of one man of good sense and good breeding, with a witty and intelligent woman; and the conclusion is a celebration of good fellowship, good food and good wine – a Platonic structure of which Peacock as a good Grecian himself must have been aware. Peacock the scholar and classicist deploys and enjoys his own pedantry in citations, learned footnotes and references for even his most fantastic ideas, most memorably in the 'character' of Sir Oran Haut-Ton in *Melincourt*, who embodies the virtue of the Rousseauist natural man, enacts the deeds of a Romantic hero, and becomes MP for the rotten borough of One-Vote, all without departing from what contemporary knowledge states are the normal abilities of the orang-outang, duly described in footnotes.

Peacock has received relatively little critical attention, though always having 'fit audience though few'. He gets a respectful mention in all histories of literature, the details varying according to the sense and wit of the writer. The first full-length biography of him was by J. B. Priestley (1927). The more serious studies of the last fifty years tend to treat with extreme solemnity the problem of his politico-ideological view, finding him reactionary, conservative, radical, liberal, and of all shades between. Similarly his comedy is characterised as like that of Aristophanes, Petronius, Juvenal, Apuleius and (by Northrop Frye) Menippus, and as used to the ends of satire, though the sense of its direction varies. The urge to make a coherent judgement of the whole output of a writer who designedly kept his own opinions largely out of his writings has as yet brought no consensus. What is general is approval; adverse judgements are few, and rest mainly on his lack of seriousness or a fixed satirical stance, a case well set out by A. E. Dyson in *The Crazy Fabric* (1965).

W. A. CRAIK

Periodicals

The Romantic period saw a great increase in the number of books and readers, and a corresponding development of literary periodicals. Between 1789 and 1834 literary articles appeared in about 100 different weeklies, monthlies and quarterlies (only the most important are discussed here). The main types were the review, dealing with new books of all kinds; and the magazine, a miscellany whose contents would include essays, poetry, criticism and sometimes fiction.

Among reviews, the monthly journals on the eighteenth-century pattern still covered as many works as possible, giving summaries, extracts and a restricted (though not negligible) amount of criticism. Such were the *Monthly Review* (1749–1845), with Whiggish and dissenting traditions; the *Critical Review* (1756–1817), long a Tory periodical but taken over by Foxite Whigs around 1792–1801, when Coleridge and Southey were contributors; the *Analytical Review* (1788–99), which published many articles by Mary Wollstonecraft and other radical and dissenting friends of the publisher Joseph Johnson; and the *British Critic* (1793–1843), founded to take the side of Church and King and, with Thomas Percy and George Ellis as contributors, particularly strong on early literature. This review press was on the whole helpful to the first Romantic poets, and in the *Monthly Review* William Taylor introduced important new German writing to British readers.

In 1802 Francis Jeffrey, with his friends Sydney Smith, Henry Brougham and Francis Horner, began the *Edinburgh Review* (1802–1929). Appearing quarterly, moderately Whiggish in tone, it dealt with far fewer books than its predecessors but those it did discuss it treated at greater length and in a more critical and entertaining fashion. Its success called forth a Tory counterpart, the *Quarterly Review* (1809–1967), founded in 1809 with the aid of Walter Scott and edited until 1824 by William Gifford. By 1818 each had a circulation of around 15,000, three times that of any earlier review. The *Westminster Review* (1824–1914), on the same pattern, was begun by James Mill to voice the views of the utilitarians and 'philosophical radicals'.

The new reviews were important channels of opinion and attracted good contributors: Hazlitt and Macaulay wrote for the *Edinburgh*, Scott and Southey for the *Quarterly*, the two Mills and

Carlyle for the *Westminster*. But they did not perform the services that might have been expected in introducing the Romantic poets to a wide public. In the *Edinburgh*, Jeffrey's derisive attack on the Lake poets (Oct 1802) was renewed each time work by Wordsworth or Coleridge appeared. Keats was pilloried by J. W. Croker in the *Quarterly* (number for Apr 1818, published Sep 1818) and lamely and belatedly defended by Jeffrey (*Edinburgh*, Aug 1820). Scott and Byron fared better, but they were already popular favourites.

Among monthly miscellanies, the *Gentleman's Magazine* (1731–1914) now specialised in antiquarian topics. A livelier work was the *Monthly Magazine* (1796–1825), published by the radical bookseller Richard Phillips and edited until 1806 by John Aikin. Contributors included Coleridge, Southey, Godwin, Hazlitt and William Taylor, whose influential translation of Bürger's 'Lenore' appeared there in March 1796. A successful rival, set up to counteract the bias towards liberal politics and dissent, was the *New Monthly Magazine* (1814–1836), edited after 1821 by the poet Thomas Campbell.

The most striking success among the newer monthlies was *Blackwood's Edinburgh Magazine* (1817–1980). In the hands of J. G. Lockhart, James Hogg and John Wilson, *Blackwood's* quickly won fame by its journalistic stunts, high-Tory politics, and scurrilous attacks (such as the campaign begun in October 1817 against 'the Cockney School', a group of 'vilest vermin' which included Leigh Hunt, Hazlitt and Keats). More creditably, *Blackwood's* published much original prose fiction, including John Galt's *Ayrshire Legatees* (serialised 1820–1). Writers in *Blackwood's* also moved away from the anonymity of earlier periodicals to exploit versions of their own personalities – especially in 'Noctes Ambrosianae' (1822–35), a series of conversations portrayed as taking place in 'Ambrose's Tavern' between 'Christopher North' (Wilson), 'the Ettrick Shepherd' (Hogg) and others. Despite much that is lamentable in its record, *Blackwood's* became a much-loved presence in the homes of distant readers such as the young Brontës.

Another miscellany offering approachability through pseudonyms was Baldwin's *London Magazine* (1820–9). Its first editor, John Scott, attracted a brilliant group of writers: Lamb's *Essays of Elia*, De Quincey's *Opium Eater*, much of Hazlitt's *Table Talk*, and poems by Keats, Clare and Hood first appeared in the *London*. But its circulation was never large (around 1600, as opposed to *Blackwood's*

6000), and, after John Scott had been killed in a duel, arising out of a quarrel with *Blackwood's*, in 1821, his place was less ably filled by John Taylor.

Among weekly magazines, *The Anti-Jacobin* (1797–8) is remembered for its parodies of Southey and others by George Canning and John Hookham Frere. *The Examiner* (1808–81), edited until 1819 by Leigh Hunt, was an outspoken reformist paper whose contributors included Hazlitt, Lamb, Wordsworth, Byron, Keats and Shelley. *The Champion* (1814–22), edited till his death by John Scott, also published work by Hazlitt and Lamb and produced some of the best literary criticism of the period. *The Literary Gazette* (1817–62), edited for thirty-three years by William Jerdan, succeeded *The Champion* as the most important literary weekly, with contributions by George Crabbe and Mary Russell Mitford.

Newspapers did not usually publish literary criticism, but most included a column of verse, which could provide an important outlet for writers: some fifty of Coleridge's poems first appeared in the *Morning Chronicle*, *Morning Post* and other papers.

DEREK ROPER

Peterloo

'Peterloo' or 'the Peterloo Massacre' was the name coined almost immediately after the event to refer to the violent dispersal of 60,000 people gathered in St Peter's Fields, Manchester, on 16 August 1819 to hear 'Orator' John Hunt. The meeting was called to support parliamentary reform in general, and to elect Hunt as 'representative' for Manchester. It formed part of a pattern of meetings and agitation, begun in the aftermath of Waterloo, demanding electoral reform, employment, higher wages and lower prices. A series of poor harvests, in the wake of the return to Britain of thousands of ex-soldiers, led to legislation to keep the price of corn high, active protests by committed radical groups, and government legislation to suppress the protests. All this formed the background to the publication of the major poems of Byron, Keats and Shelley.

The 'Peterloo Massacre' appeared at the time to be the culmination of a struggle between revolutionary and repressive movements such as had already been experienced on the Continent. In the event it was not, and, though Peterloo may have played some part in the later adoption of the Reform Act (1832), it is principally important as myth rather than event. It is as a myth that Shelley, the only great writer to have devoted energy to Peterloo, treats the subject in his 'The Mask of Anarchy' (written in 1819 and sent to Leigh Hunt to be published in *The Examiner* but not published until 1832). The essence of the myth is that a huge crowd of peaceful English people demonstrating for political reform were attacked directly by cavalry armed with sabres and that this was done, directly or indirectly, on the orders of the British government. Arthur Thistlewood, for instance, leader of the 'Cato Street conspiracy' (1820) determined to avenge the deaths in Manchester by assassinating the entire British Cabinet. He was arrested and hanged. The facts of Peterloo, sufficiently unpleasant, have more of accident and farce about them than the myth implies. The magistrates, indecisive throughout, allowed the crowd to assemble but then decided to arrest Hunt and the other speakers. Magistrates had taken a similar course of action previously at a smaller but similar meeting in St Peter's Fields (Mar 1817). The constabulary refused to make arrests without military assistance. The Manchester Yeomanry (a local body of volunteer soldiers) were successful in making the arrests but then became hemmed in by the crowd. The Riot Act had been read but scarcely heard and the Fifteenth Hussars, professional soldiers kept in reserve, were sent in to disperse the crowd and rescue the trapped Yeomanry. The Hussars were wearing the battle colours which they had won at Waterloo. In the resulting chaos, eleven people were killed and up to 500 injured. The whole operation was over in twenty minutes.

Most of the many cartoons of the incident show cavalry (the hated Yeomanry) directly slashing women and children with sabres. Shelley's poem, though it is not intended to represent the occasion directly, implies that bayonets and 'charged artillery' were used. The true horror of Peterloo was of a different kind. Central government was certainly not directly implicated in the affair but publicly supported the magistrates' action after the event. It hurriedly passed the Six Acts of December 1819 (1) to forbid arming and drilling; (2) to permit magistrates to search for arms; (3) to compel defendants to respond at once to charges; (4) to prohibit meetings of more than fifty persons without permission from a magistrate; (5) to increase penalties for blasphemy and sedition; and (6) to introduce a prohibitive stamp duty for pamphlets and papers.

Hunt, a vain and shallow figure, was subsequently imprisoned but, whilst on bail, made a triumphal entry into London. It was this event, more than the 'massacre' itself, which attracted the attention of Keats (letter to George and Georgiana Keats, 17–27 Sep 1819). Public reaction to Peterloo was unfavourable, but, except as the occasion of Shelley's brilliant poem, Peterloo has had more resonance as a political myth in the twentieth century than it had in the nineteenth.

BERNARD BEATTY

The Picturesque

Like the 'sublime', the 'picturesque' was at first a term of literary criticism. When applied to style, it was a synonym for 'graphic' or 'vivid'. After the beginning of the eighteenth century, it came also to designate a natural scenery and then it meant 'suitable for representation'. Smollett uses the word in *Humphry Clinker* (1771) to describe the mountainous landscapes of Scotland: it is then associated with 'romantic'.

The aesthetic concept was more clearly defined by three main writers: William Gilpin, Uvedale Price and Richard Payne Knight, commonly referred to collectively as 'the picturesque school'. The Revd William Gilpin (1724–1804) began his famous tours round the country in the late 1760s, sketching and taking notes as he went along. He was encouraged to publish his observations and pictures by two of his friends, the poets Thomas Gray and William Mason. His first account was entitled *Observations on the River Wye . . . Relative Chiefly to Picturesque Beauty* (1782). Others followed, with similar titles, on the Lake District (1786), the Scottish Highlands (1789) and forest scenery (1791) – six volumes in all, the author trying to understand the aesthetic potential of the landscapes he visited, the reason why they inspired painters. In addition, he developed his theory in *Three Essays* (1792). These studies – 'On Picturesque Beauty', 'On Picturesque Travel', and 'On Sketching Landscape' – had been submitted to Reynolds as early as 1776. The President of the Royal Academy commented upon them in his tenth Discourse (1780) and in a personal letter to Gilpin (1791), in both cases demonstrating that he had no high opinion of the picturesque.

Uvedale Price (1747–1829) was a gentleman farmer. Considering Gilpin's concepts too vague, he set out to supply his own definition and at the same time tried to realise his ideas on his estate at Foxley, Herefordshire. A disciple and an admirer of Burke, 'he set about converting Burke's dualism into an aesthetic trinity' (Samuel Monk) in *An Essay on the Picturesque* (1794). R. Payne Knight, in his poem *The Landscape* (1794), attracted a good deal of attention with his attack on Lancelot ('Capability') Brown's style of landscape gardening. There followed a violent quarrel between him and Brown's heir, Humphry Repton, with supporters on each side. The polemics eventually set Price and Payne Knight at loggerheads, but it helped the picturesque school to refine its ideas. Price published *A Dialogue on the Distinct Characters of the Picturesque and the Beautiful* (1801), and Payne Knight prolonged the controversy with *An Analytical Inquiry into the Principles of Taste* (1805).

What are the qualities which make objects picturesque? For Gilpin, if smoothness and neatness are beautiful, roughness and ruggedness are picturesque. For Price, beauty as defined by Burke and as practised by the 'Brownist' school of landscape gardeners lacked two basic attributes: variety and intricacy – a temple in the Grecian style is beautiful when whole, picturesque when ruinous. This concept also differs from the sublime: it has no necessary connection with dimension, infinity or uniformity. The objects that appeal most to the painter are those which offer the greatest variety of shape, colour and light. Studying art, having an artistic culture, can be useful in helping to detect and appreciate the picturesque in nature.

Like their predecessors, the theoreticians of the new aesthetic category were concerned with its psychological effects. Like Burke, Gilpin spoke of a 'pause of our judgment', a 'deliquium of the soul', but unlike him he did not favour any physiological theory: 'it has nothing to do with . . . irises and retinas'. Price followed in his footsteps: 'the effect of the picturesque is curiosity . . . it neither relaxes nor violently stretches the fibres'. Payne Knight's approach was even more radical: it did away with any psychological dimension, to limit the picturesque to a form of vision, effects on the eye being more important than emotions. He is sometimes seen as the one who opened the door for Turner's impressionistic play with shape, line, light and colour.

Thus, from the start and by definition, the picturesque was intimately connected with painting. Two varieties of it can be distinguished: the 'homely' and the 'grand' picturesque. Gainsborough was the best practitioner of the first manner: in *The Harvest Wagon* (1770), *The Watering-Place* (1777) and *The Cottage Door* (1780), for instance, he selects and arranges his elements in the best style defined by Gilpin. Ragged country folk appear among shaggy trees in well-composed landscapes exhibiting great variety of line and colour and a remarkable play of light and shade. His *Rocky Mountain Landscape with Sheep* (1783) verges on the 'grand' picturesque.

In the latter type of paintings, mountains reign supreme, their effect often enhanced by some natural cataclysm or storm: colours are dark or violently contrasted; human beings, if there are any, appear crushed by hostile nature. Masters of the grand picturesque include Joseph Wright of Derby (*The Eruption of Mount Vesuvius*, 1774), Philip James de Loutherbourg (*An Avalanche in the Alps*, 1803; and see his book *The Romantic and Picturesque Scenery of England and Wales*, 1805) and the watercolourists John Robert Cozens (*Mont Blanc and the Arve Near Sallanches*, 1776; *The Lake of Nemi, Looking towards Genzans*, 1777), and Francis Towne (*The Source of the Arveyron; Moni Blanc in the Distance*, 1781).

SUZY HALIMI

Political Thought

This period saw basic changes in English political assumptions: the word 'democracy' lost most of its pejorative bias. As always, the thinking of the time was driven by social needs and pressures. A system dominated by landowners and controlled, through patronage, by the political families who gained power in 1688 had, after nearly a century, come under attack by overlapping groups of the excluded: commercial men, dissenters, voteless inhabitants of the new industrial cities. A campaign to reform the extremely unrepresentative electoral system was already under way in the 1770s, when Wordsworth was a child, and the beginning of the French Revolution in 1789 greatly stimulated reformers. But its more violent developments and the war which followed (1793–1815) were used to discredit liberal ideas and justify a repression which continued into the hungry post-war period, most dramatically at 'Peterloo' in 1819. A limited measure of parliamentary reform was achieved only in 1832, when Wordsworth was sixty-two. Severe social problems were posed by the Industrial Revolution and the large numbers of paupers in town and country. Against this background radically diverse political theories were being canvassed.

Locke had justified the Revolution of 1688 in terms of the 'social contract'. In this theory rulers derived their authority from the governed, who surrendered part of their natural rights in exchange for the benefits of social order. When these benefits were not received, men were entitled to resume their rights and change their rulers or form of government. This theory was used to justify the American Revolution, and later by reformers such as Richard Price in his *Discourse upon the Love of our Country* (1789). It was also the point of departure for Thomas Paine in his democratic and republican *Rights of Man* (1791), immensely popular despite the attempts of government to suppress it. In *A Vindication of the Rights of Woman* (1792), Mary Wollstonecraft appealed for women to be given better education, more equal opportunities, and perhaps one day the right to vote.

For Jeremy Bentham (1748–1832) reforms needed no justification from history or myth. His *Fragment on Government* (1776) argued that all institutions could be judged simply by their effectiveness in promoting general well-being – 'the greatest happiness of the greatest number'. Given enough information this 'utility' could be measured, as Bentham indicated in his *Principles of Morals and Legislation* (1789) and other works. In a more idealistic fashion, William Godwin also applied the test of utility in his *Political Justice* (1793). He believed men were perfectible (though perfection was far distant), and that the ideal state – probably a federation of small self-governing communities – would maximise their capacity for acting in a rational and disinterested way for the common good.

Like their Whig predecessors, these writers were concerned to diminish rather than increase the power of government. But in attacking injustices and abuses they were sometimes led to propose remedies which required a strong central power. In part II of *The Rights of Man* (1792) Paine advocated free compulsory education, family allowances and old-age pensions, all to be financed by a graduated income tax. Early forms of socialism were evolved in this period, most conspicuously by the manufacturer Robert Owen (1771–1858), author of *A New View of Society* (1813–14) and *Report to the County of Lanark* (1820). Believing that character was the product of environment, he created successful industrial communities at New Lanark and elsewhere in model villages where great attention was paid to education. Owen thought competitive capitalism gave rise to false values, and would have replaced it by economic co-operation through trade unions and community workshops. Others, such as Thomas Spence (1750–1814) in *The Meridian Sun of Liberty* (1775, 1796), argued for the common ownership of land.

In the confident tradition of the Enlightenment, such writers subjected all institutions to rational inquiry. This whole tradition was opposed by Edmund Burke, who in his *Reflections on the Revolution in France* (1790) argued that the political constitution was not the result of a bargain nor a matter of practical convenience: it expressed the accumulated wisdom of an organic national community, evolved over centuries under divine guidance, with which hasty reformers tampered at their peril. Ancient usage and prescription gave a better authority than mere reasoning. Burke's line of thought was later developed by Samuel Taylor Coleridge, especially in *The Friend* (1809–10), *The Statesman's Manual* (1816) and *On the Constitution of the Church and State* (1830). A different kind of conservative was Thomas Malthus (1776–1834), who held that, since the population tended to increase much faster than the supply of food, hand-outs to the poor only hastened the day when checks such as disease and famine would be necessary to keep numbers down. The poor could be helped only by education and self-discipline (*Essay concerning the Principle of Population*, 1798).

The seeds planted in this period flowered at different times. Paine and Godwin were eclipsed by the anti-Jacobin reaction of the later 1790s, though Paine's works survived in underground publications and fed into later working-class radicalism. Godwin's thought influenced that of Percy Bysshe Shelley, some of whose poems (e.g. *Queen Mab*) also enjoyed underground prestige. The ideas of Bentham and his followers, the utilitarians (led by James Mill), were adapted to specific social and legal reforms based on a close study of facts. Their thinking was unimaginative, anti-historical, and particularly uncongenial to the Romantics. But, in combination with that of Malthus and the economic individualism of Adam Smith, it was extremely influential in the 1820s and 1830s and left its mark on the whole century. The historical conceptions of Burke and of Walter Scott helped to shape the growth of nineteenth-century nationalism. After Coleridge's death, his thoughts on Church and state, partly derived from Burke, became influential through the Broad Church and Tractarian movements. The hour of socialism came later.

At the beginning of our period political writing was almost wholly by and for the upper and middle classes. By the end, political consciousness had developed among industrial and agricultural workers, and in this process the lively popular language used by Paine and by William Cobbett played a vital role.

DEREK ROPER

Religious Thought: Wesley, Swedenborg

At the end of the eighteenth century, the established Church of England was in a state of extreme spiritual and physical decay. Rural parishes were characterised by absentee incumbents or pleasure-seeking parsons, the growing industrial towns were largely unchurched, while the cities experienced a spiritual vacuum (there were six communicants in St Paul's Cathedral, London, on Easter Day 1800). Coleridge, the son of a clergyman, had doubts in 1800 about the baptism of his sons: 'shall I suffer the Toad of Priesthood to spurt out his foul juice in this Babe's face?' (letter to William Godwin). Blake satirised the clergy as lying hypocrites.

Yet Bishop Butler of Durham, in his *Analogy of Religion* (1736), had produced a powerful argument that we act not upon certainties but upon the aggregation of probabilities. This argument profoundly influenced Coleridge and, after him, J. H. Newman in his *Grammar of Assent* (1869), one of the last great documents of English Romantic religious thought.

The old dissenters from the seventeenth century – Presbyterians, Baptists, Independents (Congregationalists) – were in a similarly torpid state. They were all, more or less, Calvinist, and had almost no effect on Romantic thought. The Unitarians, however, a tiny sect basing their teaching on scriptural authority and rejecting the doctrine of the Trinity and the divinity of Christ, formed an intellectual elite with a fine educational system. It is claimed that the Warrington Academy in the 1790s was superior to Oxford and Cambridge. Among the great Unitarian families were the Wedgwoods and the Darwins, while for some years in the early 1790s Coleridge was a Unitarian and preached in their pulpits. In the next generation James and Harriet Martineau were leading Unitarians, while in the United States Unitarian thinking deeply influenced Emerson.

But the most significant development in the religious climate was associated with the growth of Methodism. Although its theology never differed from the Anglicanism from which it emerged, its emotional sense of the drama of the human condition characterised a central element in Romanticism. On 24 May 1738 an Anglican clergyman, John Wesley (1703–91), had an experience of conversion at a reading of Luther's Preface to the Epistle to the

Romans in a meeting at Aldersgate Street, London. He had already been deeply influenced by contact with the Moravians. He took to field preaching, intending 'to promote as far as I am able vital practical religion'. Methodism stressed a conversion of emotional acceptance rather than intellectual conviction, terror of damnation, and joy at Christ's freely given gift of salvation. Other great figures in Methodism were John's brother, Charles Wesley (1707–88), a gifted and prolific hymn-writer, and the Calvinist George White-field (1714–70). Charles's hymns include 'Hark! the herald angels sing', 'Jesu, lover of my soul' and 'Love divine, all loves excelling', and his verse can achieve great poetic subtlety. F. D. Maurice in the nineteenth century wrote that 'the Methodists . . . led other men into a belief which they did not entertain themselves, they were the unconscious and unacknowledged, but not the least powerful instruments of a great change in the views of philosophers. . . . A dynamical philosophy has gradually superseded a mechanical one' (*The Kingdom of Christ*, I, 1838).

Emmanuel Swedenborg (1688–1772), originally Swedberg, was a Swedish scientist and mystic. In his many writings he stressed the fundamentally spiritual structure of the universe through a combination of pantheistic and theosophic doctrines. Kant was one of his earliest critics, although their ideas are often uncannily alike. In 1743–5 Swedenborg became conscious of direct contact with spirits and angels, and felt called by the Lord to evangelistic work, but he believed that his New Church was not to exist separately from existing churches. He died in London, where, in May 1787, five ex-Wesleyan preachers instituted the New Jerusalem Church. The following year a general conference of British Swedenborgians convened, with William and Catherine Blake as signatories. Swedenborg taught that the second coming had begun in 1757 (when Blake, coincidentally, was born), and preached a new creed based on the 'divine humanity'.

Blake was also involved in various antinomian groups in London, many surviving from the seventeenth century, including Ranters, Shakers and Muggletonians. These and other millenarian movements (believing in a future 'millennium', that a thousand years will precede the second coming of Christ, during which his reign on earth will be in a kingdom of his saints who will be taken with him to heaven) were newly inspired by the French Revolution. The most important figures in these movements were Edward Irving (1792–1834), Richard Brothers (1757–1824) and

Joanna Southcott (1750–1814). Southcott had been a Methodist, and claimed supernatural powers. To Romantics such as Wordsworth and Coleridge, such people gave mysticism a bad name. Coleridge disparaged mysticism in *Aids to Reflection* (1825).

By the end of the eighteenth century, not only Methodism, but also the evangelical revival in the Church of England, was awakening in people a new experience of personal conversion, blurring the distinction between Arminianism (salvation potentially available for all) and Calvinism (salvation for the elect). The effects of evangelical piety were profound, even on such non-evangelicals as Crabbe, Wordsworth and Coleridge; and the last two, in turn, profoundly affected John Keble, J. H. Newman and the Tractarian fathers of the Oxford Movement in the nineteenth century. (See Stephen Prickett, *Romanticism and Religion*, 1976.)

In religion the Romantic period may be characterised as a time in which many people inspired to a new and heartfelt belief were deeply suspicious of old ecclesiastical institutions, while a new and radical secularism took root through thinkers such as William Godwin and the utilitarians who followed him.

DAVID JASPER

Robespierre, Maximilien (1758–94)

Maximilien Robespierre was the son of a barrister. He was born in Arras and studied in Paris, at the Lycée Louis-le-Grand. He became a lawyer himself, and was elected to represent his native province (Artois) at the meeting of the States General in May 1789. During the years that followed, as the new constitution was being debated, he was notable as a clever and zealous supporter of the revolutionary cause. In 1791 he was nicknamed 'the Incorruptible', and it was hinted that he was the nephew of Damien, who had attempted to kill Louis XV in 1757 (and been brutally executed in consequence). Carlyle, from a misreading of Madame de Staël, described him as a man with a sea-green complexion (*The French Revolution*, 1896, I, 141): actually, the veins of his brow were this colour.

Robespierre worked his way through the various assemblies following the States General of 1789, and was a prominent figure in the Jacobin Club. He promoted the idea of electing the Convention by universal suffrage, and after 10 August 1792 (the deposition of the monarchy) he became – along with Danton – one of the leaders of the 'Montagne' in the Convention. In the following winter the King was put on trial and executed (21 Jan 1793), and after the murder of Marat by Charlotte Corday (13 July 1793) the Girondins (the more moderate rivals to the Jacobins in the struggle for power) were rounded up and guillotined (Oct 1793). The Queen, Marie-Antoinette, was executed for high treason on October 16.

During the months that followed, Robespierre had to continue to struggle with a civil war in the Vendée (the revolt there had broken out in March 1793), and other threats to the Revolution: in April 1793 he and his followers had instituted the Committee of Public Safety, which condemned not only their opponents and the former aristocrats to the guillotine, but extended its power to destroy even their former allies, including the supporters of Hébert (24 Mar 1794) and Danton (5 Apr 1794). His policy of guillotining all suspected 'traitors' to the Revolution led to his downfall, when sufficient numbers of deputies in the Convention began to fear for their own lives. Robespierre and his close followers were themselves proscribed, and he was outlawed by his own legislation, the Prairial law. He was guillotined on 10 Thermidor, in year II of the revolutionary calendar (28 July 1794).

For some English writers, Robespierre became something of a scapegoat, the individual chiefly blamed for the Terror and for the turbulent course of the Revolution in 1793 and 1794. Wordsworth rejoiced in his fall, recalling in a celebrated passage in *The Prelude* (1805 text, X. 467–539) the moment when he heard the news. However, one should read with some reservation the lines (X. 457–8) describing Robespierre as one who 'wielded the sceptre of the atheist crew'. Robespierre supported the cult of a new religion, dedicated to the Supreme Being, in whose honour a fête was held in June 1794. His ideas derived from a rational theology which had become widespread during the eighteenth century, and he presided over the burning of a cardboard statue of Atheism in the Tuileries garden.

Carlyle saw Robespierre as a man of cant and humbug, but recognised that he had a good side: 'His poor landlord, the Cabinet-maker in the Rue Saint-Honoré, loved him; his Brother died for him' (*The French Revolution*, III, 286). Historians have naturally differed over his part in the Revolution, usually condemning him from a liberal standpoint. His chief defender has been Albert Mathiez, and his reputation is still a subject for debate.

JACQUES BLONDEL

Rogers, Samuel (1763–1855)

Samuel Rogers lived a long life, from the time of Samuel Johnson to the heyday of the Pre-Raphaelite movement. As a young man, he followed in his father's footsteps and became a banker, but he also had ambitions as a poet and a man of letters. In 1792 he published *The Pleasures of Memory*, the success of which was very encouraging. On his father's death in 1793, Rogers inherited the family share in the bank, and thereafter lived the life of a prosperous London literary man: he built himself a house in St James's Street, which he filled with works of art and where he entertained freely. His celebrated breakfasts were attended by writers and politicians, and by poets as diverse as Byron and Wordsworth, Thomas Moore and Ugo Foscolo. Rogers continued to be a respected poet, publishing a small volume, *Poems* (1812); his second great literary success was the publication of the illustrated edition of his *Italy* (1830), with engravings by Turner and Stothard. Rogers was offered the laureateship when Wordsworth died in 1850, but he felt that he was by then too old.

Byron thought highly of Rogers' poems, placing him at the top of his triangular 'Gradus and Parnassum' (Scott, as the 'Monarch of Parnassus', was above all competition); below Rogers were Moore and Campbell, and below them Wordsworth, Coleridge and Southey (*Byron's Letters and Journals*, ed. Marchand, III, 220). The order shows Byron's perverse dislike of the Lake poets, but it also gives some indication of Rogers' standing and importance in 1813. His verse was conservative, both in its expression, its use of eighteenth-century forms, and in its ideas (indeed, its affinity with eighteenth-century poetry was one of the qualities that made it attractive to Byron). Rogers' heroic couplets were often described as 'polished' or 'enamelled', and *The Pleasures of Memory* was a variation on a common subject of the day: the 'pleasures' poem, following Akenside's *Pleasures of the Imagination* (1744) and Thomas Warton's *The Pleasures of Melancholy* (1774), and anticipating Campbell's *The Pleasures of Hope* (1799). Rogers' procedures owe a great deal to Hartley and to eighteenth-century philosophy of the mind, and in his 'Analysis' Rogers speaks of 'the associative principle' which 'gives exercise to every mild and generous propensity'.

As this last phrase suggests, Rogers' poems are usually benign (in contrast to his sharp tongue on social occasions). The 'Ode to Superstition', for example, is a celebration of the Enlightenment, beginning 'Hence, to the realms of Night, dire Demon, hence!' and ending with the coming of truth, which 'visits man with gleams of light and love'. Similarly the 'Verses to be Spoken by Mrs Siddons' are a charming description of role-playing and its relationship to the personality of the actress. Rogers is at his best in short or epigrammatic pieces, such as 'A Wish' ('Mine be a cot beside the hill'), or 'Written at Midnight, 1786'. The danger with Rogers' longer poetry is that its regularity and control will become boring, and its ideas will seem predictable and unexciting. 'It is in a calm and well-regulated mind that the Memory is most perfect', he wrote in the Analysis to part II of *The Pleasures of Memory*; and there are occasions when Rogers' poems seem too well-regulated for their own good. At other times, however, he is capable of writing imaginatively: *The Voyage of Columbus* (1810) is an appeal to the reader's creativity, and an experiment in fragmentary presentation that may have influenced Byron's technique in *The Giaour*. Rogers described disparate episodes of Columbus's journey, comparing them to 'shreds of old arras, or reflections from a river broken and confused by the oar; and now and then perhaps the imagination of the reader may supply more than is lost'. Similarly, the various poems which comprise *Italy* form a patchwork; together with the illustrations, the poems provide an attractive set of impressions, usefully combining narrative and description, and, although the book is conventional in both metre and thought, in other ways it looks forward to Victorian writers on Italy such as Browning and Clough.

J. R. WATSON

Rousseau, Jean-Jacques (1712–78)

Jean-Jacques Rousseau was born in Geneva, of French origin, the son of a watchmaker. His mother died when he was nine months old, and he was cared for by his father and (after his father's departure) by an uncle. He was apprenticed to an engraver, but at the age of fifteen he fled from Geneva: he was looked after by Madame de Warens, a Catholic and a secret spy of the Savoyard government, who sent him to a seminary in Turin, where he abjured his original Protestantism. He returned to France, and fell in love with Madame de Warens: the golden days spent in her presence, either at 'Les Charmettes' or at Annecy or Chambéry, were later recalled in *Les Rêveries du promeneur solitaire*. During this period, he invented a new system of music, with figures instead of notes, and for some time he was secretary to the French Ambassador to Venice. Returning to France in 1743, he moved to Paris, where he met Diderot, Grimm, d'Holbach, and Madame d'Epinay; he lived with a washerwoman, Thérèse Levasseur, and her five children, whom he sent to the foundling hospital. When visiting Diderot, imprisoned in the Bastille, Rousseau heard of the subject set for a prize essay by the Dijon Academy, on the question whether 'Progress in the arts and sciences had contributed towards the corruption or the bettering of manners'. Rousseau's prize-winning response was the *Discours sur les sciences et les arts*, published in 1750: it argued that the advancement of knowledge and culture had corrupted human beings, and that they had been superior in their primitive state. This argument was developed further in the *Discours sur l'origine de l'inégalité parmi les hommes*, which contrasted the simplicity of primitive people, who had a natural compassion ('pitié naturelle'), with the corruption of eighteenth-century life: at the root of the problem, in Rousseau's view, was the possession of private property. The *Lettre sur les spectacles* (1758) is ideologically similar, attacking the artificiality of the French theatre and the society around it.

After a period in Geneva, where he reconverted to Protestantism, Rousseau returned to Paris, where he was befriended by Madame d'Epinay: in these years he wrote *Julie, ou la nouvelle Héloïse* (1761), the novel which made him famous. It was in epistolary form, and concerned the passionate love of Julie for her tutor Saint-Preux, and her sacrifice of that love to live a virtuous life with the Baron Wolmar. A considerable part of the appeal was also in the setting (on the shores of Lake Geneva) and in Rousseau's ability to portray the Alpine landscape with an awareness of the interaction between human beings and nature.

Rousseau then turned his highly original and innovative mind to educational and social problems. *Emile, ou de l'éducation* (1762) advocated an education in which the child was to be free, was to learn by making mistakes, and was to be allowed to develop at his or her own pace. *Du contrat social* (1762) advocated a more equal justice, and by implication was a strong criticism of the current system; its ideas of the sovereignty of the citizens, and government by the 'general will', were extremely influential in creating the current of ideas that led up to the French Revolution.

The unpopularity of *Emile* led to Rousseau's banishment: he took refuge in Britain, where he was helped by David Hume; but Rousseau's quarrelsome and paranoid nature led to differences between them, and he returned to France. He led an unsettled existence until Monsieur and Madame Girardin found him a place to live at Ermenonville, near Paris, where he died. In his later years, he wrote *Les confessions* and *Les rêveries du promeneur solitaire*, both published after his death. They are exercises, as their titles imply, in reflection, remembrance and autobiography: *Les confessions*, in particular, is an early example of retrospective and self-justifying examination of an individual's development.

Rousseau's importance as a prophetic forerunner of the French Revolution was recognised when his remains were moved from Ermenonville to be placed in the Panthéon in 1794. But his contribution to eighteenth-century thought was more widespread than this. His recognition of the importance of the feelings and emotions led the way in the development of the Romantic sensibility, and his awareness of the interaction between the mind and the natural world anticipated Wordsworth. *The Prelude* was a natural sequel to Rousseau's thought, not only in its confessional and autobiographical mode, but also in its *Emile*-like approach to the bringing-up of children. Similarly, the idea that primitive man was not only uncorrupted by the artificialities and evils of civilisation, but also in closer harmony with nature, was profoundly influential. Shelley, in particular, was a fervent admirer of Rousseau. In 1816 he visited the scenes of *La nouvelle Héloise* in the company of Byron (also an admirer, though not so complete an enthusiast); and in his last poem, 'The Triumph of Life', Shelley produced a final acknowledgement of Rousseau's genius, making him say,

> If I have been extinguished, yet there rise
> A thousand beacons from the spark I bore.

JACQUES BLONDEL

Scott, Walter (1771–1832)

Sir Walter Scott (who had a baronetcy conferred upon him by
George IV in 1818) was born in Edinburgh, the son of a writer to
the signet, with an interest in church history, and of Anne Rutherford, the daughter of a professor of medicine in the University of
Edinburgh. There were, on both sides of the family, old Scottish
ancestors who had been involved in Border history, among them
Jacobites. Part of his boyhood, after the illness that crippled him,
was spent in the Tweed valley, in the Border region, where even as
a child he was receptive to the oral tradition of the area, gathering
old songs and ballads and listening to historical reminiscences of
ancient warfare between the Scots and the English; thus he began
to store essential material for his poetry and fiction. He studied
civil law at the University of Edinburgh and was for a time apprenticed to his father; he was admitted to the Bar in 1792. He became
Sheriff Depute for Selkirkshire in 1799 and, renouncing a more
distinguished legal career, obtained a post as Principal Clerk of the
Quarter Session in 1805, deriving from both positions a moderate
but steady income with enough leisure for a literary career.

Scott's first major poetical works were *The Minstrelsy of the
Scottish Border* (1802–3), a three-volume collection of songs and
ballads of the Border country which was well received, and *The Lay
of the Last Minstrel* (1805), the successful start of his career as a
narrative poet. More poems of this type followed, among them
Marmion (1808), *The Lady of the Lake* (1810, his greatest success as a
poet) and *The Lord of the Isles* (1815). His last long narrative poem
was *Harold the Dauntless* (1817). The formula – a blend of mediaevalism, supernaturalism, landscape-painting and Scottish epic –
had apparently become exhausted. Besides, with the emergence of
Byron (the first two cantos of *Childe Harold* had come out in 1812),
Scott's so-far-unrivalled fame as a poet was being effectively challenged.

Meanwhile he had turned to novel-writing, completing his first
'historical romance', published in 1814, *Waverley*. Not only was it
widely acclaimed; it was also a financial success, a weighty argument in favour of more novel-writing, in view of Scott's social
ambitions and aristocratic lifestyle as the 'laird' of Abbotsford, the
estate he had acquired in 1812 in the Tweed valley of his boyhood.
Waverley, like all the following novels until 1827, was published

anonymously, Scott hiding behind the mask of 'the Author of Waverley', the 'Great Unknown' (although many could see through the mask). Scott went on writing novels steadily at the average rate of two novels a year until the time of his death (1832). Some of the more significant novels are: *Old Mortality* (1816), *Rob Roy* (1817), *The Heart of Midlothian* (1818), *Ivanhoe* (1820) and *Redgauntlet* (1824). The later novels show signs of weakening powers, and also suffer from the strain of Scott's financial worries resulting from hazardous financial ventures going back to 1805. He had then become involved in the printing- and publishing-business through his association with the Ballantyne brothers, and then with Archibald Constable, who had taken over the Ballantyne firm; the collapse of Constable and Co. in 1826 ruined Scott, who kept on writing fiction to reimburse his creditors and maintain Abbotsford. After his death the copyrights of the Waverley Novels allowed the settlement of all his debts.

Scott's literary output also included plays, such as *Halidon Hill* (1822), *The Doom of Devorgoyl* (1830) and *Auchindrane, or, The Ayrshire Tragedy* (1830); contributions to the *Encyclopaedia Britannica* ('Chivalry', 1818; 'The Drama', 1814; 'Romance', 1822); and *Tales of a Grandfather* (1828–30), dealing with the history of Scotland and France. He edited Dryden (1808) and Swift (1814). His *Journal*, begun in 1825, was first published in 1890, and his letters were published by H. J. C. Grierson in twelve volumes (1932–7).

Scott's political stance in relation to Scottish and British politics contained built-in contradictions, which can be traced in his fiction: he was sentimentally attached to the splendours of the past, in the days when Scotland was an independent kingdom and sang of its arms and its heroes, yet was resolutely British and supported the Union with England. He posed as a moderate, advocating compromise and asserting aversion towards all kinds of extremists; yet he was perceptibly more lenient towards fanatical upholders of the royalist cause (like Claverhouse in *Old Mortality*) than towards popular preachers on the side of the Covenanters, so that, rather than a moderate, he appears as a Tory, who displayed anti-revolutionary zeal from the early 1790s, severed his connection with the *Edinburgh Review* because of its Whig politics, and was alarmed in his old age by the prospect of the 1832 Reform Bill. Scott's Toryism thus distinguishes him from such leading Romantics as Blake and Shelley, with their revolutionary fervour.

Altogether, his relation to Romanticism is problematic. In some

ways he was the most conspicuous figure in the Romantic movement. His fascination with the past, the pageantry of feudalism
and the sublime landscapes of Scotland make up a form of Romanticism. But neither the symbolic writing nor the cosmic vision of
the major Romantics can be traced in his works. His poetry is
visual rather than visionary; his interest in the supernatural is that
of a collector of old legends and superstitions, not that of a
Romantic attempting to reach out beyond the visible; his conception of history is that of a rationalist working out patterns of
change in terms of the balance of forces and causes and effects.
However, he not only invented the historical novel, but the way he
associated fiction and history shaped the nineteenth-century
novel, and his mark can be felt from Balzac to Tolstoy.

The Poems

Walter Scott's career as a poet started under the influence of
German literature, which he began to study in 1792, although
there were earlier influences. His first (anonymous) publications
(1796) were translations of Bürger's 'Lenore' and 'Der wilde Jäger'
('The Wild Huntsman'). The next stage was The Minstrelsy of the
Scottish Border (3 vols, 1802–3), a collection of ballads of the Border
country partly derived from the oral culture that had fascinated his
imagination in his boyhood. Scott was no discriminating collector,
and among pieces of authentic folklore there were some modern
imitations, including some fabrications by Scott himself, who at
times rewrote and expanded the orally transmitted poem. His
Minstrelsy, however, ranks in importance with the work of the two
other great collectors of ballads in English, Thomas Percy (Reliques
of Ancient English Poetry, 1765) and the American scholar F. J. Child
(English and Scottish Popular Ballads, 1882–98). The collection included ballads on legendary subjects ('Thomas the Rhymer', 'The
Twa Corbies') and others on historical events which later feature in
the Waverley Novels. For example, in Scott's notes to 'The Battle of
Bothwell-Bridge' –

> Alang the brae, beyond the brig,
>   Mony brave man lies cauld and still;
> But lang we'll mind, and sair we'll rue,
>   The bloody battle of Bothwell Hill. . . .

– he began to work out his treatment of the subject in his novel *Old Mortality*.

The Lay of the Last Minstrel (1805) initiated the series of long narrative poems that first made Scott famous. It tells a tale of the Border country about a feud between two aristocratic families in the sixteenth century, and the speaker is mostly the 'last minstrel':

> The last of all the bards was he
> Who sung of Border chivalry;
> For, welladay! their date was fled,
> His tuneful brethren all were dead. . . .

The prosody, of the pure stress-metre type, is borrowed from Coleridge's use of this form in 'Christabel', which, though unpublished at the time, was known to Scott. This is obvious from the first lines of canto I:

> The feast was over in Branksom tower,
> And the Ladye had gone to her secret bower;
> Her bower that was guarded by word and by spell,
> Deadly to hear and deadly to tell. . . .

The lay, like all those that were to follow, is divided into cantos, with interpolated lyrics.

Marmion, a Tale of Flodden Field followed in 1808, and, in placing a fictitious protagonist (Marmion) in a historical context (the battle of Flodden, 1513), it foreshadows the formula of the historical romances. The basic metre is the rather monotonous octosyllabic iambic couplet, which became the staple of Scott's narrative verse. *The Lady of the Lake* (1810), a Highland poem with an Arthurian title, whose scene is laid near Loch Katrine, marked the climax of Scott's poetic fame: 'its success was certainly so extraordinary as to induce me for a moment to conclude that I had at last fixed a nail on the proverbially inconstant wheel of fortune' (said by Scott, 1830; reported in J. G. Lockhart, *Life*, ch. 6). *The Vision of Don Roderick* (1811) shifts the scene to Spain, where the last Gothic king has a prophetic vision of the future which includes the recent French invasion and the arrival of British troops to free the country. In *Rokeby* (1813), the scene is in Durham and Yorkshire, and the period of the tale, which again blends history and fiction, is the days immediately following the battle of Marston Moor (1644) during the English Civil War. *The Bridal of Triermain* (1813) has an Arthurian subject; *The Lord of the Isles* (1815) reverts to Scotland in the days of Robert Bruce (1307); and *Harold the Dauntless* (1817), the last of the long narrative poems, is about a Danish Viking converted to Christianity by St Cuthbert.

Scott's poetry well exemplifies the literary tastes of the period for romantic tales of the past, with wild scenery. The Caledonian epic strain nostalgically harks back to the days when Scotland was independent, and the Ossianic minstrel boldly recites ancient stories of love and warfare. Early critics praised the narrative and descriptive powers as well as the human interest evinced in those poems. Scott was a master of easy, fluent, picturesque poetry, but the mythic and visionary powers of the Romantic imagination were never within his grasp, and he is restricted to the register of attractive fancifulness. His lays were so fashionable that they have now become period pieces – of interest to the literary historian, and for the insights they furnish into the workings of Scott's imagination, which reused some of this material in the Waverley Novels.

Waverley

Though *Waverley, or 'Tis Sixty Years Since* (1814), the first of Scott's Waverley Novels, was published anonymously, some instantly guessed its author's true identity: 'Aut Scotus aut Diabolus' ('Either Scott or the Devil'), said R. L. Edgeworth, Maria's father (*A Memoir of Maria Edgeworth*, 1867). It is the first historical novel (i.e. the first novel based on historical sources), although it may also be seen as indebted to the Irish 'national tales' of Maria Edgeworth (*Castle Rackrent, The Absentee*), as Scott gracefully acknowledged in the Postcript to *Waverley*: 'so as in some distant degree to emulate the admirable portraits drawn by Miss Edgeworth'. The first edition was sold out in five weeks, and *Waverley* was reissued three times within the year and four more times between 1815 and 1821.

The historical subject of *Waverley* is the Jacobite Rebellion of 1745, sixty years before Scott began to write the novel (1805). Charles Edward Stuart, or Prince Charlie the Young Pretender (the son of the Old Pretender or 'king over the water', the nominal James III), is seen in Holyrood, his ancestors' palace, where he formally proclaims the restoration of the Stuart family in the name of his father. The narrative includes the battle of Prestonpans, where the army of Prince Charles defeated the Hanoverians, and the Jacobites' drive southwards across Lancashire, down to Derby, where the uprising started to disintegrate as a result of internal conflicts and lack of support from the English population. The Jacobite army is then seen retreating from Derby, and after the battle of Clifton (a Hanoverian victory) the leaders of the rebellion (Highland chieftains among them) walk to their inevitable doom (some are executed in Carlisle) and the young Pretender achieves a difficult escape through the Scottish Highlands.

Waverley, a young English aristocrat whose meeting with Prince Charles in Holyrood is the pivotal point of the novel as historical fiction, is the heir to an estate (Waverley-Honour) whose present owner, his grandfather, Sir Everard, is one of the nostalgic Stuart faithfuls, whereas Waverley's father is a 'Hanoverian rat' supporting George II. The fictional hero's incipient military career, in the service of the ruling monarch, takes him to the Lowlands, where he is befriended by a Scottish laird, the Baron of Bradwardine, and is attracted towards his daughter, Rose Bradwardine, the blonde heroine. Waverley, the wavering hero, is thus involved in the world of Jacobite intrigues, and becomes acquainted with a Highland chieftain, Fergus Mac-Ivor, whose sister Flora (the dark heroine) fascinates him and draws him into the Rebellion. After the Jacobites have been routed, Waverley is granted a full pardon by George II, and marries Rose Bradwardine. Fergus is hanged in Carlisle with one of his retainers and Flora Mac-Ivor lives out her days in a convent.

Even the fictional part of *Waverley* is of historical interest beyond its romantic plot, and what Walter Scott is rendering through the fictitious characters is the process of historical change that removes the last traces of the old feudal world order and destroys the Scottish clans. The conclusion of the novel operates historical distancing in its presentation of the Castle of Bradwardine in its restored state, which makes it appear as a museum, containing among other exhibits a historical picture in the style of Raeburn: 'It was a large and spirited painting representing Fergus Mac-Ivor and Waverley in their Highland dress, the scene a wild, rocky, and mountainous pass.'

Waverley was praised by Byron ('the best & most interesting novel I have redde since – I don't know when' – *Letters and Journals*, ed. L. Marchand, 1975, IV, 146), Jane Austen ('Walter Scott has no business to write novels, especially good ones' – *Letters*, ed. R. W. Chapman, 1932, p. 404), Maria Edgeworth, Francis Jeffrey (in the *Edinburgh Review*) and the anonymous reviewer of the *Monthly Review*. Modern criticism has focused on structure, design, viewpoint and historical interest. R. A. Colby points out a similarity to the heroine of *Northanger Abbey* in the mental make-up of the protagonist: 'at the outset Edward Waverley is a victim of the same form of mental derangement as Jane Austen's Catherine Morland' (*Fiction with a Purpose*, 1967). A. O. J. Cockshut draws attention to the large spectrum of political mentalities fictionalised through character, and to the rendering of 'war in all its bearings, its large cultural implications, its splendour, its absurdity, its sadness' (*The Achievement of Walter Scott*, 1969). One of *Waverley*'s greatest merits is the scope of its historical vision, which relates the 1745 Rebellion to the larger context of Scottish history, reverberating with earlier conflicts involving Jacobites and Cameronians and pointing to a contemporary situation in which heroic values are no longer relevant. The mock-heroic mode, which in some distant way recalls Cervantes, is at times used to expose the inadequacy of the Jacobite cause even as it appears triumphant.

The Waverley Novels (1814–32)

This was the usual designation of Scott's novels and stories, or 'historical romances' as they were called on the title page, since, with the exception of *Tales of my Landlord* (described as 'collected and arranged by Jedediah Cleishbotham'), until 1827 they were published anonymously, ascribed to 'the Author of Waverley'. The Waverley Novels include *Waverley*, 1814; *Guy Mannering*, 1815; *The Antiquary*, 1816; *Tales of my Landlord* (*The Black Dwarf, Old Mortality*), *Rob Roy*, 1817; *Tales of my Landlord*, 2nd series (*The Heart of Midlothian*), 1818; *Tales of my Landlord*, 3rd series (*The Bride of Lammermoor, A Legend of Montrose*), 1819; *Ivanhoe, The Monastery, The Abbot*, 1820; *Kenilworth*, 1821; *The Pirate, The Fortunes of Nigel, Peveril of the Peak*, 1822; *Quentin Durward*, 1823; *St Ronan's Well, Redgauntlet*, 1824; *Tales of the Crusaders* (*The Betrothed, The Talisman*), 1825; *Woodstock, or The Cavalier: A Tale of 1651*, 1826; *Chronicles of the Canongate* ('The Two Drovers', 'The Highland Widow', *The Surgeon's Daughter*), 1827, 1828; *St Valentine's Day, or The Fair Maid of Perth*, 1828; *Anne of*

Geierstein, or The Maiden of the Mist, 1829; *Tales of my Landlord,* 4th series (*Count Robert of Paris, Castle Dangerous*), 1832.

In Scott's long period of novel-writing, there were two phases, with *Ivanhoe* as the turning-point. The first phase (1814–20) comprises two types of novels, though both have a seventeenth- or eighteenth-century Scottish background. One type involves stories whose plot hinges on a foundling, or a case of concealed identity: for example, the hero of *Guy Mannering,* Bertram, the son and heir of the laird of Ellengowan, was kidnapped when a child, and returns to discover his parentage and recover his property (the period is the eighteenth century); and *The Antiquary* (which, as Scott makes clear in the Advertisement, refers to the last ten years of the eighteenth century) centres on a character who is suspected of being illegitimate, and who turns out to be the legitimate son of a Scottish earl. In both cases the main focus of interest is displaced from the conventional plot towards colourful, eccentric figures such as the old gipsy Meg Merrilies (*Guy Mannering*), and Oldbuck, the laird of Monkbarns (the titular character of *The Antiquary*), who 'measured decayed entrenchments, made plans of ruined castles, read illegible inscriptions, and wrote essays on medals in the proportion of twelve pages to each letter of the legend'. Other novels of the earlier phase integrate actual historical events (as in *Waverley*): the Cameronian uprising of 1679 (*Old Mortality*), the Jacobite intrigues before the 1715 rebellion (*Rob Roy*), the Porteous riots of 1736 (*The Heart of Midlothian*), the Civil War (*A Legend of Montrose*). *The Bride of Lammermoor* stands somewhat apart as an eerie story of revenge and madness with spectral appearances.

With *Ivanhoe,* Scott's fiction moves outside Scotland and deals with a far remoter period, the scene being 'that pleasant district of merry England which is watered by the river Don' – Yorkshire at the time of Richard I and Robin Hood, who are among the prominent characters; the novel contains a good deal of mediaeval pageantry, with tournaments and the besieging and storming of a 'hoary and ancient castle'. Coeur de Lion (Richard I) also appears in *The Talisman,* together with Sultan Saladin, in a romantic tale of the Holy Land at the time of the Third Crusade. In *Count Robert of Paris,* Scott went even further back in time, the scene being Constantinople at the time of the First Crusade. *Quentin Durward,* a story of fifteenth-century France, includes Louis XI together with the Wild Boar of Ardennes and Charles the Bold of Burgundy (who also appears in *Anne of Geierstein*). Other novels of the later phase deal with English and Scottish history at less remote times: the days of Mary, Queen of Scots, and Elizabeth (*The Abbot, Kenilworth*); London at the time of James I of England (*The Fortunes of Nigel*), and the Civil War (*Woodstock*). *Redgauntlet*, one of Scott's most interesting novels of the 1820s, reverts to his first historical hero, Prince Charles Edward Stuart, and to the theme of *Waverley*, once again telling the tale of a Jacobite attempt, for which however there is in this case no historical basis. Within *Redgauntlet*, 'Wandering Willie's Tale' is an inset ghostly tale echoing *Old Mortality* through the image of General Claverhouse, the scourge of the Covenanters, seen among the demons of Hell.

In the 1842 preface to the *Comédie humaine*, Balzac praised Walter Scott for bringing to the novel the philosophical dimension of history. Despite Carlyle's adverse comments, there is evidence that the Waverley Novels were widely read in the nineteenth century, and the brief, inexplicit references to *The Pirate* and *Waverley* in, for example, George Eliot's *The Mill on the Floss* simply assume a knowledge of Scott's fiction. Scott has been largely ignored by the twentieth century and is still insufficiently read in spite of Georg Lukács reassessment of the Waverley Novels in *The Historical Novel* (1936; English tr. 1962), and more recent criticism by A. O. J. Cockshut, David Daiches, Edgar Johnson and others. The novels of the earlier phase, written in the wake of *Waverley*, are the more important: *Old Mortality* is remarkable for the historical grasp with which it renders the process that was to lead to the establishment of moderate Presbyterianism in Scotland after a bitter civil war opposing Royalists and Covenanters; *The Heart of Midlothian*, whose title refers to the old Edinburgh Tolbooth (prison), explores the conflicting relations between Scots and English after the Act of Union (1707), and probes the iniquity of the judicial system in a way that foreshadows *Bleak House* and *Adam Bede*. In the development of the English novel, Scott is a key figure whose fictional inventiveness and understanding of historic change led to Dickens, Thackeray, Elizabeth Gaskell and George Eliot. A major new edition, the 'Edinburgh Edition', is in progress.

HUBERT TEYSSANDIER

The Self

Romantic poetry might be defined as a poetry of the self. For example, in Coleridge's 'Frost at Midnight' and Wordsworth's 'Tintern Abbey' – the finest examples of the 'conversation poem', which in many respects is characteristic of Romantic poetics – we observe the presence of the self on all levels of the text: as speaking *subject* or 'I' who generates the message; as *object* or theme of the message (whether the poem deals with a moment of emotion or with the total history of the self); and most of all as 'mirrored' in the very *form* of the message, for the poem works itself out as a self-bounded verbal space to contain or reflect the self, thus establishing a basically narcissistic relationship between the text and the 'I'. Hence the natural circularity of the 'conversation poem', which thus offers itself as a verbal icon of the microcosm; hence also the plastic architecture of the 'Ode to Psyche', Keats's own 'conversation' poem, wrought as a 'fane' for the image of the loved psyche. In such threefold investment of the self in the poetic message – enunciation, theme and verbal lay out – lies the essence of Romantic 'lyricism'.

While thus measuring the intensity of the Romantic commitment to subjectivity, it is essential to realise the tension it creates in the artist. For the Romantic has at the same time a high notion of his role as a poet: he hears the call 'of zeal and just ambition' which urges him to achieve 'some work / Of glory' (*The Prelude*, 1850 text, I.256, 78); and as such he longs to relate to the great poetic tradition, to the 'great verse' of the past (Keats's 'Ode to Maia') and most of all to the grand genres – dramatic and epic poetry – supremely exemplified in the English language by Shakespeare and Milton in works characterised by impersonality of approach (the infinite variety of Shakespeare's characters) and universality of theme (Milton's display of the 'ways of God to men' in *Paradise Lost*): works whose grandeur goes together with the poet's self-effacement, and where the creator remains (as Joyce said in *The Portrait of the Artist*, ch. 5) 'invisible, refined out of existence, indifferent . . . like the God of creation'. The Romantic poet is thus, even while creating, caught in a latent self-conflict – sometimes fully articulated as in the first 300 lines of *The Prelude*, where Wordsworth grapples with the problem in a poetic *agon*; torn between two seemingly irreconcilable longings: for Miltonic grandeur (which prompts the choice of some 'noble theme' of universal scope) and for lyrical self-expression (involving the renunciation of epic amplitude). He has – in the words of Keats assessing the achievement of the older Wordsworth and implicitly voicing his own poetic anxiety twenty years after 'Tintern Abbey', in 1818 – 'epic passion [but] martyrs himself to the human heart, the main region of his song'. A Romantic poet's development can thus be seen as, in a large measure, a struggle to transcend the epic – lyric dilemma; and his poetic achievement evaluated as an attempt to work out a voice adjusted to the self's intimacy while retaining (or re-creating) the grandeur of epic utterance.

The Romantic solution lies in the intuition – central to Romanticism – of vast areas of the self stretching beyond the frontiers of consciousness, in the sense of 'unknown modes of being' like those discovered by the boy of *The Prelude* (1850, I.357–400). In his nightly adventure he catches sight of 'a huge peak, black and huge / As if with voluntary power instinct' beyond 'The horizon's utmost boundary' – metaphorically outside the range of consciousness – in a passage which may be read as emblematic of the new lyric epic of Romanticism. Here the old 'objective' amplitude of epic poetry is redefined in terms of the subject's own depth, and the traditional feat of the epic hero (the journey into the Underworld, a central theme *in* the poem) turns into the artist's adventurous voyage into the self's unknown territories in the very genesis *of* the poem. The self-expression of Romantic lyricism thus deepens into self-exploration and, ultimately, into a reconstruction of the self as a much vaster subject: as the true Psyche, celebrated by Keats, with her 'untrodden region[s]' sought by the loving poet.

The privileged locus for such re-creation of the self is often the 'nature poem', because nature's images provide a symbolic language for that hitherto speechless intimacy to be spoken or objectified, while, as external, vast and remote, they register the paradoxical sense of an internal *otherness*.

> In looking at objects of Nature while I am thinking, as at yonder moon dimglimmering thro' the dewy window-pane, I seem rather to be seeking, as it were asking, a symbolical language for something within me that already and forever exists, than observing anything new. Even when that latter is the case, yet still I have always an obscure feeling as if that new phaenomenon

were the dim awaking of a forgotten or hidden Truth of my inner
Nature.

This remarkable entry in Coleridge's notebooks (ed. K. Coburn,
1957– ; II, 2546), with its musing subject no longer anchored in
the first person, but hovering between an uncertain 'I' and a
fascinating cosmic 'It', is probably the best key to the movement
and structure of the Romantic lyrical discourse as it might be
observed by comparing textual units such as the opening sequence
of Wordsworth's *Prelude* (I.1–58), the first section of Coleridge's
'Frost at Midnight' (II.1–23) and the second part of Shelley's 'Mont
Blanc' (II.12–48). Each poem's imagery moves from metonymy (as 'I'
and 'It', or landscape, are introduced in mere contiguity) towards
metaphor as a sense of secret correspondences is expressed
(Wordsworth feeling 'within / A correspondent breeze'; Coleridge
perceiving 'a companionable form' in the pulsating fire; Shelley,
gazing at the torrent, feeling as if he was beholding his 'own
separate fantasy'). Chiefly this occurs as the identity between self and
scenery is poetically realised in the psycho-cosmic space worked out
in the text (Shelley's 'dark, deep ravine', with its stream and
clouds, indistinguishable from the 'still cave' of the mind), a space
which becomes, by the end of the sequence, the new referent of 'I'.
The internal 'sublime' to which the Romantic self-exploration leads
finds two major objective correlatives: the 'inward immensity'
(Gaston Bachelard's *immensité intime*) that the Romantic subject
experiences in introversion as he sinks into 'the deep and labyrinthine soul' (*Prometheus Unbound*, I.805) or explores the psyche's
'wild ridged mountains steep by steep' ('Ode to Psyche', l. 55);
and the stream of (un)consciousness, realistically described in
Coleridge's 'Eolian Harp' as the poet watches 'Full many a
thought uncall'd and undetain'd, / And many idle flitting phanta-
sies, / Traverse [his] indolent and passive brain' (ll. 39–41), or
objectified in the 'unresting sound' of Shelley's mountain stream
flowing through the valley of the mind. Beyond the self-
transparent, self-knowing ego of the Cartesian *cogito* ('I think'), the
Romantic mind thus experiences a much deeper and mightier
cogitor ('I am thought') and, through it, an 'It thinks' whose echoes
or emergences are contemplated. Hence the Romantic reconstruction of the classical 'I think' as an 'I' watching a stream of thought
or of imagery in the making and in the welling from a self below
the self: the Romantic *mindscape*, supremely exemplified in Coler-

idge's 'Kubla Khan', whose ego ('I' or the Khan) watches in wonder a 'deep romantic chasm' from which 'a mighty fountain momently [is] forced' (l. 19).

Romantic self-exploration thus tends to lead to epiphany, to the discovery of 'underpresences' (Wordsworth's word) within the self. Such epiphanic experience is also played out in the second act of Shelley's *Prometheus Unbound*, where the hero (impersonated by Asia) explores the depth of his 'labyrinthine soul' and discovers a 'mighty darkness' – the 'It' figure of Demogorgon – in the ultimate cavern. This appears also in the first section of 'Tintern Abbey', where the Wordsworthian 'I' sinks, as the description unfolds, into circle after circle of mindscape down to a 'me–it' figure coalescing with the 'her' of nature, retired in the cave 'where by his fire / The hermit sits alone'. And a comparable experience might be traced in Byron's poetry in the context of its major myth, that of the sea wanderer. The moralist–narrator of Don Juan's adventures plays the part of 'some Columbus of the moral seas' setting out on a voyage of exploration to reveal 'icebergs in the hearts of mighty men' (*Don Juan*, XIV.ci–cii), thus promising unsuspected cartographies of the soul; and as the emotionally involved narrator of *Childe Harold's Pilgrimage* (e.g. III.i–vii), the Byronic subject drifts in his text – more than he travels on seas – in the movements of his uncertain syntax and in the space of his ambiguous imagery, between a rather passive 'I' and an energetic 'It' – wind or wave – whose manifestation might be the true revelation of the journey.

Romanticism thus redefines the subject as structurally dual or plural in a vision which remarkably anticipates, in several respects, our modern post-Freudian view of the psyche. Ultimately this sense of a multiple self generates the need, in the Romantic artist, for a new literary mode: a plurivocal speech fit to give utterance to the plural subject in its urge towards self-expression. Paradoxically, therefore, this supreme form of lyricism will no longer be a lyricism of the first person, for the subject, in the grammar of the new idiom, overlaps the traditional 'I' (which is no larger than the Cartesian *cogito* or smaller consciousness) and invests the other 'persons' as channels for the other forces or aspects of itself to enter the text. In the process it goes beyond the self-conscious 'I' (or ego) of Shelley's Prometheus, the paralysing 'he' of Jupiter (an inhibiting and restricting psychical agency), the 'she' of Asia (a power of integration at work in the self) and the secret 'It' of Demogorgon (a figure of the potentially creative unconscious).

That idiom which turns the structures of intersubjective communication into the intra-subjective grammar of a modern lyricism is no other than *myth*, as re-created by Romanticism as language of the total self, whether in narrative or dramatic form. Narrative works such as Coleridge's 'Ancient Mariner', Keats's *Fall of Hyperion*, and Blake's 'prophecies' and epics, and dramatic works such as Byron's plays and 'mysteries', and Shelley's 'play for voices', *Prometheus Unbound*, all ultimately read like projections of an 'inner scene' from which the artist's 'I' may be absent or invisible because coextensive with the stage and split up into the mythic figures which embody forces or agencies in the self. The 'impersonality' of myth – an ironical response, after all, to the challenge of Shakespearean and Miltonic art – thus turns out to be the deepest and, we may suggest, happiest expression of Romantic subjectivity. Thanks to the construction of myth, the Romantic artist is able to articulate the newly discovered plurality of the self while retaining its unity, to objectify the 'underpresences' while holding them together in an imaginative system. The security of myth conjures in the creator the danger of neurotic dissociation: it may be highly significant that Blake, whose ever-recurring theme is the 'fall' of man into 'Urizenic' division (a disintegration which finds its fullest expression in the great epic of *The Four Zoas* [1797–1803], Blake's most elaborate myth of the self, with the Zoas – 'living creatures' or 'Energies' – as figures within a latent psyche where the drama unfolds) should always play out his theme in a strongly designed mythic system which involves, as such, the potential reintegration of the fragmented self. This happens indeed in the final 'night' of the epic, which projects a state of wholeness and of creative interplay between the four Energies – Urizen, Luvah, Tharmas and Urthona (or Los) – traditionally interpreted as the four faculties of reason, passion, instinct and imagination, and which a modern (and more interesting) reading can construe (in view of how the figures function and dynamically relate to each other in the myth) as psychical agencies, in a configuration which anticipates the views of twentieth-century depth psychology on the structure of the psyche.

When myth no longer keeps up the unified image of the self, or when the writer turns away from it to convey another quality of experience, Romantic lyricism turns into another 'impersonal' mode of self-expression: that of the 'tale of terror' or, in modern terms, of fantastic literature. The 'fantastic' is myth in a broken form. It projects an irretrievably dissociated self where some vital element which has grown alienated from the conscious personality (through being repressed into the unconscious id or disowned and projected into the outer world or the other) returns and confronts the ego which perceives it as 'strange', frightening or monstrous, while being helpless because what it faces is indeed the 'stranger within' or the 'double'. The double, embodying the self's repressed intimacy, generates that peculiar kind of 'terror' – or, better, anxiety – which will be later defined by Freud as *das Unheimliche* ('the uncanny'), felt when 'something which ought to have remained hidden . . . comes to light'. The uncanny (self-) meeting, occasionally projected in the major poetry of Romanticism when myth momentarily modulates into the 'fantastic' – as in moments of 'soft alarm' in *The Prelude* or in Prometheus's encounter with the Furies in *Prometheus Unbound* (I.445–51) – finds its full expression in Romantic fiction. It is basic to the poetic narrative of Coleridge's 'Christabel', whose fascinating opening presents the meeting of Christabel with Geraldine, an id-figure that steps in like the Shadow or mirror image of the heroine and binds her up in a nightmarish symbiosis; but it is, most of all, a generative cell in the 'Gothic' and (more generally) the Romantic novel. M. G. Lewis's *The Monk* (1796), with its sexually unbound Ambrosio as potential double of Lorenzo, the 'good' hero; Mary Shelley's *Frankenstein* (1817), with the monster as virtual Shadow of his creator; and Hogg's *Confessions of a Justified Sinner* (1823), with the mysterious Gil-Martin as an id-figure objectifying the Puritan hero's repressed sexuality and sadism – all these remarkably exemplify this 'fantastic' mode. It emerges as the alternative or the 'shady' side of unifying myth and thus constitutes, together with myth, the final avatar of Romantic lyricism and the outcome of the Romantic attempt at creating a new language of the self.

CHRISTIAN LA CASSAGNÈRE

Sensibility

The concept of sensibility underwent a deep change in the course of the eighteenth century, although the word only came into fashion in the 1750s. It first meant the power of sensation or perception and was closely related to Locke's tenet that sense perception is the 'inlet of all knowledge in our minds'. The revaluation of sense impressions played a major role in the development of nature poetry. Sensibility is also the power or faculty of feeling, a delicate 'sensitiveness of taste', and it became in the 1750s 'the capacity for refined emotion', the 'readiness to feel compassion for suffering and to be moved by the pathetic in literature or art' (*Oxford English Dictionary*).

Lord Shaftesbury, whose *Characteristicks* (1711) were widely influential, contributed to the shift of meaning and came to be regarded as 'a sort of unofficial official philosopher of the movement of sensibility' (L. I. Bredvold). He was the first to build a philosophic system founded on benevolence and not on abstract, rational principles.

Influenced by the Cambridge Platonists, he believed in the essential goodness of human nature. Man has, he thought, an intuitive sense of right and wrong and is normally incapable of performing a rude or brutal action. Shaftesbury held that benevolence springs from the tender emotions of pity and compassion; man takes a real delight in doing good and virtue is its own reward. The 'friend of mankind' is also an enthusiastic lover of nature, for his 'moral sense' (or 'inward eye') is an aspect of the 'taste' that enables him to perceive and enjoy the beauty and harmony of the universe. Wherever he discusses happiness and virtue, Shaftesbury stresses the importance of the heart, the value of 'passions and affections'.

As the century wore on, the notion of judgement gradually faded away in favour of an ethics of feeling. David Hume's ethical system is based on unselfish sympathy or benevolence. Adam Smith argued in his *Theory of Moral Sentiments* (1759) that sympathy precedes moral approbation or disapprobation. The philosophers helped define the psychology of the man of feeling. Their ideas met a favourable reception among the rising middle classes and the new female reading public.

Yet the term 'sentimental' followed the fluctuations of sensibility. In the 1740s and 1750s it still referred to a virtuous way of thinking and acting. It meant 'highly moral' or 'engaged in moral reflection'. It was gradually associated with the 'sympathetic imagination', coloured with emotion and 'characterized or exhibiting refined and elevated feeling' (*Oxford English Dictionary*). In later use it acquired derogatory connotations and meant 'addicted to indulgence in superficial emotion' (ibid.).

The literature of the eighteenth century offered ample occasion to shed the sympathetic tear over the sufferings of the unfortunate. Poetry of nature and solitude, meditations on death and the vanity of human wishes encouraged by the religious revival, midnight walks in country churchyards or ruined abbeys – all favoured moral elevation as well as the sweet 'Pleasures of Melancholy' (the title of a poem by Thomas Warton, 1747). The man of feeling developed a real cult of melancholy that often verged on the morbid and indulged in tender sorrow or luxurious grief. Melancholy became the hallmark of an elegant soul and a refined sensibility. In the 1770s it gave way to darker moods and stronger emotions. The man (or woman) of taste relished the exquisite terror and horror of Gothic fiction.

Sentimental comedy meant to teach the audience how to think and feel. Ever since Steele's *Conscious Lovers* (1722), it had been brimming over with moral sentiment, pity and sympathy. Goldsmith declared in *The Good-Natured Man* (1768), 'I love to see a gentleman with a tender heart.'

The sentimental novel initiated by Richardson had a tremendous impact on the middle classes. The general tone of *Pamela* (1740), the situations discussed and the system of values advocated in the book reflect the current of sensibility that was to sweep over Europe in the second part of the century. Yet 'sentimental' was still linked with 'moral'. But *Sir Charles Grandison* (1753–4) presented a hero who is a paradigm of virtue and tries 'to conceal from others those sensibilities which they cannot relieve'. Richardson, Sarah Fielding and a long line of female novelists examined the more elusive passions, and traced 'the Mazes, Windings and Labyrinths which perplex the Heart of Man' (Henry Fielding). Sterne's *Sentimental Journey* (1768) marked the dissociation between morals and sentiment, the frankly avowed pursuit of altruistic emotions for egoistic ends, and exalted the cult of sentiment as a source of hedonistic pleasure. Sterne hails 'Dear Sensibility! source unexhausted of all that's precious in our joys, or costly in our sorrows!'

A young Scot, Henry Mackenzie (1745–1831), published in 1771 a short novel, *The Man of Feeling*, which attained a wide popularity. Educated at Edinburgh University, Mackenzie became a practising lawyer and frequented the city's literary circles, which then included Hume, Robertson, Adam Smith and Hugh Blair. A disciple of Adam Smith, he was also a fervent admirer of the English and French novelists Fielding, Richardson, Rousseau and l'Abbé Prévost, and an imitator of Sterne in sentiment, pathos and style.

The novel is presented as a fragmentary manuscript set in order by a mysterious 'editor'. Harley, the hero of the story, is a mild eccentric who, like Sterne's Yorick, is gifted with an acute sensibility. His bashfulness rises from 'a consciousness which the most delicate feelings produce'. His generosity and warm sympathies do not fit him to achieve worldly success (or wisdom), since his main pleasure is to assist those who suffer from poverty or injustice.

The disconnected episodes all illustrate the fine texture of Harley's sensibility. He gives money to a cheerful beggar on the road to London; visits Bedlam, where he feels compassion for a beautiful young woman in distress; saves a prostitute from starvation and moral despair; and offers his protection and a roof to a discharged soldier. Mackenzie lays the emphasis on Harley's modesty, benevolence and melancholy refinement. Deeply in love with a charming heiress whose 'humanity was a feeling, not a principle', his scruples prevent him from expressing his passion except on his deathbed. He pines away and dies from an excess of delicacy. Gestures, sighs, tears and broken expressions convey emotional intensity.

But the Man of Feeling, like the poet evoked in the stagecoach conversation (ch. 33), proves ineffectual in a selfish and cruel society, and may only have his function as a 'pattern of perfection'. Harley exemplifies the moral qualities and virtues advocated by the philosophers and philanthropists of the time, while experiencing a vivid pleasure in doing good. The story, however, points to Mackenzie's ambivalent attitude, 'a puzzling combination of idealistic sensibility and hard-hearted prudence' (E. Tuveson). He feels deep sympathy for his sensitive hero, and yet seems implicitly to condemn his 'unfitness for the world' which is at the root of his tragic end.

The Man of the World (1773), Mackenzie's next novel, shows the influence of Richardson. Sir Thomas Sindall is a stereotyped villain–hero who persecutes innocent victims, whose distress and sufferings are painstakingly carried to the limit. The rake's deathbed sensibility and penitence are out of keeping with the general tenor of the character.

Julia de Roubigné (1777) was Mackenzie's last extended fictional work and was inspired by *Julie, ou la nouvelle Héloïse*. It is often considered as a further step towards the Gothic novel on account of its melancholy mood and gloomy atmosphere. There is no flicker of hope to relieve the overwhelming misery and despair of the characters involved in a series of catastrophes. Like *The Man of Feeling*, it contains a fervent anti-slavery plea, which places Mackenzie among the 'literary abolitionists'.

Henry Mackenzie intensifies the cult of emotion outlined in Sterne's *Sentimental Journey*, explores abysses of sensibility and suffering resulting from an excess of delicacy and scruples, while preserving a keen sense of propriety (love is more strongly felt when it remains at the stage of mere yearning) and pursuing a didactic intention.

If he contributed to the success of the novel of sensibility, Mackenzie did not fail to warn the reader of its dangers. In *The Lounger* (no. 20) he denounced the 'refined sentimentalists who are contented with talking of virtues which they never practise' and considered as pernicious 'this separation of conscience from feeling'. These words echo Dr Johnson's warning in his *Life of Savage* and *Rambler* (no. 28), which proves that, however close Mackenzie was to the Romantic movement, he also remained within the tradition of the great eighteenth-century moralists. His successor, Jane Austen (who also greatly admired Dr Johnson) also portrayed the dangers of emotional attachment in *Sense and Sensibility* (begun c.1797, published 1811).

MICHÈLE PLAISANT

Shelley, Mary (1797–1851)

Mary Shelley was born Mary Wollstonecraft Godwin on 30 August 1797, the daughter of William Godwin and Mary Wollstonecraft; her mother died of post-natal septicaemia ten days later. She was brought up by her father and stepmother, Mrs Clairmont; when Shelley became friendly with Godwin in 1812 she made his acquaintance, and in July 1814 she eloped with him. They travelled on the Continent, together with her stepsister, Claire Clairmont, and returned to Switzerland in 1816 for a celebrated stay near Geneva, during which they saw much of Byron and his companion Dr Polidori. *Frankenstein* was begun during the visit, in June 1816, and continued during a tour with Shelley of the Mont Blanc region in July.

In December 1816 Shelley's wife Harriet committed suicide, and Shelley and Mary were married (30 Dec). *Frankenstein* was completed in May 1817 and published in 1818. Later in that year the Shelleys moved to Italy, living in various places during the next four years (Venice, Rome, Naples, Florence, Pisa). Mary published a second novel, *Mathilda*, in 1819. In May 1822 they settled at Lerici, and Shelley was drowned in July 1822 when sailing from Leghorn to Lerici in an unstable boat. Mary Shelley subsequently published *Valperga* (1823), *The Last Man* (1826), *Perkin Warbeck* (1830), *Lodore* (1835) and *Falkner* (1837). In 1839 she supervised the publication of Shelley's *Poetical Works*, with extensive biographical and critical annotations. She travelled widely in later years, dying in London in February 1851.

Frankenstein, the novel by which Mary Shelley is chiefly known, has remarkable technical skill and psychological insight. Its subtitle, 'The Modern Prometheus' concerns two legends of Prometheus – that he stole fire from heaven, and that he made a man from clay and used fire to give it life. The narrative, which deals with the moral and spiritual consequences of making such a creature, was greatly influenced by conversations during the wet summer of 1816 between Byron, Shelley and Dr Polidori on science, electricity, galvanism and 'the principle of life': each member of the party agreed to write a tale of the supernatural, of which *Frankenstein* is the best known.

The novel is told from three points of view, which makes it interesting technically. The first is that of Robert Walton, in letters to his sister which describe his journey towards the North Pole and his meeting with Victor Frankenstein (who is in search of the monster across the icy wastes). The second is Frankenstein's story of how he created the monster, after discovering the secret of imparting life to matter. Horrified by the monster's appearance, he abandons it, and the monster then turns to murder and revenge, killing Frankenstein's friend Clerval, his brother, and his bride Elizabeth. The innermost narrative is the story of the monster itself, in which the creature tells of his own misery at being rejected by human beings: 'Everywhere I see bliss, from which I alone am irrevocably excluded. I was benevolent and good; misery made me a fiend. Make me happy, and I shall again be virtuous.' In such Godwinian statements the monster shows the effect of loneliness, isolation and friendlessness on his character.

As a kind of monstrous Adam, he demands an Eve from his creator, promising to go and live with her in the South American jungle; but Frankenstein, after starting to build one, destroys it, and the monster is thereafter doomed to a life of solitude and (in consequence) crime. The monster finally destroys Frankenstein, though not without self-reproaches: in admitting that 'evil . . . became my good', the monster shows himself to be a fallen angel that has not yet lost all his original brightness, and through the preservation of his moral sense (even though he acts wickedly) becomes an object of sympathy.

Walton's matter-of-fact letters and his journey towards the Pole hold these improbable narratives within a frame of realism, although Walton's dangerous journey is itself emblematic of a desire for knowledge and travel beyond the normal (this part of the novel may well have been influenced by Coleridge's 'The Ancient Mariner'). The moral implications of a science taken further than normal are explored with an acute awareness of the dangers involved, and a profound sympathy with human beings whose aspirations overreach themselves and lead to tragic consequences. The mixture of sympathy and horror with which the monster is portrayed is particularly effective.

The Last Man, Mary Shelley's other remarkable novel, describes the extermination of the human race by plague in the twenty-first century, and contains a portrayal of Shelley in the character of Adrian, and of Byron as Lord Raymond; while Lionel Verney, who remains alone at the end, may be a reflection of Mary's own feelings of bereavement and loneliness after Shelley's death in 1822 and Byron's in 1824. The novel is also, like *Frankenstein*, a study of solitude and isolation, and of the inescapable self. Against the terrible forces of the plague, Mary Shelley sets the love of Adrian; but, when Adrian is drowned and Lionel is left alone, the novel explores the ultimate desolation of the last man left on earth (a theme also treated by Thomas Campbell). Lionel wanders from town to town, through Ravenna, Forli and Rome, finding each of them dark and silent, and seeing only himself in a mirror, 'without love, without sympathy, without communion with any'. Yet he journeys on, sailing to other lands, a figure who is tragic, absurd and heroic.

J. R. WATSON

Shelley, Percy Bysshe (1792–1822)

Percy Bysshe Shelley, the eldest child and only son of a baronet, was born on 4 August 1792 at Horsham, Sussex. He soon started learning Latin and Greek, and at Eton (1804–10) had to face the first hardships of a thenceforth difficult and on the whole unhappy life. He was ruthlessly persecuted by his fellow pupils, because of his feminine sensitivity, love of reading, and aversion to the brutal tradition of fagging, and as a result his already fragile nervous balance was badly disturbed. He then spent a few months at University College, Oxford (April 1810–March 1811), where he asserted his personality by challenging society and traditions. He evinced a predilection for physics and chemistry, but was expelled for publishing a pamphlet entitled *The Necessity of Atheism*. This was the beginning of a lifelong disagreement with his father and of a wandering existence in Great Britain and abroad.

Soon after his expulsion from Oxford, he eloped with Harriet Westbrook to 'save' her from her boarding-school. Together they travelled first to the north of England, then to Ireland, where Shelley actively supported the Irish in their struggle for independence by preaching non-violence and publishing a pamphlet, *An Address to the Irish People*, which he distributed in the streets. Disenchanted, he came back to Wales, where the famous Tan yr'allt incident took place: Shelley declared that he had been shot at by an unknown assailant, whom he saw indistinctly. The drawing that he later made of his attacker bears a strange likeness to the Devil. It is commonly thought that he was the victim of a delusion, or possibly a plot to get rid of him because of his radical sympathies.

While taking part in various humanitarian actions, Shelley started a correspondence with Godwin, and in 1813 published his first important poem, *Queen Mab*. In 1814, with his marriage to Harriet failing, he met Godwin's daughter Mary and eloped with her (in the company of her stepsister, Claire Clairmont) to the Continent for two months. They took a second trip two years later, joining Byron in Switzerland; this marked the beginning of a very profitable friendship between the two poets. It was during this trip that Shelley paid a visit to Mont Blanc, and wrote his famous philosophical poem inspired by the mountain.

In December 1816, after Harriet had committed suicide, Shelley married Mary. By March 1818 life in England was becoming extremely difficult and his health was deteriorating, so he decided to leave the country for Italy with Mary, their children and Claire. He was never to return. They led a wandering life, successively taking up residence at Milan, Leghorn, Venice, Rome, Naples, Leghorn again, Florence and Pisa. During this period he suffered several severe blows (especially the death of his two children) and wrote most of his main works, including *Prometheus Unbound*, his drama *The Cenci*, and his odes and lyrics. His death has become legendary: he was drowned in a storm in the bay of La Spezia, on 8 July 1822, as he was sailing in the *Don Juan*. His body was burnt a few days later on the beach, in the presence of his friends Byron, Leigh Hunt and Edward Trelawny.

'His mental hunger for knowledge was insatiable – no one ever saw him without a book in his hand or pocket' (Trelawny, *Records of Shelley, Byron and the Author*, 1878). A prodigious reader, like Coleridge, Shelley shared with him the privilege of being one of the most learned men of his time. Poems such as *Queen Mab* and *Prometheus Unbound* contain a most impressive store of knowledge and reflections in various fields: science (astronomy, electricity, chemistry, magnetism), history, philosophy, poetry. Shelley, a natural Platonist, was also very much influenced by English empiricism and eighteenth-century philosophers (Volney, d'Holbach, Rousseau, Godwin, Paine). Like other great Romantics, he attempted a synthesis of their different, even antithetical theories; hence inevitable difficulties of interpretation for his readers. But he was not only one of the greatest masters of harmonious verse in English literature (although his own speaking voice was shrill and discordant); he also had a natural bent for philosophy, as his best poems and his prose works both attest. His prose writings, dominated by his famous *Defence of Poetry* (1821), include essays on love, marriage, vegetarianism, philanthropy and Christianity, and a criticism of religion.

Shelley's influence may easily be inferred from the names of the leading members of the Shelley circle: Godwin, Leigh Hunt, Horace Smith, the Lambs, Keats, Peacock, and later Byron, Trelawny and Mrs Gisborne. Many were fascinated by his keen intelligence, culture, earnestness and powerful idealism, and many benefited from his almost wild generosity – not only anonymous paupers, but also famous contemporaries such as Leigh Hunt and his father-in-law William Godwin, who sponged on him. Shelley gave away much that he had to help others and kept very little for himself, which caused endless financial difficulties.

The key aspects of his personality – wildness and earnestness, benevolence, generosity and revolt, mysticism and opposition to Christianity, unhappiness and optimism – give an idea of the paradoxes and contradictions he had to cope with during his life. A born rebel with a gentle temperament, he devoted all his efforts to fighting tyranny in all its forms; *The Revolt of Islam* (1817) was his first poetic contribution to the struggle. But what set him apart from all his contemporaries was the importance of ideas to him, and the total absence of a gap between principles and actions. He was often impractical in the affairs of everyday life, which led Matthew Arnold to describe him as 'a beautiful *and ineffectual* angel' ('Shelley', *Essays in Criticism*, Second Series, 1888).

Dogged by death, misfortune, ill health, financial difficulties, an object of scandal in his own country, hated for the strength of his convictions by those who did not know him, loved for his generosity and benevolence by his friends, Shelley was one of the leading figures of the Romantic movement owing to the power and depth of his main works, and because, to quote Trelawny (who knew him well), 'he was the ideal of what a poet should be'.

Adonais

In April 1821, Shelley heard that Keats had just died in Rome. They had met at Leigh Hunt's in 1817, and if Keats did not take to Shelley, probably because his own humbler origins made him feel uneasy in the presence of a young aristocrat, Shelley had a real liking and admiration for the author of *Endymion*. When it became known that Keats was sick in Rome, Shelley kindly invited him to his own home, an offer which was proudly refused. On hearing the sad news of Keats's death he immediately decided to write an elegy, which was completed in early June 1821 and forthwith printed at Pisa. A number of copies were sent to the brothers Ollier for sale in London.

The construction of the poem, written in Spenserian stanzas, is simple. It falls into two main parts: first, thirty-seven stanzas of 'mythological narrative' (Carlos Baker, *Shelley's Major Poetry*, 1948, p. 246), written in the elegiac style of the Greek poets Bion and Moschus, which express the desolation of the author, of natural forces, of the Muse Urania and of brother poets; and then eighteen stanzas of philosophical, consolatory speculation which develop the theme:

> Peace, peace! he is not dead, he doth not sleep –
> He hath awakened from the dream of life –
> . . .
> He has outsoared the shadow of our night;
> . . .
> He lives, he wakes – 'tis Death is dead, not he;
> Mourn not for Adonais.

The range of interpretations is, however, not so simple. Of course, all critics agree that Shelley made use of the Greek myth of Venus and Adonis but somewhat altered it to suit his purpose. E. B. Hungerford distinguishes four levels of interpretation: mythological, literal, polemical and philosophical (*Shores of Darkness*, 1941).

The polemical level is stressed by all analysts: Shelley readily accepted the common belief that Keats's death had been caused by a malevolent, anonymous attack on *Endymion* in the *Quarterly Review*. As he himself was often the victim of harsh, adverse criticism, not only did he sympathise with the dead poet, but he even instinctively projected himself into the case, and the description of stanza xxxii has all the characteristics of a self-portrait.

Earl R. Wasserman's is probably the best and most complete investigation of the mythological level. In it he suggests the double etymology of 'Adonais' (Adonis and Adonai, the Hebrew God) and the various connotations of Urania, muse but also star (Lucifer–Vesper): 'the star image functions most significantly in terms of the Platonism of the poem, since it is the means of establishing a conjunction between Adonais and the light, which eventually symbolizes here the One, the ultimate reality' (*The Subtler Language*, 1959).

The ultimate conjunction of the mythological and philosophical levels takes place in the famous stanza lii:

> The One remains, the many change and pass;
> Heaven's light for ever shines, Earth's shadows fly;
> Life, like a dome of many-coloured glass,
> Stains the white radiance of Eternity,
> Until Death tramples it to fragments. . . .

N. I. White stresses the influence of Petrarch here (*Shelley*, 1947).

In his letters Shelley expressed great satisfaction with this poem: 'It is a highly wrought *piece of art*, perhaps better in point of composition than anything I have written' (letter to the Gisbornes, 5 June 1821), but the publication did not meet with the success he had hoped for. Twentieth-century criticism has borne out Shelley's judgement.

'Alastor'

'Alastor' was written in the autumn of 1815 at Bishopsgate Heath, near Windsor Park, where the poet was then living with Mary and Claire; it was published in March 1816 as the title piece of a small volume containing other poems. Shelley's second long poem, it was 'one that could hardly have been credited to the author of such a vigorous, objective call to arms as *Queen Mab*' (White, *Shelley*). After stressing the materialistic law of necessity, Shelley now felt the need 'to dramatize a conflict of allegiance between what might be called the law for things (natural law) and the law for man (the law of love)' (Baker, *Shelley's Major Poetry*).

He states his intentions in the Preface: 'It represents a youth of uncorrupted feelings and adventurous genius led forth by an imagination inflamed and purified through familiarity with all that is excellent and majestic, to the contemplation of the universe. . . . His mind is at length suddenly awakened and thirsts for intercourse with an intelligence similar to itself.' In a dream the poet has a vision of Ideal Beauty embodied in a woman whom, on waking, he starts seeking in the world, a solitary pursuit and a hopeless quest which leads him on until 'blasted by his disappointment, he descends to an untimely grave'. The idea is that man is not meant to exist without human sympathy, and so 'the poet's self-centred seclusion was avenged by the furies of an irresistible passion pursuing him to a speedy ruin'. Alastor, in Greek mythology, is the name of this avenging spirit. The poem describes continuous wanderings in settings and landscapes now hostile, now milder, becoming thoroughly pleasant at the end. 'Alastor' is Shelley's first attempt at writing pure symbolic poetry, not unlike *Prometheus Unbound*; what must be understood is that space here is a poetic representation of time and that the meaning is to be grasped through the connotations of the imagery.

'All three of the reviewers who deigned to notice the volume in 1816 agreed in vigorous condemnation of its obscurity' (White). Modern critics all agree on the sense and the poetic value of this work, though of course most note the discrepancy between the intentions stated in the Preface, to drive the hero to an early death as a punishment for his sin of isolation and selfishness, and for his anti-Platonic delusion; and the actual pleasantness of the death described as a gentle loss of consciousness in a 'silent nook', 'a tranquil spot that seemed to smile / Even in the lap of horror'. In other words, it is a typically regressive death.

It is the first appearance in Shelley's poetry of the 'epipsyche' theme and the theme of Ideal Beauty embodied in woman, which will again be central in *Prometheus Unbound*, *The Witch of Atlas* and *Epipsychidion*.

It is also to be noted that this quest for the ideal woman remained Shelley's own lifelong obsession, a source of endless hopes, illusions and disappointments.

The Odes and Lyrics

Shelley was the author of a number of odes and lyrics, the most famous of which were composed in 1819–20 and published with *Prometheus Unbound* in 1820.

His greatest and best-known shorter poem is certainly the 'Ode to the West Wind', written in *terza rima* and

conceived and chiefly written in a wood that skirts the Arno, near Florence, and on a day when that tempestuous wind, whose temperature is at once mild and animating, was collecting the vapours which pour down the autumnal rains. They began . . . at sunset with a violent tempest of hail and rain, attended by that magnificent thunder and lightning, peculiar to the Cisalpine regions.   (Shelley's note)

Attacked by J. C. Ransom, Allen Tate, Cleanth Brooks and F. R. Leavis, it has been defended by critics with a better insight into its inspiration, symbolism and significance: R. H. Fogle, I. J. Kapstein, Francis Berry, Edmund Blunden, Harold Bloom and S. C. Wilcox.

In the development of its five parts (five sonnets), Berry makes out four stages: an invocation, a listing of the wind's attributes, a confession and a petition (*Poets' Grammar*, 1958). Both 'destroyer and preserver' (the Siva–Vishnu of Hindu mythology), the West Wind is the great informing force at work in nature – a force which the poet implores ('O lift me as a wave, a leaf, a cloud') and to which he assimilates himself in the history of mankind ('One too like thee: tameless, and swift, and proud'):

> Be thou, Spirit fierce,
> My spirit! Be thou me, impetuous one!
>
> Drive my dead thoughts over the universe
> Like withered leaves to quicken a new birth!

Harold Bloom contends that 'the *Ode to the West Wind* is actually a poem about this process of making myths, a poem whose subject is the nature and function of the nabi [Hebrew for "prophet"] and his relation to his own prophecies. Specifically we can take it as being a poem about Shelley's relationship to *Prometheus Unbound*' (*Shelley's Mythmaking*, 1959). And it ends on one of the most striking expressions of Shelley's optimism: 'If Winter comes, can Spring be far behind?'

This optimistic conception of the poet's role was to be echoed, a few months later, in 'To a Skylark', in which the bird soaring up into the light of the sun becomes at once a disembodied voice, a pure emanation of the divinity, and, as such, a symbol of the poet in a state of creative ecstasy ('the Trumpet of a prophecy' celebrated in the 'Ode to the West Wind'):

> Like a Poet hidden
>   In the light of thought
> Singing hymns unbidden,
>   Till the world is wrought
> To sympathy with hopes and fears it heeded not. . . .

Also written at the very moment of the experience, on an Italian plain, this poem is another good example of the flashing immediacy of Shelley's inspiration.

In 'The Cloud', as in the two previous poems, the poet projects himself into a natural phenomenon or creature. The mood is one of exalted, exalting and exulting freedom that sweeps the reader along and makes him participate in the sublime joy of cosmic power. Such statements as 'I change but I cannot die' and 'I silently laugh at my own cenotaph' could be accepted as 'evidences of Shelley's belief in some kind of immortality, as well as his yearning towards a supernal status. . . . These lyrics would suggest, more forthrightly than *The Sensitive Plant*, that he found what he was seeking: an emblem of permanence in a world of change' (Baker, *Shelley's Major Poetry*).

'The Cloud' succeeds where 'The Sensitive Plant' fails, wrote Benjamin P. Kurtz (*The Pursuit of Death*, 1933). Kurtz thought that the latter poem was more a superficial exercise lightly undertaken by Shelley, to see how well he could describe living and dead flowers, than a true expression of the moods that were deepest in him at this time. Earl R. Wasserman, on the other hand, made a long and subtle analysis of this pretty and touching poem with Platonic overtones (asserting that 'in some general sense the material world is an illusion and the ideal world is real' – *The Subtler Language*, 1959); but he acknowledged and demonstrated that 'the moral does not in fact follow from the narrative'. However, the gist of his investigation was to present 'the linguistic triumph' of this important poem:

> with this subtler language within language he has expressed the illusory nature of "worldly" change and decay, the probable reality of the only metaphorically true that inheres within the structure of this new language, and the identity of the sensorily experienced world in its perfection with the metaphorically revealed reality – an identity in some ultimate and immortal oneness that forever eludes all language.

To express what eludes all languages certainly was what Shelley never ceased to attempt in all his poems.

Mention must also be made of more solemn, political odes, such as 'Ode to Liberty' and 'Ode to Naples' (both written in 1820) – interesting works with a very classical construction, but not so moving or original as the poems discussed above.

Prometheus Unbound

It seems that Shelley had the initial inspiration for *Prometheus Unbound* in the Gorges des Echelles, near Chambéry, on his way to Italy in the spring of 1818. The first act was written at Este in September and October of the same year. The composition was interrupted by the death of his daughter Clara, which nearly drove him to suicide. It was resumed in Rome, under more favourable circumstances, in March 1819. Acts II and III were written in the ruins of the Baths of Caracalla. Finally Shelley wrote Act IV at Florence, several months later, as a sort of afterthought and triumphal crowning-piece.

The author gives a good account of his purpose in the Preface, in which he explains that he has indeed borrowed the subject from Aeschylus, but 'was averse from a catastrophe so feeble as that of reconciling the Champion with the oppressor of mankind'. We know that in Aeschylus's *Prometheus Bound* Prometheus becomes reconciled to Jupiter by disclosing to him the danger threatened to his empire by the consummation of his marriage with Thetis. In the first act of Shelley's drama, Prometheus, bound to his rock, laments and is tortured by Furies and tempted by Mercury, Jupiter's messenger. But he refuses to yield up his secret. The main action, though, is his recantation of his curse of Jupiter, an act of forgiveness and repentance. The second act is essentially feminine, with Asia and Panthea, the two Oceanides. In scene IV they descend into the cave of Demogorgon (described as 'a mighty darkness'), a character invented by Shelley and something like the voice of Ultimate Reality. The act ends on the famous lyric 'My soul is an enchanted boat', sung by Asia. Act III opens with the fall of the tyrant Jupiter, followed by the unbinding of Prometheus and the resurrection of nature. The last act, full of musical exuberance and rhythmic variety, is a celebration of the Golden Age.

Prometheus Unbound is one of the richest and most powerful

Romantic works. Shelley was conscious, when he wrote this un-stageable drama, that it could not be understood, then, by more than five or six people. Contemporary critics were certainly at a loss, and it has taken many decades of careful study to elucidate its most obscure passages. All principal Shelley critics have discussed it at length, and the variorum edition of L. J. Zillman, with its important introduction and abundant notes, gives a very complete account of the various approaches until 1959. There is no need to insist on their diversity, since this poem is the best example of Shelley's mythopoetic power. Like all true mythical works it can be approached with equal relevance from a number of points of view: biographical, sociological, historical, scientific, philosophical, an-thropological. All, naturally, must first take into account Shelley's own declaration of his intentions in the Preface, in which he compares his hero to Satan, without 'the taints of ambition, envy, revenge, and a desire for personal aggrandisement, which, in the Hero of *Paradise Lost*, interferes with the interest. . . . But Pro-metheus is, as it were, the type of the highest perfection of moral and intellectual nature, impelled by the purest and the truest motives to the best and noblest ends.' In very Shelleyan fashion this ideal character fights against Jupiter and defeats him without resorting to violence. Prometheus is usually considered to symbol-ise mankind, or the mind or soul of man; Jupiter all tyrannical powers (political, religious, paternal and so on), i.e. evil; Demogor-gon the primal power of the world, or destiny or necessity; Asia love, nature or beauty; Panthea faith or insight; Ione possibly hope or perfection. Carl Grabo (*A Newton among Poets*, 1930) has shown how thickly stored the poem is with scientific notations (especially in astronomy, chemistry and magnetism), and indeed several passages make sense only if this is known. Christian La Cassagnère has given a very different, but extremely consistent, and just as convincing interpretation of *Prometheus Unbound* as a neo-Platonic initiation (La mystique de 'Prometheus Unbound'). The contemporary approach afforded by anthropological structur-alism helps assess the great imaginative consistency of the poem, which Shelley rightly considered to be 'the most perfect of my productions'.

'The Triumph of Life'

'The Triumph of Life' is Shelley's last poem, written in *terza rima*. Composed at Lerici, on the Gulf of Spezia, in the spring and early summer of 1822, it was interrupted by the poet's death, and the unfinished draft was published by Mrs Shelley in the *Posthumous Poems* of 1824.

The first part is devoted to the description of a vision: the pageant of the car of earthly life with its victims, in 'a cold glare, intenser than the noon'. The chariot is guided by 'a Janus-visaged Shadow', with four faces and his 'eyes handed'. In front of the procession maidens and youths, 'tortured by their agonizing pleasure, . . . fling their wild arms in air', while the rear is brought up by 'old men and women foully disarrayed' who 'wheel', 'with impotence of will', until 'they sink, and corruption veils them as they lie'. In the second part Rousseau appears in the shape of 'an old root which grew / To strange distortion out of the hill side' and helps identify some historical characters among the victims of earthly life: Napoleon, Plato, Aristotle, Alexander the Great, Catherine the Great, Frederick the Great, Gregory the Great.

'The Triumph of Life' has received little critical attention in comparison with Shelley's other major poems. A number of commentators consider that it is the poet's 'palinode', although this is denied by the greatest names of Shelley criticism (especially Carlos Baker and Harold Bloom): 'The poem is Rousseau's palinode, not Shelley's' (Bloom, *Shelley's Mythmaking*). This fundamental disagreement between critics shows how difficult some passages are to understand and interpret. Probably because the poem is unfinished, the moot point is the interpretation of the charioteer ('Human Intelligence', according to Todhunter; 'Destiny' or 'Necessity', for Locock, Stawell and Bradley; the 'calculating faculty', for Hughes; 'the human soul', for Barnard; 'the type of the poet who has been hoodwinked by the worldly life', according to Baker. Similar problems are raised by the 'Shape all light' (l. 352), the 'fierce Spirit' (l. 145), and even the car of life; although all agree on the influence of Petrarch's *Trionfi* on the poem's conception and on that of the pageant.

The most complete and subtle analysis of the poem is probably that provided by *Shelley's Mythmaking* by Harold Bloom, who differs from many of his predecessors on a number of important points and agrees with Hazlitt in understanding the title as 'bitterly paradoxical': 'it is in fact a new and terrific dance of Death'. He stresses the probable influence of Ezekiel and the important connections with the Blake of *Milton* and *Jerusalem*, and the Wordsworth of 'Ode: Intimations of Immortality' (as pointed out by Carlos Baker), and admits the influence of Dante (noted by Bradley), but doubts that of Calderón and Goethe (suggested by Stawell). He differs from Knight and Baker in his interpretation of the meaning of Rousseau – 'that natural religion is the enemy within himself, in his selfhood proper. . . . There is no natural religion'; and from Bradley and Stawell in considering the 'Shape all light' to be 'a parody of the Witch of Atlas'. His conclusion is that 'the final aspect of Shelley's mythopoeia is that the myth, and the myth's maker, are fully conscious of the myth's necessary defeat. There are no Thou's of relationship in *The Triumph of Life*; the poem commemorates the triumph of the "It" of experience'.

Carlos Baker's judgement in *Shelley's Major Poetry* can be considered as a fit expression of the modern general positive assessment of Shelley's last unfinished work:

> the likelihood is that, if he had been able to finish what he had begun, the result would have been a fourth great poem to add to the *Prometheus*, the *Epipsychidion*, and the *Adonais*. The *Triumph* is filled with solemn music, charged with deep melancholy, more nearly mature in its inward control and majestically dignified in its quiet outward demeanor than anything Shelley had done before.

JEAN PERRIN

Smart, Christopher (1722–71)

Christopher Smart was born in Kent and brought up in the north of England, where he seems to have fallen in love with Anne Vane, daughter of his family patron, Lord Barnard; she later became a visionary figure (like those in John Clare's poems three quarters of a century later) under such guises as Hope and Constancy in his poem *Jubilate Agno*, written in the 1760s but not published until 1939.

Smart's excitable temperament and volatile imagination allied personal experiences of places and people with the language of Scripture from the start. He wrote poetry during his years at Pembroke College, Cambridge, and continued to do so after he gave up his fellowship on marrying in 1752. By this time he was already spending much time in London as a hack writer; he had also (as Thomas Gray's letters make clear) gained a reputation for drunkenness and hopeless eccentricity.

Smart was a cheerful, companionable man, but, as his early-contracted religious enthusiasm grew into mania, he gradually lost contact with other people. He suffered a total breakdown and was confined in an asylum for much of the period 1754-63. There he produced his finest religious poem, the ecstatic, rhapsodising *A Song to David* (1763), in which the language of the Apocalypse is drawn upon for an immense litany of praise for all created things. The poem possesses both charm and real power, but it seemed too hopelessly eccentric to be taken seriously in its own day (the parallel with Blake's fortunes a generation later need hardly be stressed).

A Song to David began to be republished and taken notice of in Victorian times, but both it and the long-unknown *Jubilate Agno* had to wait until recent years for real appreciation. In 1770 Smart also published *Hymns for the Amusement of Children*, combining his characteristic sharpness of eye and feeling with a greater-than-usual control of tone and form.

Smart was a small, neat, sociable man and his minor poems are often marked by genuine high spirits and a sense of humour. But his religious mania, which gave his verse imaginative power, also trapped him within a framework of personal obsessions and a private series of codes and values. There is, however, an enduring appeal (akin to that in certain poems by Cowper and Clare) to be found in Smart's tenderness for small and helpless living creatures. Smart's capacity for drawing on contemporary knowledge in his religious poems is remarkable: even in states of mental disturbance he was able to draw upon his vast store of reading. Smart points the way forward towards Blake's poems in his capacity for finding the simple metaphor which can illuminate the affections vividly for the reader. His children's hymns also look towards

Blake in their ability to combine the approachable style of Isaac Watts with genuine visionary power.

WILLIAM RUDDICK

Southey, Robert (1774-1843)

Robert Southey was born in Bristol on 12 August 1774. The son of a linen-draper, he was handed over, from two to six years of age, to an eccentric and dictatorial maiden aunt, Miss Elizabeth Tyler, who 'laboured under perpetual *dusto-phobia*' and inculcated on him an almost morbid sense of physical and moral cleanliness. At the age of six the child was removed from Miss Tyler's household in Bath. After attending schools at Bristol and Corston (1780–88), he entered Westminster School (Apr 1788) where he formed lifelong friendships with C. W. W. Wynn and Grosvenor Charles Bedford, with whom he exchanged letters for forty years. Expelled from Westminster (Apr 1792) for writing a pamphlet entitled *The Flagellant*, in which corporal punishment was stigmatised as inspired by the Devil, he was a prey to severe dejection for a few months. His sympathy for the French Revolution was no obstacle to his entering Balliol College, Oxford (Nov 1792), where he never felt happy. There he read Godwin's *Political Justice* and wrote a revolutionary epic, *Joan of Arc*, in which he denounced all the representatives of the establishment. With Coleridge, whom he first met in Oxford (June 1794), he planned a utopian scheme of emigration to America called 'Pantisocracy' (the idea was to establish an egalitarian community of six families 'on the banks of the Susquehanna'), which came to nothing for lack of money. No sooner had he secretly married Edith Fricker (thus becoming Coleridge's brother-in-law) on 14 November 1795, than he left for Portugal alone. He travelled through Spain and Portugal (Dec 1795 – Apr 1796) with an uncle of his, the Rev Herbert Hill, chaplain to the British Factory at Lisbon, who wanted his nephew to become a clergyman. With a head full of Rousseau and *Werther* and his religious principles undermined by Gibbon's *History*, Southey felt unable to comply with his uncle's wish. The publication of *Joan of Arc* (1796), of two volumes of *Poems*

(1797, 1799) and of *Letters Written during a Short Residence in Spain and Portugal* (1797) persuaded him that settling into the life of a professional man of letters was his true vocation. A second stay in Portugal (May 1800 – June 1801), whose climate and landscapes suited him so well (Cintra was another Eden to him), turned him into an authority on Portuguese and Spanish history and literature.

In September 1803, Southey, deeply grieved by the loss of his daughter Margaret, settled with the Coleridges at Greta Hall, Keswick, in the heart of the Lake District. There he spent the last forty years of a life remarkable for unremitting literary activities. By 1810 the former radical supporter of the French Revolution had become a staunch Conservative branded as a 'turncoat' by his political enemies. His appointment to the office of Poet Laureate (1813) was hailed by the latter as the symbol of his ideological apostasy. The publication of *A Vision of Judgement* (1821), a high-falutin poem in praise of George III, and its preface containing a scathing attack against 'the Satanic School' of poetry, led to a quarrel with Byron which reached a climax in 1822. From his seclusion in Keswick, only interrupted by a few occasional journeys (to London, Wales, Scotland, the Low Countries and France), he pointed out the means of reforming British society to avoid violent revolution. A warm advocate of public order and a leading contributor to the *Quarterly Review*, he opposed Catholic emancipation and parliamentary reform but resented social inequality. Owen's socialist plans he considered with sympathy and he unhesitatingly encouraged Lord Ashley's interest in factory reform. Neither the offer he received to write for *The Times* (1817) nor his election as MP for Downton (1826) could tear him away from his retreat at Keswick, where he stuck to his motto 'In labore quies' ('In work quiet'). Of all the domestic misfortunes that befell him, the most terrible was the death (in 1816) of his favourite child, Herbert, at the age of nine – a loss from which he never fully recovered. After the death of his wife (1837), he married Caroline Bowles, a poet with whom he had been corresponding since 1818. Shortly afterwards his mind and memory failed. The last four years of his life were melancholy ones: incapable of reading, he could only sit and pat the books on the shelves of his superb library at Greta Hall. He died on 21 March 1843.

Southey's literary career was a long crusade against evil. In a letter of July 1796 he explained that he once saw half a dozen men

stoning a dog to death and that he wished he had been the
Exterminating Angel. 'Oh, how I delight to see him trample on his
enemies!' Mrs Piozzi said in 1817. The pride verging on haughti-
ness which he displayed was probably a mask hiding extreme
uneasiness and anxiety. A very emotional man, he insisted on
repressing the sensibility which formed the deeper aspect of his
personality. 'In Southey the *frein vital* curbed the *élan vital*, almost
from the first' (H. N. Fairchild *The Noble Savage*, 1928). Southey's
rejection of heartfelt lyricism may be related to a characteristic
process of repression in which the deeper self is overcome by the
superego. 'Whenever anything distresses me', he once wrote, 'I fly
to hard employment, as many fly to the bottle'. He had made it a
rule to beware of the imagination, which he called 'a kind of
mettled horse that will most probably break the rider's neck, where
a donkey would have carried slow and sure to the end of the
journey'. In this sense the author of such long narrative poems as
Madoc, *Thalaba* and *The Curse of Kehama* appears as a Romantic poet
who lost his way. Yet Southey's influence in the development of
English Romanticism was far greater than is commonly admitted.
He, and not Coleridge or Wordsworth, was regarded as the leading champion of a new spirit in late eighteenth-century poetry.
With his *Joan of Arc*, his *Botany Bay Eclogues*, his *English Eclogues* and
some of his ballads, Southey actively participated in the regener-
ation of poetry, paving the way for greater poets than himself.
Thalaba and *The Curse of Kehama*, despite their flashy exoticism,
were seminal poems which appealed to the Shelley of *Queen Mab*,
Alastor and *Prometheus Unbound*. Lake man of letters rather than
Lake poet, Southey stood halfway between rationalism and Romanticism – participating in both, possessed by neither.

The poems

Southey's poetical works, of which Kenneth Curry rightly said that
'they do not glow with that white incandescence which one finds
in the best works of his contemporaries', readily fall into three
categories: the early poems, mostly influenced by the French
Revolution; the long narrative poems, remarkable for their exotic
qualities; and the Laureate verse.

His revolutionary views Southey expressed as early as 1794 not
only in the *Botany Bay Eclogues*, where convicts were depicted as

victims of the oppressive justice of the state, but also in *The Fall of Robespierre* (the first act of which was written by Coleridge) and *Wat Tyler*, two very crude radical 'dramas' bearing the stamp of Thomas Paine's influence. *Joan of Arc* (1796), an epic whose heroine is 'a Tom Paine in petticoats' (Coleridge's words), is a provocative political manifesto interspersed with invectives against kings, aristocrats and churchmen in which mediaeval history is but an excuse for extolling the French revolutionary ideal of liberty, equality and fraternity. The pieces in Southey's *Poems* (1795, 1797) are imbued with a kind of humanitarianism linked to his early revolutionary ideology (see for example 'The Widow', 'The Soldier's Wife', 'The Pauper's Funeral', 'The Soldier's Funeral', the 1794 African slave-trade sonnets, and 'The Battle of Blenheim'). His *English Eclogues* (composed between 1797 and 1803) offers a realistic picture of country life. 'Pastoral poetry', Southey wrote in his common place book, 'must be made interesting by story. The characters must be such as are to be found in nature; these must be sought in an age or country of simple manners.' The resemblance to Wordsworth's own theory of poetry is striking. The language of the *English Eclogues*, though not exactly a reproduction of the rustic idiom, aims at conveying much of its raciness and colloquial vigour. The early Southey was a nature devotee. The 1798 revised edition of *Joan of Arc*, which appeared a few months before the *Lyrical Ballads*, offered a Wordsworthian vision of nature: like the heroine of the 'Lucy poems', Southey's Joan is a child of nature whose soul was nursed 'in solitude'. The ballads in *Metrical Tales, and Other Poems* (1805), where the threefold influence of Percy's *Reliques*, German Romantic poetry and Gothic fiction can be traced, do show commendable vigour in their treatment of the supernatural, the terrifying and the gruesome. 'Mary, the Maid of the Inn', 'Donica', 'Rudiger', 'Jaspar', 'Roprecht the Robber' and 'Lord William' are still worth reading. The terrifying vein is usually associated with a simple moral message – villains will be punished in this or another world – as in 'God's Judgement on a Wicked Bishop', 'The Old Woman of Berkeley' and 'The Inchcape Rock'. Some of Southey's ballads ('The Surgeon's Warning', 'St Michael's Chair', 'The March to Moscow') are not devoid of humour, the Devil being made a laughing-stock in 'St Romuald' and 'The Pious Painter'. Humour was one of the paths which Southey followed to elude the hold of the supernatural upon him. Whether humorous or serious, the ballads remain the most living part of Southey's poetical output.

It is not at all certain that Southey's long narrative poems should be called epics. Exoticism artficially coupled with Christian pre-occupations is the dominant feature of these imposing structures. *Madoc* (1805) tells the adventures of a twelfth-century Welsh prince fighting the Aztecs in the New World, while *Thalaba* (1801) takes the reader to Arabia, where the eponymous hero, a young Muslim, finally destroys, after a long quest, the occupants of the Domdaniel cavern under the sea. *The Curse of Kehama* (1810), which is based on Hindu mythology, can be considered a by-product of the craze for oriental tales in the second half of the eighteenth century and the early part of the nineteenth. To the oriental images of earth, heaven and hell in *The Curse of Kehama* the modern reader will prefer the more familiar Spanish setting of *Roderick, the Last of the Goths* (1814), unquestionably a better poem than Walter Savage Landor's *Count Julian* (1812) and Scott's *The Vision of Don Roderick* (1811). Southey's *Roderick* is very different from the protagonists of *Thalaba*, *Madoc* and *The Curse of Kehama*, who are exceptional and immaculate characters more like abstract allegories than creatures of flesh and blood. Also, the plot of *Roderick, the Last of the Goths* is unencumbered with the marvellous, or the demonic and angelic figures, magic rings, oaths, curses and other such paraphernalia found in *Thalaba* or *The Curse of Kehama*. Unlike Ladurlad or Thalaba, Roderick is a guilty hero, whose 'original' sin, the rape of Florinda, caused the ruin of a whole nation. A man 'more sinned against than sinning', Roderick is by far the most human of Southey's heroes, precisely because he is not the embodiment of virtue and perfection. He is close to us in so far as he illustrates the biblical aphorism 'The spirit indeed is willing, but the flesh is weak.' Deprived of his throne through his own fault, he is doomed to a life of expiation during which he has ample time to meditate on his own tragic fate: 'What's done cannot be undone'. The subtitle, 'A Tragic Poem', which Southey gave to *Roderick*, is certainly not undeserved.

Of Southey's Laureate verse little need be said. Byron had good reason to sneer at the author of such dull poems as *The Poet's Pilgrimage to Waterloo* (1816) or *A Vision of Judgement* (1821), whose only merit lay in giving rise to Byron's devastating parody entitled *The Vision of Judgment* (1822). That Southey took his office very seriously is evidenced by what he said in 1820: 'Without reckoning the annual odes, I have written more upon public occasions (on none of which I should otherwise have composed a line) than has been written by any person who held the office before, with the single exception of Ben Jonson, if his Masques are taken into account.' Of all the Laureate odes, only 'Ode Written during the Negotiations with Buonaparte, in January, 1814' is worth preserving from oblivion: inspired by an indomitable hatred of Napoleon, it is an impressive sample of political invective.

The prose

Southey, who was almost entirely dependent upon his pen for his livelihood, was a versatile and prolific prose-writer. Much of his writing, especially his review articles for the *Monthly Magazine*, the *Critical Review*, the *Annual Review* and above all the *Quarterly Review*, he regarded as task work. His *Essays, Moral and Political* (1832), in which some of his articles on the poor were reprinted, represent only a very small portion of the numerous essays – usually on general topics – which he kept writing to the end of his working life. His editorial labours and translations, pot-boilers as they were, nevertheless helped to gain readers for English, Spanish and Portuguese literature: his *Select Works of the British Poets, from Chaucer to Jonson*, his editions of *The Works of Chatterton*, *The Pilgrim's Progress* and *The Works of William Cowper*, and his translations *Palmerin of England* (from the Portuguese), *Amadis of Gaul* and *Chronicle of the Cid* (from the Spanish) all served a useful function.

Letters from England, by Don Manuel Alvarez Espriella* (1807) arose from a sincere love of Spain combined with a genuine interest in social and political questions. Through seventy-six letters supposedly written by a young Spaniard spending eighteen months in England, Southey expressed his views on a wide variety of subjects – notably the evils of the Industrial Revolution, whose 'inevitable tendency' was 'to multiply the number of the poor'. Similar topics are discussed at greater length in *Colloquies on the Progress and Prospects of Society* (1829) where the ghost of Thomas More, encountering Montesinos, alias Southey, on several occasions, debates with him major social and political problems (Catholic emancipation, poverty, education, the status of women, the manufacturing system). T. B. Macaulay's review of the *Colloquies* (*Edinburgh Review*, Jan 1830) did much to bolster the notion that Southey was an incorrigible reactionary.

Southey's bitter opposition to Catholic emancipation could not but bias *The Book of the Church* (1824), a would-be popular history of the Church of England. The Laureate's achievements in the field of historical research were not up to his ambitions. Neither his *History of the Peninsular War* (1823–32) nor his *History of Brazil* (1810–19) – three huge quartos resulting from his interest in Portugal – could make him the Gibbon of the Romantic age. Prolixity, overemphasis of details, and a chronic lack of broad, synthetic views of his subject mar the *History of Brazil*, which has nevertheless been translated into Portuguese and is treasured by the Brazilians.

Southey was more successful with biography. The *Life of Wesley* (1820), which includes a study of 'the Rise and Progress of Methodism', is neatly written, but it is doubtful whether its author conveys the spirit of Methodism. Southey's biography of Cowper (1835–6), two thirds of which consists of excerpts from the poet's correspondence, suffers from his propensity to digress. No such defect mars the *Life of Nelson* (1813), unanimously acknowledged as one of the best biographies in English. The main reason for the uninterrupted popularity of this clear and concise biography, divided into nine chapters, is to be found in Southey's sympathy and admiration for Nelson, a national hero whose death at Trafalgar, in the hour of victory, corresponded to the triumph of good over evil. With the shadow of Napoleon, 'that bloody Corsican', lurking in the background, the little book, in which the great admiral's charisma (the famous 'Nelson touch') is remarkably conveyed, actually takes on an epic quality. To resist the Devil (the French Emperor), England needed no less than an archangel. Nelson was that archangel. It is no small tribute to Southey's *Life of Nelson* that it made the hero of Trafalgar a legendary figure. Besides, the Laureate may have obscurely felt that his own life and code of conduct had certain parallels with Nelson's.

Southey's personality can be traced in most of his prose writings. But his travel journals, his correspondence and *The Doctor*, his only attempt at fiction, are plainly autobiographical. Southey proves a close and often humorous observer of men and manners in *Letters Written during a Short Residence in Spain and Portugal* (1797), *Journal of a Residence in Portugal, 1800–1801, Journal of a Tour in the Netherlands in the Autumn of 1815* and *Journal of a Tour in Scotland in 1819*. Eccentricity is the dominant note of *The Doctor*, an unclassifiable work first published anonymously (1834–8). 'I see in the work', its author said, 'a little of Rabelais, but not much; more of

Tristram Shandy, somewhat of Burton, and perhaps more of Montaigne; but methinks the *quintum quid* predominates.' That the *quintum quid* is Southey himself cannot be doubted. A hybrid book with hardly any plot – the story of Dr Daniel Dove of Doncaster is only a slender thread – *The Doctor* amounts to a collection of endless digressions upon an infinite variety of topics, teeming with quotations from innumerable authors. Reading such a queer volume means entering Southey's home at Keswick, a cosy world of books, children and cats. 'The Story of the Three Bears', told in chapter 129, evidences Southey's eagerness to amuse the children at Greta Hall: it has remained a classic of the nursery ever since. *The Doctor* is no less noteworthy for the sheer ease and raciness of its style. Southey's vigorous, direct, pellucid prose, utterly devoid of mannerism, is the very opposite of Johnson's Latinised style. Nowhere will it be better enjoyed than in his voluminous correspondence, where Southey is most completely himself. As the *Edinburgh Review* (Apr 1848) remarked, in Southey's correspondence 'the lovers of pleasant English prose may make sure of having as agreeable a specimen of unconscious autobiography, in the form of letters, as any in the language'.

JEAN RAIMOND

The Sublime

Deep and fruitful reflection on aesthetics characterised the eighteenth century in England. It crystallised progressively around the three concepts of 'the beautiful', 'the sublime' and 'the picturesque', and provided for a harmonious transition between the neo-classical tradition, on the one hand, with its deep attachment to rules, and to the notions of order, balance and harmony associated with beauty, and Romanticism, on the other, fascinated by the sublime, and giving prime importance to emotions.

The father of the sublime was Longinus, whose treatise *Peri Hupsous* (*On the Sublime*) was translated into French by Boileau (1674), but also into English in 1652 and again in 1739. The sublime was at first a critical concept relating to literature, and as late as

1755 Dr Johnson in his *Dictionary* defined it as 'the grand and lofty style'. Then its meaning widened, and it became a mode of aesthetic experience associated with the emotion felt before nature, the more majestic landscapes in particular. John Dennis (*The Grounds of Criticism in Poetry*, 1704), Joseph Addison (*The Spectator*, 1712), Lord Shaftesbury (*Characteristicks of Men, Manners, Opinions, Times*, 1711), John Beattie (*Essay on the Sublime*, 1747) and William Hogarth (*Analysis of Beauty*, 1753) all made their own contribution to the definition of the sublime. It was left to Edmund Burke to clarify contemporary analyses on the subject, in his *Philosophical Enquiry into the Origin of our Ideas of the Sublime and Beautiful* (1757), and it is indicative of the evolution of taste that the sublime was given precedence over the beautiful in the title of the book.

Both concepts are analysed on two levels: the material causes, found in the world, and the psychological effects produced on the spectator. Obscurity, power, privation (vacuity, solitude, silence, for instance), vastness (what Addison called 'the great'), succession and uniformity (a rotunda, for example), magnificence, extreme light, or 'sad and fuscous colours' – all help to make an object sublime. As all senses are involved in the experience, Burke also included sections on 'loud sounds', 'bitter tastes', 'stenches' and 'intense feelings'. For that reason Burke has been accused of sponsoring the horrors of the Gothic novel, set in grandiose landscapes, with hair-raising noises in the dead of night, and involving exaggerated passions.

As far as its effects are concerned, the sublime inspires a mixture of horror and pleasure: 'a delightful horror' (Dennis), 'an agreeable kind of horror' (Addison). These writers are obviously struggling for adequate words to describe new feelings, alien to the quiet satisfaction raised by the contemplation of beauty. Burke uses the term 'delight' for that 'negative pleasure' born of the removal of pain once one realises one has escaped from some great danger; the passion that fills the soul under such circumstances is 'astonishment', with 'all its motions suspended with some degree of horror'. Believing in the physiological basis of all aesthetic experience, Burke then analyses the physical effects of the sublime: 'teeth set . . . eyebrows violently contracted, forehead wrinkled . . . eyes dragged inwards . . . hair on end . . . short shrieks and groans' (*Philosophical Enquiry*, IV.iii). This was going too far, and even his warmest supporters, such as Uvedale Price, thought it better to ignore that questionable part of his demonstration, while paying homage to the coherence of the rest.

Poetry responded to the sublime very early, with the 'excursion poets'. These gave an important place in their verse to spacious panoramas, majestic nature and geographical catastrophes, as in David Mallet's *The Excursion* (1728), James Thomson's *The Seasons* (1726–30) and Mark Akenside's *The Pleasures of Imagination*, which was written in imitation of Addison and revised in 1757 after the publication of Burke's *Enquiry*. All betray the same attraction to the new 'aesthetics of the infinite' (Majorie Hope Nicolson, *Mountain Gloom and Mountain Glory*, 1959). So does Edward Young, whose *Night Thoughts* (1742–5) carry the reader into interstellar space. The novel 'caught' the sublime much later, with the Gothic tales of Horace Walpole (*The Castle of Otranto*, 1764) and Clara Reeve (*The Old English Baron*, 1778). The fashion reached its culmination in the last decades of the century, with the novels of Mrs Radcliffe (*The Mysteries of Udolpho*, 1794; *The Italian*, 1797) and her imitators. (See under the GOTHIC NOVEL.)

In the painting of sublime landscapes, an early connection appeared between the 'sister arts', painters illustrating scenes from contemporary or past poets: Milton, Ossian and Shakespeare, offered rich sources of inspiration to James Barralet, Alexander Runciman and Paul Sandby (for instance, in his *A Storm from the Winter's Tale'*, 1774). But under the growing influence of Salvator Rosa, whom English tourists discovered on the Continent and whose works they brought back to England, majestic scenes attracted more and more artists and their customers. (See under LANDSCAPE and the PICTURESQUE.)

This many-sided and progressive reflection on the sublime, and the literary and pictorial illustrations of that aesthetic category, were certainly among the main achievements of the eighteenth century. Taste was changing: if beauty remained inspired more or less by classical canons, the sublime introduced, or rather gave expression to, a new form of sensibility that had been latent in English consciousness, art and literature since the days of Shakespeare and Milton, and that was to blossom in Romantic poetry.

SUZY HALIMI

Thomson, James (1700–48)

James Thomson was the outstanding descriptive poet of the early eighteenth century. Born at Ednam, in the south of Scotland, he was educated at Edinburgh University, where he studied for the ministry and acquired some reputation as a poet. He set out for London in 1725 and wrote 'Winter', the first part of *The Seasons*, which appeared piecemeal in 1726–30 and was published in collected form in 1730 (including 'Autumn' and 'A Hymn on the Seasons'). 'A Poem Sacred to the Memory of Sir Isaac Newton' appeared in 1727; 'Britannia. A Poem', in 1729. The literary world admired Thomson's frankness and genial temper and was impressed by his talent. The success of *The Seasons* won him the friendship of Pope, Gay, Young and the protection of such famous patrons as Lord Lyttelton and Frederick, Prince of Wales. In November 1730 he began the Grand Tour of Europe as tutor to Charles Richard Talbot, son of the Solicitor-General, and on his return received a sinecure that enabled him to live in easy circumstances. In 1735–6 he published *Liberty*, a long philosophic and patriotic poem in which he traces the progress of liberty through the ages in Greece, Rome and Britain. Thomson also produced a series of tragedies and a masque, *Alfred* (1740). In May 1748, a few weeks before his death, appeared *The Castle of Indolence*, an allegorical and humorous poem in Spenserian stanzas, originally intended to amuse his friends and considered as his most delightful and polished work. William Collins, the latest of his friends, who had taken lodgings in Richmond, near the poet, bemoaned 'the woodland pilgrim', 'meek Nature's child' in a moving ode, 'In yonder Grave a Druid lies' (1749).

The Seasons, written in flexible Miltonic blank verse, proved to be one of the most widely influential didactic–descriptive nature poems in English literature. 'Winter', in its first, 1726 version (405 lines), is a short, spontaneous, meditative poem on the 'Il Penseroso' model. The sadness of the season is suited to the poet's melancholy mood. The 'gloomy' sight of the stormy elements exalts 'the soul to solemn thought' and incites a 'pleasing Dread' that testifies to the poet's subjective response to nature. 'Winter' was composed in England, when Thomson was longing for 'the chapt mountain', 'the mossy wild' and 'rocks abrupt' of his native Borders. His sensitiveness to light and 'feel', his attention to the dynamic, ever-changing aspects of nature, his racy and vivid notations, his immediacy of impression and discreet appeal to emotion offer a new departure from the static and colourless topographical poems following the tradition of Denham's *Cooper's Hill*.

The *Seasons* underwent a continuous process of enlargement and revision. It swelled from 4470 lines in the complete 1730 edition to 5541 lines in the final 1746 edition. Some critics have taxed it with 'unevenness' on account of its increasing length, and digressions – either moral, historical, philosophic or scientific – that detract from the unity of atmosphere. Dr Johnson blamed Thomson's 'want of method' but pointed to the revisions as an extension of book-knowledge and actual experience. *The Seasons* reflects the poet's taste for classical and travel literature, and his deep interest in the intellectual and scientific currents of his time. A. D. McKillop (*The Background of Thomson's 'Seasons'*, 1942) and Marjorie Hope Nicol-son (*Newton Demands the Muse*, 1946) have shown his indebtedness to Thomas Burnet, the astro-and physico-theologians, Newton, Locke and Shaftesbury. Throughout the stages of his work, the 'philosopher–poet–virtuoso' explores the workings of creative nature, while evincing a landscape-painter's eye for light and colour.

Thomson's descriptions reveal a keen perception of the 'beauty', 'variety' and 'magnificence' (or sublimity) of nature – the three aesthetic categories defined by Addison in his 'Pleasures of the Imagination'. They cover a wide range of landscapes and pros-pects, from English country gardens, orchards and fields to Cale-donia's 'airy mountains', 'the wonders of the torrid zone' and arctic wastes. They include idyllic pictures suffused with Claudian light, as well as excursions into cosmic nature. Pastoral vignettes alternate with scenes of violence and horror in which the raging elements unfurl with destructive energy and 'burst into chaos', stressing the insignificance of man in front of hostile forces. The all-embracing poet introduces human activities, juxtaposing peace and war, retirement and social life, Virgilian Arcadia and com-merce, rural shades and harbours bristling with 'groves of masts', primitivism and progress.

If Bonamy Dobrée describes *The Seasons* as a 'most extraordinary hotchpotch' (*English Literature in the Early Eighteenth Century, 1700–1740*, 1959), Patricia Meyer Spacks traces the organic relation between nature and ideas (*The Poetry of Vision*, 1967). She suggests that Thomson is deeply concerned with the total pattern of nature and controls his descriptions through a 'unifying vision' rather than through visual details. The poet wants to convey the order and harmony he perceives in the universe behind the apparent disorder, although we may feel at times, in the description of natural phenomena, that he is more fascinated by diversity, exuberance, plenitude (and vital, chaotic forces) than by the abstract principle of unity.

Indeed, the 'book of nature' and man's responsiveness to its 'grand Works' disclose the secret-working hand of the remote and mildly pervasive 'varied God'. He, in the sentimental and optimistic 'Hymn', which belies the darker tenor of the rest of the poem, reconciles all conflicting elements and assumptions.

Recent criticism has emphasised the constant change governing Thomson's world and focused attention on the cyclical rhythm of the seasons with their blending or opposing forces, shifting perspectives, and fragmentary views and perceptions, so that 'even organically unified parts are merely another aspect of fragmentation' (Ralph Cohen, The Unfolding of 'The Seasons', 1970). The poet's original conception of time and space emerges from a close study of nature's transformations, while a symbolic and/or psychoanalytic approach to elemental imagery leads to a revaluation of Thomson's artistry.

Most critics have praised the vigour of Thomson's imagination, which is both visual, penetrating the mysteries of the universe, and creative. Thomson often contrasts man's 'bounded view' with 'fancy's eye' or 'the mind's creative eye', whose strange visionary power is akin to Coleridge's conception of the secondary imagination. The primary sense becomes transformed into an imaginary power. It can soar from pole to pole, survey infinitive space, or exalt thought 'beyond dim earth' and anticipate 'those scenes of happiness' to be found in afterlife.

The wandering narrator alternates motion and rest. He likes to 'woo lone Quiet' in the autumnal forests. The 'haunts of meditation', 'twilight groves' or 'weeping grottoes' favour contemplation, pleasing melancholy, dream or vision. The poet's many voices prove that he is not merely writing descriptive verse but relating an inner experience.

The poem suffers from an excess of scientific material and abstract terms. Yet Thomson's never-failing sense of wonder and personal responsiveness to the least vibrations of nature, his impressionistic rendering of light and colour (effects that Turner greatly admired) and his subtle variations of tone all struck a chord in Wordsworth, who hailed him as the first poet since Milton to present new images of external nature.

MICHÈLE PLAISANT

Turner, Joseph William Mallord (1775–1851)

J. M. W. Turner was born on 23 April 1775 in Covent Garden, London, the son of a barber and wigmaker. When a child, he showed great promise as a painter; and he was admitted as a student of the Royal Academy as early as December 1789, when he was fourteen. He exhibited his first watercolour a year later. He continued to make progress, especially in topographical art, and was employed by Dr Thomas Monro, Surgeon to the King and a patron and connoisseur, to copy drawings by J. R. Cozens and others. Monro had an eye for youthful promise: one fellow copyist was Thomas Girtin, who died young but who was always described by Turner as the finer artist of the two of them: 'if Tom Girtin had lived, I should have starved'. During these years Turner made a number of tours in England, painting in the West Country and (in 1797) the Lake District. He was elected an Associate of the Royal Academy in 1799, at the earliest possible age, and became a full member three years later. In the same year, 1802, he made his first visit to France and Switzerland, taking advantage of the brief peace which interrupted the long war with France.

Turner was a devoted member of the Royal Academy, becoming Professor of Perspective in 1807, and Professor of Painting in 1811. He was also fortunate in his patrons, who included Walter Fawkes, of Farnley Hall, Yorkshire, and Lord Egremont, of Petworth, Sussex. He visited Farnley Hall every year from 1810 to 1824 (Fawkes died in the following year); but his love of travel and fine scenery (and towns) is chiefly seen in the visits which he made to parts of Europe. During many summers after the end of the Napoleonic wars, he visited European countries, including Belgium and Holland (1817), Italy (1819–20) and France (1821); he returned to many of these places in the following years, particularly Switzerland and the Alps, and the Rhine valley.

In 1840 he met John Ruskin for the first time; Ruskin bought some of Turner's pictures, and became an enthusiastic advocate of his work. *Modern Painters*, the first volume of which was published in 1843, was, and still is, a classic exposition of Turner's qualities as a landscape-painter, with specific reference to a number of notable pictures; and Ruskin, who thought of Turner as 'my earthly Master', continued to work on Turner after the painter's death in 1851.

Turner was in great demand as an illustrator: his topographical accuracy, and his ability to capture the atmosphere and 'feel' of a place were highly prized by poets and their publishers. His superb command of line and tone contributed greatly to the success of Samuel Rogers' *Italy* (1830) – the book which first attracted the attention of the young Ruskin – and he also illustrated the great popular poets of the day, Scott and Byron. Other topographical work included *The Southern Coast* (1811) and work on English seaports, which after his death became *The Harbours of England* (1856, with text by Ruskin). Throughout his life, Turner had strong associations with literary texts: like Constable, he was an admirer of James Thomson's *The Seasons*, and of the poetry of Thomas Gray. Turner was unusual, however, in that he also wrote his own fragmentary poems to accompany his pictures when they were exhibited. One of these is described (in the catalogue entry for *Snow Storm: Hannibal and his Army Crossing the Alps*, exhibited at the Royal Academy, 1812) as being from 'MS P Fallacies of Hope', where 'P' has been taken to stand for 'Poem'; the title 'Fallacies of Hope' indicates something of Turner's shrewd and sometimes sardonic literary imagination. His paintings often demonstrate a moment of hope, as in *Hannibal*, or in *Dido Building Carthage; or the Rise of the Carthaginian Empire* (exhibited 1815), together with an underlying sense of foreboding. This sense of impending doom is found in the verses attached to *Hannibal*, which forecast his downfall, or in the companion picture to *Dido*, *Decline of the Carthaginian Empire* (exhibited 1817). Paintings such as *Dido and Aeneas* (1814) similarly suggest a fragile happiness, and later pictures, such as *The Fighting 'Téméraire' Tugged to her Last Berth to be Broken Up* (1838, exhibited 1839), give an elegiac impression of a former glory. At other times, Turner suggests the power of nature and the littleness of human beings, as in *The Shipwreck* (1805), where small boats toss in a huge sea, and in *The Fall of an Avalanche in the Grisons* (exhibited

1810, and influenced by Thomson's 'Winter'). Sometimes the human beings are made to seem comic in their ineffectual struggles, as in *Calais Pier, with French Poissards Preparing for Sea: an English Packet Arriving* (exhibited 1803) or *Messieurs les Voyageurs on their Return from Italy (par la Diligence) in a Snow Drift upon Mount Tarrar* (watercolour, 1829). At the opposite extreme are paintings demonstrating human cruelty, the finest example of which is *Slavers Throwing Overboard the Dead and Dying – Typhoon Coming On* (exhibited 1840). The painting shows slaves being thrown overboard and being eaten by sharks, against a background of an angry sky: it is one of Turner's most powerful statements about humanity, cruelty and nature.

Turner loved to paint spectacular effects, such as the typhoon in this picture, or the storm in *Hannibal* (based on a storm he had witnessed on the Yorkshire moors): he depicted avalanches, volcanic eruptions (*The Eruption of the Souffrier Mountains*, exhibited 1815), and the burning of the Houses of Parliament in 1834, which he painted in oil and in watercolour. Other pictures, such as *Snow-storm, Avalanche and Inundation – a Scene in the Upper Part of Val d'Aouste, Piedmont* (exhibited 1837) reflect Turner's delight in sublime and terrific Alpine effects. This is linked with his literary interests, which included such epic subjects as *Ulysses Deriding Polyphemus – Homer's Odyssey* (exhibited 1829) or biblical ones such as *The Fifth Plague of Egypt* (1800) and *The Tenth Plague of Egypt* (1802).

In addition to these sublime and spectacular subjects, Turner could also paint tranquil compositions, initially in the manner of Claude and of Richard Wilson (*Aeneas and the Sibyl, Lake Avernus, c. 1798*), but later in his own experimental and inimitable manner. He was particularly interested in the effects of light and colour in different places and at different times: he anticipated the Impressionists in his later years by painting the Rigi (a mountain on the side of Lake Lucerne) at many different hours of day and in many lights and colourings. His landscapes are luminous in their treatment of light and water; Turner uses watercolour to great effect to convey the misty and reflected light of Italian scenes, especially Venice, which he visited in 1819. The same kind of technique is found in British scenes, such as *Somer-Hill* (1811) or the later versions of Norham Castle on the river Tweed, which he painted on a number of occasions, principally in watercolour but once in oil.

Turner's paintings often have figures at work in the foreground, with beautiful effects behind, as in *Sun Rising through Vapour; Fishermen Cleaning and Selling Fish* (exhibited 1807) or *Ploughing up Turnips, near Slough* (1809), which has a radiant Windsor Castle in the background. Other strong contrasts occur in Turner's use of black and flame in *Keelmen Heaving in Coals by Night* (1835), painted on the river Tyne, and his tribute to his friend David Wilkie, *Peace – Burial at Sea* (exhibited 1842). Such paintings are great because of their wonderful use of light and colour, their daring experiments with different effects; but Turner's pictures are also great because of their humanity. His portrayal of grandeur, and of the sublime, is part of a tremendous variety and energy in his work, which is found in his endless curiosity and his interest in many different kinds of landscape; but it is also linked with human emotion felt with a deep amusement, honesty and compassion.

J. R. WATSON

Warton, Joseph (1722–1800)

The son of Thomas Warton, the elder, who was Professor of Poetry at Oxford (1718–28), Joseph Warton was educated at Winchester and at Oriel College, Oxford. His several offices in the Church included that of Canon of St Paul's Cathedral, and he was headmaster of Winchester from 1766 to 1793. His two volumes of *Odes* (1744 and 1746) show an interest in nature and the country traceable in part to classical sources, especially the *Eclogues* and *Georgics* of Virgil, which he translated for an edition of 1753. Like the richer and more accomplished examples published by his friend William Collins in 1747, however, Warton's odes are important primarily for their cultivation of feeling and imagination, primitive in kind though polished in expression; albeit in a minor key, they helped to develop that rediscovery of lyrical modes and private sensibility which has been viewed historically as a crucial aspect of 'pre-Romanticism' and the mid-century challenge to the supremacy of Pope's social vision and technical sophistication.

This undermining of Augustan tradition and emplacement of new values is apparent also in Warton's major critical work, *An Essay on the Genius and Writings of Pope* (vol. I, 1756; vol. II, 1782), which J. R. Lowell, the American essayist and poet, called 'the earliest official declaration of war against the reigning mode'. Warton does see much to admire in Pope, designating him, for example, 'the first of ethical authors in verse'; but in the course of his long seriatim commentary on the whole of Pope's poems he polemically celebrates, at Pope's expense, not only Spenser and Milton but also the poetry of nature and 'passion' to be found in such moderns as Thomson ('scenes . . . as wild and romantic as those of Salvator Rosa'), Akenside, Gray, Collins and Shenstone. He argues for the superiority of blank verse over the rhymed couplet, and recommends the study of Miltonic sublimity rather than the 'correctness' of the 'French writers and their followers'. Above all, he asserts that 'the most solid observations on human life, expressed with the utmost elegance and brevity, are morality and not poetry', for it is 'a creative and glowing imagination' that alone makes the true poet. His own poetic imagination was hardly 'glowing', though it was 'creative' in its way; and his knowledge and independence as a critic had a substantial influence on the direction of eighteenth-century taste – as well as representing a methodological advance, alongside the even more ambitious work of his brother, Thomas Warton the younger, in the scholarly study of native, as opposed to classical, literature. When, in 1797, a year before the appearance of *Lyrical Ballads*, he published a full-scale edition of Pope's works (followed in 1800 by one of Dryden) he may have been seeking in some degree to redress a balance that had swung decisively to the side of his own earlier critique and aesthetic principles. He was, with his brother, a member of Samuel Johnson's 'Literary Club', and wrote for both *The Idler* and *The Adventurer*.

VINCENT NEWEY

Warton, Thomas, the younger (1728–90)

The brother of Joseph Warton, Thomas the younger was educated at Trinity College, Oxford, and, like his father, became Professor of Poetry (1757–67). He was later Camden Professor of History, and held the office of Poet Laureate from 1785. Some of his poetry, notably the early *Pleasures of Melancholy*, is similar to that of his brother in both quality and bias; but he produced a substantial and varied body of verse during his career, including important revivals of the sonnet, historical romances on mediaeval and other themes, and such lively humorous lyrics as the once-popular 'Panegyric on Oxford Ale'. His *Poems* appeared in 1777. He edited several classical texts and miscellanies of verse, and in 1785 published an excellent edition of Milton's long-neglected shorter poems. Among his copious prose writings there are biographies, topographical works, and an interesting contribution to the debate over the pseudo-mediaeval 'Rowley' poems of Thomas Chatterton.

Warton's most significant achievements, however, are as an innovative and vastly informed critic and literary historian. His *Observations on the Faerie Queene of Spenser* (1754) established an authoritative model for the study and appreciation of literature, indicating the importance of cultural context, sources and influence, manuscripts, comparison, evaluative response, and close consideration of authorial techniques as they operate within given stylistic and generic forms. This method, in effect highly original though derived in part from the field of classical studies, is carried forward with even greater scope and erudition in his famous *History of English Poetry* (3 vols, 1774, 1778, 1781), which covers the period from the thirteenth century to the death of Queen Elizabeth and marks a comprehensive breakthrough in the rehabilitation of mediaeval poetry and drama at a time when it was widely thought to be merely barbarous. The *History* also contains seminal accounts of the thematic patterns of myth and romance. It was republished in elaborate annotated editions throughout the nineteenth century, beginning with that by Ritson and others in 1824. This, together with Mant's edition of the *Poetical Works* (1802), bears witness to the continuing presence of a writer who had given signal impetus to emergent Romanticism itself and to the understanding of literary movements in general.

VINCENT NEWEY

Waterloo

Waterloo is a small village near to the famous battlefield. It became the British name for the battle itself. The French name is 'Mont Saint-Jean' after the site of the battle. Here Napoleon was defeated (18 June 1815) by British, Netherlanders, Hanoverians and Nassauers under the Duke of Wellington and by Prussians under General Blücher. On 16 June, Napoleon had divided Blücher's forces (whom he defeated at Ligny) from Wellington's (who defeated Marshal Ney at Quatre Bras). At Mont Saint-Jean itself Wellington's troops resisted a series of attacks by Napoleon's army, including, finally, the Imperial Guard, and, with the arrival of Blücher's reconstituted Prussian forces, the French were defeated. Over a quarter of a million soldiers were involved and more than 50,000 died. Napoleon's power, and twenty-two years of nearly continuous warfare, were over.

Everyone, Wellington included, was appalled by the scale of the carnage and the battle's chaotic ferocity. The result was viewed differently according to political conviction. Hazlitt went into formal mourning and thought it the victory of kings over mankind. Hobhouse, Godwin, Leigh Hunt, Cobbett and other 'Napoleonists' were downcast by the apparent end of the events begun in 1789. Wordsworth, Coleridge, Southey and Scott shared British popular rejoicing in the British victory. It was the first time that British troops had been in the forefront of a battle against Napoleon himself. Shelley, no admirer of Napoleon, seems to allude to it as a temporary defeat for liberty in 'The Masque of Anarchy' (ll. 240–45). Thomas Moore, in the persona of Phelim Connor in letter XI of *The Fudge Family in Paris* (1818), presents a similar view of Napoleon's demise as also that of liberty.

There were a host of minor poems on the occasion, but Scott's *Field of Waterloo* (1815), Southey's *The Poet's Pilgrimage to Waterloo* (1816) and Byron's stanzas in canto III of *Childe Harold* (1816), in *The Vision of Judgment* (1822), in cantos VIII and IX of *Don Juan* (1823) and in *The Age of Bronze* (1823) are the most important. Southey and Scott celebrate Waterloo but not without some elegiac disquiet. Southey journeyed to the battlefield in the autumn of 1815 specifically to collect material for a major Laureate poem on the occasion: we are as much aware of the poet's struggle to win some affirmation out of Waterloo as of the battle itself.

Byron's magnificent rhetoric in *Childe Harold* (III.xvii–xliv) became the definitive poetic account of the battle. He registers its scale, erotic glamour, carnage and waste in lines which express both the desperate energy and tainted elegiac sentiment disclosed by it. Scott was indignant (*Quarterly Review*, XVI, 1816) that Byron did not drop 'even one leaf of Laurel on the head of Wellington'. Byron is more caustic still in *The Vision of Judgment* ('the crowning carnage Waterloo' – l. 38), *The Age of Bronze* and *Don Juan*. In canto XI he wryly presents his own poetic career as 'The grand Napoleon of the realms of rhyme'. Like Napoleon his earlier successes have declined and his later works are unread:

> But Juan was my Moscow, and Faliero
> My Leipsic, and my Mont Saint Jean seems Cain.

(XI.lvi)

BERNARD BEATTY

Wellington, Arthur Wellesley, first Duke of (1769–1852)

Arthur Wellesley was born in Ireland and educated at Eton. He entered the army in 1787, and rose to become a lieutenant-colonel of the Thirty-third of Foot (later known as the Duke of Wellington's Regiment) in 1793. In 1796 the regiment was sent to India, and Wellington became a major-general in Mysore. It was there that he learned the art of commanding in battle, particularly with the remarkable victory at Assaye (1803). He returned to Europe in 1805, and in 1808 became a subordinate commander in Portugal, winning the battle of Vimiero in 1808. This was followed by the Convention of Cintra, in which the defeated French armies were allowed to return to France by sea: this feeble diplomacy, which undid the heroic work of the campaign, was attacked by Wordsworth in his pamphlet *On the Convention of Cintra* (1809).

Wellington returned to Portugal in 1809, and fought a series of brilliant actions at Talavera (1809), the lines of Torres Vedras, near Lisbon (1810), Albuera (1811), Ciudad Rodrigo (1812), Badajoz (1812) and Salamanca (1812). His attention to detail (particularly in relation to transport and supplies), his mastery of the tactics of defence, and his mobile warfare were the secrets of his success. By 1814 he was across the Pyrenees, had captured Toulouse, and was installed at Bordeaux.

Following Napoleon's abdication in 1814, Wellington became British Ambassador in Paris, but he was then sent to the Congress of Vienna in 1815. It was at Vienna that he learned the news of Napoleon's return from Elba: he and the Prussian marshal Blücher were put in charge of the allied armies in the north.

Waterloo (18 June 1815) was Wellington's most famous victory. After it he returned to Paris, and thence to London: he was appointed Commander-in-Chief of the Army in 1827 (on the death of the Duke of York!), and became Prime Minister in 1828. His opposition to the Reform Bill of 1832 led to his removal from power, but he served under Sir Robert Peel in the Tory administrations of 1834 and 1841. He retired from active public service in 1846, after Peel's defeat following the repeal of the Corn Laws.

Tennyson's *Ode on the Death of the Duke of Wellington* was a fine tribute to Wellington, and perhaps the best of Tennyson's public poems as Poet Laureate. In earlier years, Wellington was less popular, and was the object of some nasty sniping from Byron. Byron, punning on the French pronunciation, called him 'Villain-ton', and asked (*Don Juan*, IX.iv),

> If you have acted *once* a generous part,
> The world, not the world's masters, will decide,
> And I shall be delighted to learn who,
> Save you and yours, have gain'd by Waterloo?

Byron regarded Wellington as the servant of the oppressive European governments, and regretted the part that he played in restoring the *status quo* after 1815 (Byron and Shelley also hated Castlereagh, the chief civil agent of this restoration, and Britain's representative at the Congress of Vienna). The battle of Waterloo, however, became famous as the place of Napoleon's final defeat (see WATERLOO).

J. R. WATSON

Wollstonecraft, Mary (1759–97)

Mary Wollstonecraft was, and is, best known as the author of *A Vindication of the Rights of Woman* (1792). She was an active member of the radical circle of intellectuals centred around the publisher Joseph Johnson's bookshop (which included at various times, Richard Price, Joseph Priestley, Horne Tooke, Tom Paine, Godwin, Blake, Wordsworth and Coleridge). In many ways, the *Vindication* extends the radical arguments of that group to the cause of women and shows that 'the rights of men' necessarily include the rights of women. Wollstonecraft's book can therefore be seen as offering a critique of the radical movement in its own terms: although her political thought is grounded in an admiration of Rousseau's egalitarian politics, one of the main targets of the *Vindication* is the discussion of women in *Emile* (1762). Wollstonecraft sets out to disprove the contemporary view that women were 'naturally' weak, vain and frivolous by pointing out that society systematically educates and conditions them to be so. She turns against itself Rousseau's argument – which was shared by conservatives who would have repudiated his egalitarian politics – that women would lose their 'power' over men if they became more rational and robust: 'This is the very point I aim at. I do not wish them to have power over men; but over themselves' (*Vindication*, p. 154). If women were to receive the same physical and intellectual education as men, then they would benefit not only themselves but the whole of society: 'It is time to effect a revolution in female manners – time to restore to them their lost dignity – and make them, as a part of the human species, labour by reforming themselves to reform the world' (p. 132). But, in addition to exhorting women towards exercise and effort, and devising schemes for equal education of the sexes, Wollstonecraft takes the argument of radical intellectuals a step further by showing that women will only gain true social and economic independence through a revolution of the social, political and economic structures which systematically repressed the majority of both sexes.

Wollstonecraft's life was very much an embodiment of the self-sufficiency and resourcefulness she urged upon women. In a society which systematically induced women's dependence on men, Wollstonecraft left home at nineteen years of age in order to gain economic independence. In 1783 she established a school at Newington Green with the evident aim of using its income to create a self-sustaining community of women (including her sisters and a friend). It was at Newington that she met the aging radical dissenter Dr Richard Price, who profoundly influenced her. After the failure of the school and an instructive period as a governess, Wollstonecraft became a professional writer – an almost unprecedented thing for a woman – publishing journalism and fiction for Joseph Johnson. In 1790 Edmund Burke's *Reflections on the Revolution in France* attacked the French Revolution in general and Richard Price's celebration of it in particular, and Wollstonecraft indignantly defended both in *Vindication of the Rights of Men* (1790). In the process, she set the tone and the terms for the many replies to Burke that followed, and developed the insights which would later mature in the *Vindication of the Rights of Woman*.

Wollstonecraft went to Paris in 1792 in order to see the Revolution at first hand. The result of this – *An Historical and Moral View of the French Revolution* (1794) – confronted the problem that faced many radicals: how to maintain the principles that had inspired the

Revolution in face of the bloodbath that it was becoming, and how to explain that bloodbath within radical theory. Wollstonecraft met and fell in love with an American in Paris; flaunting the moral code of the period, she lived openly with Gilbert Imlay and gave birth to a daughter in 1794. Although the relationship was already under strain, Wollstonecraft (accompanied only by her daughter and a maid) undertook an arduous business trip to Scandinavia for Imlay in 1795, writing a series of letters which would later be published as *Letters Written during a Short Residence in Sweden, Norway and Denmark* (1796).

Having returned to London and found that Imlay had begun a new relationship, Wollstonecraft attempted suicide. She subsequently rejoined the radical group which met above Johnson's shop, and began a relationship with William Godwin in which the two philosophers and professional writers attempted to maintain the mutual independence which each of their philosophies insisted upon. When Wollstonecraft became pregnant, however, the couple embarrassedly underwent the ceremony that they had each condemned in their writings. Even after marriage, though, 'they did not entirely cohabit' and 'were both careful not to intrude on one another's friendships' (Miriam Brody Kramnick, Introduction to *Vindication of the Rights of Woman*, 1975). Mary Wollstonecraft died in August 1797 from complications following the birth of the daughter who would later write *Frankenstein* and marry Percy Bysshe Shelley.

The publication of Godwin's *Memoirs of the Author of 'A Vindication of the Rights of Woman'* (1798), which candidly dwells on her love affair, illegitimate child and attempted suicide, provided excuses for an aggressive backlash against the woman as well as the book. Walpole had already called her a 'hyena in petticoats', and in the general reaction against radicalism after 1793 the *Vindication* became subject to a series of violent attacks by men and women: 'Wollstonecraft's name, in the years [of political repression and fear of French radicalism] following her death, . . . became virtually synonymous with free love and Jacobinism' (Kramnick). Anti-feminist readers have attacked Wollstonecraft ever since: if the *Historical Magazine* of 1799 (I, 34) suggests that properly delicate females would read the *Vindication* with 'disgust . . . detestation . . . and indignation' (quoted by Kramnick), Ferdinand Lundberg and Marynia F. Farnham, in the late 1940s, called it the 'single fateful book' behind the 'deep illness' of feminism (*Modern Women: The*

Lost Sex, 1947; quoted by Kramnick). Although feminists of the nineteenth and early twentieth centuries generally shunned Wollstonecraft as much for her 'immoral' life as for the radically sweeping scope of the *Vindication*, her work continued to influence more radical feminist thought – particularly that of John Stuart Mill's *The Subjection of Women* (1869) – and since the late 1960s and early 1970s, when Emma Goldman, Kate Millett, and Germaine Greer fully endorsed the *Vindication*'s more radical arguments, 'Mary Wollstonecraft is being read all over again' (Kramnick). New editions of Wollstonecraft's works have followed: Kramnick's 1975 edition of the *Vindication*; Gary Kelly's 1976 combined edition of Wollstonecraft's early and late novels, *Mary, a Fiction* (1788) and *The Wrongs of Woman* (1798); and Richard Holmes's 1987 edition of *A Short Residence in Sweden, Norway, and Denmark* (together with Godwin's *Memoirs*). In addition, there have been at least four biographies of Mary Wollstonecraft published since 1970.

In order to understand the radical scope of Wollstonecraft's pioneering feminism it is important to read her influential contributions to the defence of the French Revolution. Her *Vindication of the Rights of Men* was one of the first of the many replies to Edmund Burke's *Reflections on the Revolution in France* which constitute the 'Revolution controversy'. Wollstonecraft's pamphlet quickly went through two editions and made her at once famous and notorious. But, unlike Paine's *The Rights of Man* (1791–2) Wollstonecraft's *Rights of Men* quickly became neglected (it has been out of print since 1791 and is only available today in a facsimile reproduction of 1960, edited by Eleanor Louise Nicholes). This is unfortunate since, although J. T. Boulton, in *The Language of Politics in the Age of Wilkes and Burke* (1963), is dismissive of the first *Vindication* because of its lack of control and organisation, Ronald Paulson, in *Representations of Revolution* (1983, pp. 79–87), shows how her text underlines the relationship between Burke's particular representation of women and the repressive structures of the traditional order he defends. If Burke laments that 'the age of chivalry is gone' (*Reflections on the French Revolution*, ed. Conor Cruise O'Brien, 1968, p. 170), Wollstonecraft shows how the codes and structures of chivalric society render both men and women 'effeminate', and corrupt those in power as well as those subjected to that power (*Vindication of the Rights of Men*, pp. 47–55, and pp. 111–17). Her reply to *Reflections* shows how Burke's aesthetics of the sublime and the beautiful are inextricably bound up with his politics, and her insight is thus to see the urgency not only of countering Burke's arguments but also of reworking those aesthetics for a radical programme: 'truth, in morals, has ever appeared to me the essence of the sublime; and, in taste, simplicity the only criterion of the beautiful' (p. 2). In many ways, then, Wollstonecraft's attempt to transvalue aesthetic values and rhetorical practices seems to anticipate the poetics that Wordsworth develops in the Preface to *Lyrical Ballads* (1800). At the same time, the fact that Wordsworth's 'radical' poetics are in some way a withdrawal from his earlier *political* radicalism is pointed up through Wollstonecraft's insistence that such an aesthetics can only be fully realised through political transformation: 'Such a glorious change can only be produced by liberty. Inequality of rank must ever impede the growth of virtue, by vitiating the mind that submits or domineers' (p. 116).

TOM FURNISS

Wordsworth, Dorothy (1771–1855)

Dorothy Wordsworth, who was born on Christmas Day 1771, was the middle child of the Wordsworth family, and closest both in age and temperament to her older brother William. At her mother's death in 1778, Dorothy went to live with her second cousin at Halifax: she was happy there, but (following her father's death in 1783) she was removed by her grandparents in 1787 to live with them at Penrith, where she met her brothers during their holidays from school at Hawkshead. In 1788 she went to live with her uncle, the Revd William Cookson, and his wife, at Forncett in Norfolk, where she remained until 1794: she kept in touch with William by letter during these years, interesting herself greatly in his progress at Cambridge, his walking-tour of 1790, his adventures in France and his relationship with Annette Vallon. There were short meetings during these years, but in the spring of 1794 they stayed together for the first time in the Lake District at Windy Brow, near Keswick; and in 1795 Raisley Calvert's legacy enabled them to live frugally but independently, first at Racedown in Dorset (where she encouraged William to become a poet, and supported him during a period of anxiety and uncertainty) and then at Alfoxden. There the friendship with Coleridge, who described Dorothy as 'a most exquisite young woman in her mind and heart', deepened daily, and became a fruitful poetic collaboration between William and Coleridge, with active support from Dorothy. The brilliant year 1797–8, which saw the composition of *Lyrical Ballads*, was also the period of Dorothy's first journal, the 'Alfoxden' journal. Later in 1798 she accompanied William and Coleridge to Germany, and lived with William at Goslar during the coldest winter of the century, when William began to write the blank-verse passages on his early life which later became *The Prelude*. In 1799 they returned to England, and in December of that year they settled in Grasmere at Dove Cottage. Dorothy's 'Grasmere' journal was begun on 14 May 1800, and continued until 1803: in it may be found episodes which subsequently became the foundation of a number of William's poems, most notably 'Resolution and Independence' and 'I wandered lonely as a cloud'.

Dorothy accompanied William on the visit to Calais in August 1802, and continued to live at Dove Cottage after William's marriage to Mary Hutchinson on 4 October of that year. She went with William and Coleridge on a tour of Scotland in 1803 (which Coleridge left after two weeks) and wrote a journal recording it; accounts by her of visits to Ullswater and of a climb of Scafell appear in William's *Guide to the Lakes*. Dorothy continued to live with the family during the moves in Grasmere and to Rydal Mount in 1813, and lived at Rydal until her death, though suffering after 1835 from pre-senile dementia.

There seems to have been a passionate and deeply felt affection between brother and sister, though there is no evidence to support the charge that the relationship was incestuous: the language of affection between members of the same family was probably much stronger then than it is now. William paid tribute to Dorothy in many places, most notably in 'Tintern Abbey' and in *The Prelude* (1805, XII; 1850, XIII). In 'The Sparrow's Nest' he wrote,

> She gave me eyes, she gave me ears;
> And humble cares, and innocent fears;
> A heart, the fountain of sweet tears;
> And love, and thought, and joy.

But Dorothy Wordsworth deserves to be known for her own

writings, as well as for being the sister of a great poet. Her journals are full of a direct, startlingly beautiful observation of nature (as in the 'red leaf' image used by Coleridge in 'Christabel'); they are also full of a delicate sensibility, an exquisite emotion, in relation to nature and to other people; and they are also acute in their human curiosity, their perception of character, and in their compassion for vagrants and for the poor. The journals are therefore much more than a record of daily life in a rural household: they are the reflections and observations of an extremely talented, perceptive and sensitive woman.

J. R. WATSON

Wordsworth, William (1770–1850)

William Wordsworth was born on 7 April 1770, at Cockermouth, Cumberland, the son of a law agent. After the death of his mother in 1778, he was sent to the Grammar School at Hawkshead, where he lodged with a cottager called Ann Tyson, and was allowed a considerable amount of freedom (described in book I of *The Prelude*). His father died in 1783, leaving the five children – William, three brothers, and his much-loved sister Dorothy – in the guardianship of their relatives: William's rebellious behaviour at this time is evidence of his emotional distress at losing his mother and then father – a distress which is not referred to in *The Prelude*, which concentrates on the happier side of his childhood at Hawkshead.

Probably when he was seventeen, Wordsworth wrote 'The Vale of Esthwaite', a poem in which several of his later interests, notably the idea of 'retrospect', are found amid the Gothic paraphernalia. He left Hawkshead in 1787 to go to St John's College, Cambridge, where he disliked the academic course. During the last summer vacation (1790), just before his final examinations, he went to France and the Alps on an extended walking-tour with a college friend, Robert Jones. After graduating, and living in London for a short time, he returned to France (Nov 1791), where he fell in love with Annette Vallon, who bore him a daughter, Caroline, born on 15 December 1792. During 1792 he had also met a French officer, Michel Beaupuy, who roused his enthusiasm for the French Revolution.

After his return to England at the end of 1792, Wordsworth published two poems, 'An Evening Walk' and 'Descriptive Sketches', written in the tradition of the eighteenth century and in heroic couplets, but with passages of an intensely personal vision. He also wrote (without publishing it) a 'Letter to the Bishop of Llandaff', against rank and privilege and in support of the French Revolution. The declaration of war between England and France shocked him: he continued to support the Revolution, though his faith in it was shaken by the Terror and by his dislike of Robespierre. After the fall of Robespierre, his hopes in the Revolution turned into hopes of reform by education and reason; he was greatly influenced at this time by the writings of William Godwin, whom he met in 1795.

A legacy of £900 allowed him to settle with his sister Dorothy, first at Racedown in Dorset (Sep 1795), where they were visited by Coleridge; and subsequently to move to Alfoxden, in Somerset (July 1797) in order to be near him. Coleridge and Dorothy helped him to recover from his disenchantment with the Revolution and with Godwin's rationalism. The year 1797–8 was a period of intense creativity, which produced the joint volume *Lyrical Ballads*, published in September 1798; in the same month Dorothy, William, and Coleridge left for Germany. Coleridge studied at Göttingen, while Wordsworth and his sister spent the coldest winter of the century at Goslar, where he began parts of what was to become *The Prelude*, and wrote some of the 'Lucy poems'. They returned to England in 1799, and in December of that year settled in Grasmere, at Dove Cottage. In 1800 a second edition of *Lyrical Ballads* appeared, containing the celebrated Preface; a further remarkable period of poetic composition took place in 1802. In that year, taking advantage of the Peace of Amiens, which brought about a brief lull in the war with France, William and Dorothy visited Calais, where they met Annette Vallon and the child Caroline; he must have explained to Annette that he did not intend to marry her, because in October of that year he married Mary Hutchinson, from Sockburn-on-Tees in County Durham. Their first child was born in June 1803.

During the following years many of Wordsworth's most notable poems were begun or completed, including 'Michael'. 'Resolution and Independence', and the 'Ode: Intimations of Immortality from Recollections of Early Childhood'. Many of them were published in *Poems in Two Volumes* (1807). Some of the poems show an increasing sense of fortitude and stoicism after the death at sea of his brother John (Feb 1805), and (in the absence of Coleridge in Malta) Wordsworth turned increasingly for support to friends such as Sir George Beaumont and Charles Lamb. As his family grew, he moved house in Grasmere (although two of his children, Catherine and Thomas, died in 1812), until in 1813 his appointment as Distributor of Stamps for Westmorland enabled him to move to the comfort of Rydal Mount, where he lived for the remainder of his life.

In August 1814 he published a major long poem, *The Excursion*, mocked by Byron for its length, but much read by other contemporaries; in March 1815 a collected edition of his poems with a preface; in 1819 'Peter Bell' (originally composed in 1798); in 1820 a series of sonnets on the River Duddon, and in 1822 the first separate edition of the *Guide to the Lakes*. A major edition of his poems appeared in 1835. He was given an honorary degree by the University of Durham in 1838, and by Oxford in 1839; on the death of Southey in 1843 he became Poet Laureate. *The Prelude*, the account of 'the growth of a poet's mind', and intended as the 'ante-chapel' to his great philosophical work (hence its title) was published by his wife shortly after his death in 1850.

Wordsworth was both the poet of humble beings and the poet of the sublime. He saw 'love unutterable' in the eyes of a London father eyeing his poor babe, and he also read 'unutterable love' in the silent faces of the clouds (*The Prelude*, 1850, VII.618; *The Excursion*, I.205). His world was 'the actual world of our familiar days' (*The Prelude*, 1850, XIII.357), a world upon which he could exert the power of his mind, and which he felt was 'exquisitely' fitted to the mind (Preface to *The Excursion*, ll. 66–8). He believed in the love of nature leading to the love of mankind, but he also had a great tragic vision, and peopled his poetry with many poor and suffering people. He also found expression for his great visionary experiences, often involving a loss of selfhood in the world of the imagination. He is known as the poet of nature, but he insisted that 'The Mind of Man' was 'My haunt, and the main region of my song' (Preface to *The Excursion*, ll. 40–1). A yellow primrose never became to him what it was to Peter Bell, a mere yellow primrose and nothing more: it remained the source of silent raptures, and testified to the presence of wonder and mystery, originating in Wordsworth's awareness and intimation of the depths of the human mind.

The Excursion

The history of the composition of this poem is complex. It began during the years of Wordsworth's close friendship with Coleridge (1797–8), when Wordsworth, stimulated by Coleridge, thought he would write a great philosophical poem, *The Recluse*. He 'was to treat man as man, – a subject of eye, ear, touch and taste, in contact with external nature, and informing the senses from the mind, and not compounding a mind out of the senses' (Coleridge's *Table Talk*, 1832). Wordsworth made many attempts. The first positive result was the writing of what was to become *The Prelude*. In 1800 he began writing 'Home at Grasmere', which he completed in 1806 (known as *The Recluse*, part I, book I). In 1797 he had written 'The Ruined Cottage' and in 1798 had begun 'The Pedlar', both of which were later included in book I of *The Excursion*. Part of the beginning of book IX also belongs to 1798, and quite a few early passages were thus later included in the poem. Most of book II–IV was composed in 1809; and books V–IX in 1811, although work on the poem went on through 1813. A study of the dates of composition will be found in Mark Reed's *Chronology, 1800–1815*. The poem was published in August 1814, with a preface, and offered, as a 'Prospectus', a fragment composed between 1798 and 1800 previously used as the conclusion to 'Home at Grasmere'. There was a subtitle: 'Being a Portion of The RECLUSE. A Poem'. This was a reminder of the unity of the poet's work: in the Preface he compared his whole work to a Gothic church and his various poems to parts of it, but a great part remained unbuilt in 1814, and was never completed. The Preface also emphasised that the author did not 'announce a system' but suggested that the reader would 'have no difficulty in extracting the system for himself'. It was a poem, not a treatise.

The title finds its justification in the excursion which takes the main characters through the mountains and vales of Lakeland. It also points to the possibilities of 'the mind's excursive power'. The poet travels with the Wanderer, who tells the sad story of Margaret. In the second book they meet the despondent Solitary, who has suffered much from life, and in the following book they attempt to assuage his sorrows and bring him comfort by gentle argument and apt illustrations from other people's lives. Most of these are afforded by a third character, the Pastor, who is met in book IV. In the last two books the social situation in England is criticised and reforms hoped for.

Hazlitt regretted that 'an intense egotism swallows everything. . . . The Recluse, the Pastor and the Pedlar are three persons in one poet.' Other contemporaries responded variously to the poem: Francis Jeffrey's rejection is well known. So are Shelley's and Byron's, but Keats found the fourth book delightful; Charles Lamb praised the poem and it did contribute to Wordsworth's popularity. Modern criticism has lately found more interest in the poem than was granted it in the first half of the twentieth century. J. S. Lyon wrote a full study of it in 1950, showing its connection with loco-descriptive poems (Thomson, Cowper), with philosophical dialogues (Richard Savage, Bishop Berkeley), rustic tales (Shenstone, Langhorne, Crabbe). Its relationship with *The Georgics* has also been seen. Russell Noyes (*The Wordsworth Circle*, 1973), sums up the qualities of the work: the utility of its philosophical optimism, the revelation of Wordsworth's inner mind, the beauty of the description of the Lake country, the pictures of enduring virtues by the elegiac poet, the realism of the stories of the Pastor (based as they are on those of Grasmere people Wordsworth knew) and the interest of the social criticism.

But the recourse to Providence, instead of creative love as in the earlier works, has been interpreted as the mark of a decline, of a 'flight from vision' (Hartman). The Pastor's prayer, in book IX, has been said to sum up all that Wordsworth had to say from 1814 to the end of his life. But Kenneth R. Johnston qualified those views: 'Having gone as far as he dared as a "Prophet of Nature", giving himself up to the influx of her spirit on his feelings, Wordsworth turned to a sacramental, priestly style of natural interpretation which enabled him to complete *The Excursion*, but it finished *The Recluse*,' ('Wordsworth's Reckless Recluse', *The Wordsworth Circle*, 1978). In *Wordsworth and The Recluse'* (1984), the same critic has written the best history to date of the difficult relations between Wordsworth and his never-completed philosophical poem.

Guide to the Lakes

In 1809 Wordsworth accepted an invitation to write an introduction to Joseph Wilkinson's collection of scenes drawn in the Lake District. *Select Views in Cumberland, Westmoreland, and Lancashire* (1810) was privately published in twelve monthly instalments; Wordsworth's name did not appear. In 1820 he published *The River Duddon, a Series of Sonnets: Vaudracour and Julia: and Other Poems*, and appended, to 'illustrate the poems', a *Topographical Description of the Country of the Lakes, in the North of England*, which is the second edition of the *Guide* (without references to Wilkinson's drawings and with additions). The third edition was *A Description of the Scenery of the Lakes in the North of England* (1822), the first separate edition (500 copies). A fourth edition followed in 1823 (1000 copies) with the same title and more additions. The present title appeared with the fifth edition (1500 copies): *A Guide through the District of the Lakes in the North of England* (1835). A very different version was published in 1842, including Wordsworth's advice, but without his direct contribution, and with further directions for the tourist.

The fifth edition contains 'Directions and Information for the Tourist', a description of the scenery of the lakes in three parts, 'Miscellaneous Observations', the narration of two excursions, the 'Ode: The Pass of Kirkstone', and an 'Itinerary' giving distances from town to town and the various excursions.

From the very beginning the reader is told that the 'Manual' is meant to guide 'the Minds of persons of taste, and feeling for landscape', which underlines its difference from other guide books previously published, in spite of its new title (*A Guide . . .*). Besides topographical directions, Wordsworth comments on the beauty of the Lakes. He starts by taking his reader to an imaginary position on a cloud above the hills and lakes, and shows him how the valleys diverge 'like spokes from the nave of a wheel'. He then studies the forms, the surfaces and the colours of the mountains, occasionally illustrating his prose with excerpts of his poems. The landscape is shown as a stimulus to the imagination: thus the still surface of lakes can carry the imagination 'into recesses of feeling otherwise impenetrable'. The presence of man is not neglected: his intercourse with nature illustrates Wordsworth's faith in its benignant influence, and he celebrates a perfect Republic of Shepherds, 'a pure Commonwealth'.

The aesthetics of landscape are more specially dealt with in the third part. The essential harmony between houses and natural landscape has been broken (in Wordsworth's opinion) by some new buildings which stand out on the tops of the hills, because their owners have yielded to the fashion in their search for 'prospects', instead of letting their houses be 'gently incorporated with the works of nature'. This desire for harmony exemplifies in a practical way Wordsworth's deep desire for oneness between man and nature.

The visitor, Wordsworth says, should visit the Lakes in September or October (for colours) or at the end of May and the beginning of June (long days, fine weather and variety of impressions). He ought to come with the proper mind,' a disposition to be pleased'.

Many passages may be read as indirect comments on the poetical works of the author: his descriptions (judged 'exquisitely faithful' by Dr Arnold) give the reader a parallel vision to that of the poems. The scenery is studied in terms that recall eighteenth-century discussions of the sublime and the beautiful, but with a pervading sense of the unity of things. Among the texts connected with the *Guide* is part of an essay on the sublime and beautiful (*Prose Works*, ed. W. J. B. Owen and Jane Worthington Smyser, 1974, II, 349–60).

'Lines Written a Few Miles above Tintern Abbey, on Revisiting the Banks of the Wye during a Tour, July 13, 1798'

The poem was composed by Wordsworth during a four-day ramble with his sister Dorothy in the Wye Valley. Possibly begun on the evening of the 10th, it was probably completed by 13 July 1798 (Mark L. Reed, *William Wordsworth: The Chronology of the Early Years, 1770–1799*, 1967). The poem was sent to Cottle, who was having *Lyrical Ballads* printed in Bristol, when the other poems were already at the press; and it was printed at the end of the 1798 edition. In the collected edition of 1815 it was classified under 'Poems of the Imagination'. In the note on the poem dictated to Isabella Fenwick, Wordsworth insisted that 'I began it upon leaving Tintern, . . . and concluded it just as I was entering Bristol in the evening. . . . Not a line of it was altered, and not any part of it written down till I reached Bristol.' The poem has indeed on the whole remained unrevised except for a few minor changes. Wordsworth had first visited the Wye in 1793, when, after a carriage accident in Salisbury Plain, he had pursued his journey to Wales with the sole help of his 'firm friends, a pair of stout legs' (Dorothy Wordsworth).

'Tintern Abbey' is written in blank verse in a series of verse paragraphs. In *Lyrical Ballads*, 1800, Wordsworth added a note to say that he hoped 'the transitions, and the impassioned music of the versification' might recall the manner of the ode. Though Tintern Abbey is neither seen nor mentioned in the poem, it is partly about place (the green landscape of the Wye Valley), and is a 'revisit poem' in the eighteenth-century tradition.

After recalling his first visit and thus introducing the theme of time, the poet describes the setting: a green landscape seemingly linked to the sky by lofty cliffs (ll. 1–22). Then he illustrates the restoring power of memory and shows how, when recollected in tranquillity, former perceptions gather a power so intense that we may 'see into the life of things' (ll. 23–49). Lines 50–66 reassert his faith that there is 'life and food' in his recollections of the Wye. Lines 66–99 recapitulate his evolution from childhood (ll. 73–4), through youth, to the time of his new visit. Lines 99–112 rejoice that what he may have lost has been replaced by an awareness of transcendence, though he still cherishes the world of the senses. In the remaining verse paragraph (ll. 113–60) he turns to his 'dearest friend', his sister Dorothy, who, having been long associated with nature and with him, seems to be a living reassertion that nature betrays those who love her.

Critics have shown that the influence of Coleridge's 'conversation poems' on the language and the rhythm of the poem, and its own rhetoric of persuasion (its repetitions, its turns and counter-turns) have much contributed to its success. The way Wordsworth uses the past, present and future exemplifies the central role of memory in rescuing experience from the wear-and-tear of time and giving it power to lift us up when fallen. The poem is not only a poem of time; it is also a poem of nature, and of man's possible insight into the life of things, as it is experienced in trance-like moments. This experience of transcendence has been seen as inwardly and downwardly directed into the depths of the ecstatic self, in a sort of 'transdescendence'. 'Tintern Abbey' is also a shorter *Prelude*. It is autobiographical and sums up the evolution of the imagination of the poet from the days of the 'coarser pleasures' of childhood to his mature love of nature, chastened and subdued after hearing the music of humanity. Finally it is his tribute to his sister Dorothy's part in the making of the poet.

Lyrical Ballads, with a Few Other Poems

This joint volume of poems was printed by Cottle in Bristol in the summer of 1798 and published in London in October 1798, anonymously, and with a short 'Advertisement'. This one-volume edition contained twenty-three poems: four by Coleridge and nineteen by Wordsworth. In January 1801, a second edition came out, dated 1800, in two volumes, with the famous Preface. In 1802 was issued a third edition, with a revised Preface and an appendix on poetic diction. The fourth edition, in 1805, *Lyrical Ballads, with Pastoral and Other Poems*, contained sixty-five poems.

Lyrical Ballads, 1798, may be said to have originated in the two authors' need of money. The book, they hoped, would pay for the expenses of a journey to Germany. The poem first planned was 'The Ancient Mariner', partly composed in November 1797; then in the spring of 1798 they probably drew up the plan described by Coleridge in *Biographia Literaria*. Two sorts of poems were to be composed: in some the incidents 'were to be, in part at least, supernatural'; in others they would be 'chosen from ordinary life'. The former would have been mostly written by Coleridge, the latter by Wordsworth. But Wordsworth never confirmed this view and dwelt rather on the psychological interest of the poems, which should trace 'the primary laws of our nature: chiefly as far as regards the manner in which we associate ideas in a state of excitement'. He also thought of them as 'experiments' to see how 'the language of conversation' may be used in poetry.

The 1802 version of the Preface made it clear that the language of poetry should be 'a selection of the language really spoken by men'. The word 'selection' strongly qualifies the idea that there should be no essential difference between the language of prose and that of poetry. Hasty readings have thus often oversimplified the views of the Prefaces. An example is the famous statement that 'all good poetry is the spontaneous overflow of powerful feelings'. What immediately follows ought to be quoted, too: 'but though this be true, Poems to which any value can be attached, were never produced . . . but by a man who being possessed of more than usual organic sensibility had also thought long and deeply'. A proper appreciation of the Prefaces requires a comparison of the 1800 version with that of 1802. The former leans more on the social side, the latter on language; hence Wordsworth's appendix on poetic diction, commenting on his rejection of eighteenth-century 'adulterated phraseology' ('finny tribe' for 'fish', and the like).

Most of the poems deal with men and women who 'do not wear fine clothes and can feel deeply'. But some are narratives and some are personal poems. 'The Thorn' is a dramatic monologue in which the narrator's part is as important as that of the thorn and the deserted girl who weeps near it. Several poems are modern pastorals. 'Goody Blake and Harry Gill, a True Story' and 'Simon Lee, the Old Huntsman with an Incident in which he was Concerned' are poems about old age. In 'The Idiot Boy' Johnny is sent to fetch a doctor, fails to reach the town and causes great anxiety to his mother; readers are made to feel for her and for other mothers, as also in 'The Mad Mother'. 'The Female Vagrant' and 'The Convict' (as well as Coleridge's 'The Dungeon') show compassion for social outcasts. Some poems are lyrics or personal effusions or meditations: their titles often draw attention to the particular circumstances of their composition, as in 'Lines Written at a Small Distance from my House, and Sent by my Little Boy to the Person to whom they were Addressed' (subsequently called 'To my Sister') and 'Lines Written a Few Miles above Tintern Abbey. . . .' Coleridge's 'The Rime of the Ancyent Marinere' (1798 title) was the main purveyor of supernatural elements. Many more poems were added in 1800 and after: 'Michael' , 'The Brothers', the 'Lucy poems', the 'Matthew poems' and blank-verse passages such as 'There was a boy' and 'Nutting'.

The terms 'lyrical' and 'ballads' have been seen as contradictory. Ballads in a narrow sense are impersonal narratives, and lyrics personal non-narrative effusions. But the choice of 'lyrical' may have emphasised the fairly frequent subjective stance of the poems, as well as their musical quality. Besides, ballads of many sorts were known in the second half of the century: traditional ballads collected in Percy's *Reliques of Ancient English Poetry*, translations of German ballads, various poems published as 'ballads' by the magazines, and the 'broadside ballads' that were hawked in the streets. The term, then, was a vague one, and Wordsworth and Coleridge followed Burns and others in writing subjective 'ballads'. Moreover, they were mainly using as a basic metre that of mediaeval ballads, trying to make it 'more impressive'.

Long considered as the starting-point of a new era, the 1798 *Lyrical Ballads* is now, after the debate which followed Robert Mayo's article (*PMLA*, 1954), seen in a new perspective. Mayo showed that many of the themes and forms of the supposedly revolutionary poems were familiar to contemporary readers of magazines and volumes of minor poetry. Indeed, not only minor authors, but Southey in his *Botany Bay Eclogues* had defended convicts; and from 1797, in his *English Eclogues*, he had turned his attention to characters of rural life not unlike those of *Lyrical Ballads*. But the forceful simplicity of the language and the intensity of the feelings of a poet who based his own rural tales on his personal experience of country life added much to what might be derived from tradition, as did the tender pathos of some poems, the visionary elements of 'Tintern Abbey' and the mystery of 'The Ancient Mariner'. The work represents no clear-cut change in literary history, but as Mary Jacobus's *Tradition and Innovation in the 'Lyrical Ballads'* has shown, marks a sort of crossroads, where eighteenth-century ways meet a new beginning.

Another moot point is the relationship of Wordsworth and Coleridge. Who influenced whom, and how? Neither of them seems to have been deeply altered by their friendship. But both of them felt, and probably were, the richer for it during the days in Somerset spent discussing the poems and their philosophy. Wordsworth, while he was recovering from his Godwinian and revolutionary disenchantment, needed the love and understanding of Coleridge. Coleridge also needed friendship and comfort. Critics disagree as to the respective amount of inspiration their creative parnership brought them. Some emphasise the role of Coleridge's philosophical knowledge; others belittle it and make Wordsworth's heart the great giver. The truth probably lies midway between the two.

'Ode: Intimations of Immortality from Recollections of Early Childhood'

The dates and circumstances of this famous ode are not altogether clear. The first four stanzas were probably composed on 27 March 1802. Stanzas v–viii were perhaps written before June 1802, but it is not likely (Mark L. Reed, *William Wordsworth: The Chronology of the Middle Years, 1800–1815*, 1975). Stanzas ix–xi may well have been written in early 1804, thus completing the ode. The two years' interval corresponds to that mentioned by Wordsworth in the note he dictated to Isabella Fenwick: 'Two years at least passed between the writing of the four first stanzas and the remaining part.' This interval may sway the interpretations. The poem was published in *Poems in Two Volumes* (1807), at the end of the second volume, under the simple title 'ode', with the epigraph 'Paulo majora canamus' ('Let us sing a nobler song'). The present title appeared in the collected edition of 1815. Several similarities of phrasing have induced most critics to consider Coleridge's 'Dejection' (written 4 Apr 1802) as a counterpart to the first four stanzas of Wordsworth.

In these four stanzas he ponders over a sense of loss of vision. Stanzas v–viii are about the growth of the child, whose powers are gradually weakening. The last three stanzas strive to give an optimistic view of the process of growing and to offer compensations.

In stanzas i and ii the poet feels he has lost his visionary power; the qualities of past and present vision are contrasted. Stanza iii opposes the poet's grief and nature's joy, but he speaks of a 'timely utterance' that has brought him relief. In iv, he feels the universal joy before his sense of loss returns. Stanza v introduces an account of the evolution of the child after pre-existence; and vi shows Earth trying to make the child forget the glories he has known in pre-existence. In vii, the child, impatient to grow up, is seen imitating all the parts played by adults on the world-stage. Stanza viii shows the child as the best philosopher, a seer, unafraid of death yet, who longs to become an adult (and thus fall under the weight of 'custom'). Stanza ix introduces memory as a compensation which endows man with a sort of immortality. In stanza x, the imagery and the structure of the first stanzas are echoed; but the sense of lost power is compensated for by the poet's resolution to find strength in what life brings, 'the philosophic mind'. The last stanza celebrates his new humanised love of nature and his sedate acceptance of his new condition.

Critics generally praise the poem (some think it is one of Wordsworth's greatest poems, if not the greatest), but they often disagree as to its meaning. Some details have been much discussed (the 'timely utterance', the 'fields of sleep . . .'), and some key passages have determined overall interpretations ('obstinate questionings . . .'). The difficulties of the chronology make biographical references uncertain. The acceptance of the date of March 1802 for stanzas i–iv and 1804 for the remaining part has enabled Alan Grob to study the parallel between the changes in the poem and the changes in Wordsworth's thought. The value of the Platonist myth of the fall into the cave of the sense has also been debated, and Wordsworth in his note dictated to Isabella Fenwick, saying he used the myth 'as a Poet', may discourage hasty generalisations about possible changes in his philosophy. The nature of the 'immortality' mentioned in the title (from 1815) has also made readers pause. It has been found in the permanence of self-conscious identity, or thought to be the consequence of the visionary experiences and the 'shadowy recollections'. The main debate probably focuses on the balance of loss and gain, and its paradoxes – the losses growing into gains. Arthur Beatty saw the triumph of life in the advent of maturity; and the idea that it was mainly about loss of power has been rejected by Lionel Trilling, who also thinks it is not about growing old, but about growing up and reaching philosophical maturity. The union of self and nature is an illusion that the child has to reject. Poetic power comes with the sense of personal identity. But, in spite of the optimism of the last stanzas, there remains, underlying it, a sort of obscure anguish. Jared R. Curtis, after a careful study of the design of the ode, concludes that 'the human heart has not frozen, but neither has it been fully warmed in the sun. Growing up and growing old are the same'.

The Prelude

An autobiographical poem in blank verse addressed to Coleridge, *The Prelude* was originally intended as an introduction to *The Recluse*, the vast philosophical work that Wordsworth never completed (see *The Excursion*).

The Prelude was the work of a lifetime, and only a brief summary of its complicated textual history can be given here. From 1798 to 1839 it was constantly revised and several times remodelled, eventually to be published after the poet's death. Three different *Preludes* have now been published. In 1850 his wife published the much revised work now known as the '1850 *Prelude*'; in 1926 Ernest de Selincourt edited the thirteen-book version that Wordsworth had completed in 1805 (the '1805 *Prelude*'); and in 1974 Jonathan Wordsworth and Stephen Gill published in *The Norton Anthology of*

English Literature the two-book version now known as the 'two-part *Prelude*' of 1799).

In October–December 1798, while at Goslar in Germany and failing to get on with *The Recluse*, Wordsworth wrote fragments to assess how far he was equipped to write the great poem. Thus he wrote the childhood episodes of book I, and, as he tried to find how they helped his creative power, composed the 'spots of time' section and went on recalling his adolescence. 978 lines were transcribed in 1799 (the 'two-part *Prelude*').

In 1804 he started to work again to give Coleridge a copy before his departure for Malta. This version in five books is sometimes alluded to as the 'five-book *Prelude*'. Soon afterwards, he extended it to thirteen books, completed in 1805 (the '1805 *Prelude*'). These books were then revised again and again. Fair copies were made in 1816/1819, 1832 and 1839. The 1839 text, in fourteen books, was the one published in 1850 by the poet's wife, who gave it the title *The Prelude, or Growth of a Poet's Mind*. Wordsworth himself had never given it a title and used to refer to it as 'the poem on my own life'. The 1850 *Prelude* is more elaborate in style than the 1805 version, and occasionally attempts to tone down too-radical political views or too-pantheistic visions of nature. Several modern editions present the two versions on facing pages for easy comparison. No final text can of course incorporate the valuable variants given in the notes of the best editions.

The overall structure is chronological. The poet takes the reader from his childhood to the days of his friendship with Coleridge. But several events are described out of their proper chronological frame.

The first two books are devoted to his childhood; they relate many famous episodes (bird-snaring, skating, boat-stealing, and so on). Book III tells of his first year at Cambridge and discusses ideas on education. In book IV he returns to Hawkshead during the summer vacation (July–Oct 1788) and reports how he became a dedicated spirit. Book V, 'Books', is mainly about children's books, and about the dream of the Arab, with his symbolic stone and shell (geometric truth and poetry). Book VI is mainly devoted to Wordsworth's 1790 journey to France and the Alps in the days of celebration of the French Revolution. Book VII is about the three and a half months spent in London in 1791. In book VIII the poet returns to the Lakes and surveys his past, praising his native hills. In book IX he goes to France again and meets Beaupuy, who converts him to the cause of the Revolution. Book X shows him at the height of his enthusiasm for the Revolution. In book XI (book X continued, in the 1805 *Prelude*) he describes his declining faith in the Revolution and his acquaintance with Godwin's theories. Book XII (XI in 1805) praises nature and the senses and introduces the 'spots of time'. Book XIII (XII in 1805) covers the period of his recovery from his disappointment with rationalism, and book XIV (XIII in 1805) uses the ascent of Snowdon to illustrate the workings of the imagination. The poem ends on an assertion of the greatness of the human mind.

Wordsworth speaks of his theme as being 'no other than the very heart of man'. It is also about nature, and about the relationship of man and nature; it is about time, and memory, and the building of the self. It has been approached from all sorts of perspectives and subjected to all sorts of inquiries. The poem's form has been analysed, together with its genre, its rhetoric, its structure, and its narrative voices. Its contents have been weighed in scales of all kinds: its imagination (never moving the poet directly, but in the guise of nature), its composition in response to a specific trauma, its democratic feeling and its redeeming power are but a few of the many aspects dealt with in the maze of *Prelude* criticism.

Wordsworth and the 'I' of the Prelude are not to be too readily identified. Wordsworth explores himself by means of this 'I'. And the 'I' of the child is not exactly the 'I' of the poet who writes *The Prelude*. The distance between what he was and what he is is worth studying. Memory and imagination combine but do not coincide. One 'I' explores another and in this exploration alters it; and the poet himself is creating a new poet as he writes, the richer for the poem he has just written, a poet closer to the ideal poet whose realisation is Wordsworth's final purpose. Thus the poet of one moment, as he lives and writes of his past again through memory, attends the birth of the poet of the next moment, who becomes his new self. And so it goes on: self-exploring, self-discovering and self-building. The text is the palimpsest of those experiences and literally the record of 'the growth of a poet's mind'.

An awareness of the complexity of meaning of the poem has long superseded the Byronic notion of a 'simple' Wordsworth. Discussions of the philosophical position of *The Prelude* oppose those who stress the associationist, Hartleian aspects to those who emphasise of the transcendental view. Either the mind is a *tabula rasa* in Locke's tradition or it is, in the words of Wordsworth, 'Lord and Master'; in other words, the mind is either a passive receiver, or it is gifted with divine creative powers. Yet another perspective is offered to the student of *The Prelude*. According to Wordsworth 'the appropriate business of poetry' is 'to treat of things not as they *are*, but as they *appear*; not as they exist in themselves, but as they *seem* to exist to the *senses*, and to the *passions*' ('Essay Supplementary to the Preface', *Poems*, 1815; Wordsworth's emphasis). Thus noumena do not concern the poet; only phenomena do. This is a definition of phenomenology, the description of what appears to the conscious subject, of 'states of mind as they are experienced' (Karl Jaspers). Thus critics, much as Hartman, who speak of Wordsworth's poetry as 'phenomenological' may well be pointing to the true position of the poem and to the main reason for its modernity.

MARCEL ISNARD

Young, Edward (1683–1765)

Edward Young was born at Upham, near Winchester, son of the rector (later Dean of Salisbury); he was educated at Winchester College and at New College and Corpus Christi College, Oxford. He was nominated to a law fellowship at All Souls', Oxford, in 1708 (Bachelor of Civil Law, 1714, and Doctor, 1719). Though 'a ready celebrator' of the great (including Robert Walpole and George I) in his early writings, he was disappointed in his hopes of professional advancement and took Holy Orders in 1727, becoming one of the royal chaplains and, from 1730, incumbent in the college living of Welwyn, Hertfordshire. Despite his renown as a man of letters and real expectations of a bishopric, he gained no further promotion in the Church. In 1731 he married Lady Elizabeth Lee, daughter of the Earl of Lichfield; her death is said to have occasioned his most famous work, *The Complaint, or Night Thoughts on Life, Death, and Immortality* (1742–5), which may also contain allusions, in the characters of Narcissa and Philander, to his stepdaughter and her husband, both of whom had died just previously.

Young is one of those eighteenth-century authors whose compositions cover what now seems a surprising variety of genres. His lyrics, panegyrics and poems on historical subjects illuminate contemporary taste and affairs but have little intrinsic merit. He wrote declamatory tragedies of passion and violence, including *Busiris* (1719) and, on the French model (but using the plot of *Othello*), *The Revenge* (1721), both of which were successful in the theatre. It is prophetic of his later individualistic approach to his art that these two plays, in which the heroines die by their own hands, should have epilogues that cast an ironic shadow over the elevated feelings supposedly excited by the representation of tragic events, and thus call in question the status and assumed 'purity' of the dramatic from itself. *The Universal Passion: The Love of Fame* (1725–8), a series of seven general satires in heroic couplets, is original and often witty, terse and hard-hitting, not least in the fourth and fifth sections, which deal with the female personality. This work brought Young not only reputation but the considerable sum of £3000. It has the distinction of predating Pope's treatment of the 'ruling passion' and the 'characters of women' in the *Moral Essays*, and would no doubt be more widely read were it not for Pope's immeasurably greater achievement.

Young's standing, and also his significance within the context of Romanticism, firmly depend, however, upon *Night Thoughts* and his important prose essay, *Conjectures on Original Composition* (1759). The latter, written as an epistle to Samuel Richardson, signals, along with such other treatises as Edmund Burke's *Philosophical Enquiry into the Origin of our Ideas of the Sublime and Beautiful* (1757) and Alexander Gerard's *Essay on Taste* (1759), a ground-swell of mid-century critical opinion, springing above all from a response to Longinus's *On the Sublime*, in favour of 'inspiration' and 'genius' as opposed to the neo-classical emphasis on form and the 'kinds' of literary creation. Lamenting the 'fetters of easy imitation', but stressing the value of taking nourishment from the spirit and vital force of the ancients and particularly of Homer, Young proclaims the liberty of poetic imagination to eschew 'the beaten road' and range at will in the realms of nature and the supra-natural, discovering worlds unrealised and giving life to its own autonomous conceptions. Though couched in a somewhat extravagant rhetoric, and plainly reductive in its polemical assault on Dryden and Pope, the *Conjectures* is a landmark in aesthetic theory, a singularly enthusiastic anticipation, indeed emplacement, of the Romantic upvaluing of the subjective origins of poetry – poetry as 'process' rather than 'product', an outgrowth of the mind rather than a deliberate construct. It had notable influence on the Continent, especially in Germany.

One salient feature of the *Conjectures* is a plea for the use of blank verse, as opposed to the rhymed couplet, for 'high' subjects, blank verse being 'verse reclaimed and re-enthroned in the language of the gods'. Young had chosen this medium for *Night Thoughts*, which extends to nine books diversely combining moralism, feeling, melancholy, devout reflections, and engagement with the sublime wonders of the universal creation. The poet moves, in a series of soliloquies, from thoughts on life's vicissitudes, death and the survival of the soul to sustained declarations (addressed to the worldly non-believer Lorenzo) of the blessings of faith and virtue, and finally, in 'The Consolation', to a preview of eternity, an account of the profound mystery and beauties of the nocturnal heavens, and a supplication to the Deity. While keeping the 'Christian triumph' steadily in view (it is specifically expounded in the fourth book), the work develops a broad religio-philosophic sweep which encompasses celebration not only of the living reality of a divine universe but also of the spiritual and psychological rewards of contemplation and experiential encounter with that reality. It achieved immediate popularity and remained a classic for more than a century, running through copious editions in English and in European translations. Together with Robert Blair's contemporaneous meditation *The Grave* (1743), it established the so-called Graveyard School of poetic writing. Its influence on the Romantics – which was both direct and through a prevailing current of sensibility – is reflected in contexts as distinct as those of Wordsworth's *Prelude* and *Excursion* and, less predictably, Byron's *Childe Harold's Pilgrimage*, where, in canto III, there are manifest reworkings of Young's extempore, yet also aphoristic and self-consciously impressive, handling of the themes of mortal bondage and transcendental aspiration involving intimations of the Divine in the presence of nature.

VINCENT NEWEY

Index

Primary entries to individual authors and subjects appear first, with subordinate entries following.

Abolition of the Slave Trade, 1–4, 84
Adam, Robert, architect, 12
Addington, Henry, Lord (Viscount Sidmouth), Prime Minister, 67
Addison, Joseph, eighteenth-century essayist, 4, 168, 178, 270
Aikin, John, periodical editor, 210
Akenside, Mark, 4–5
  influenced Collins, 78
  and Samuel Rogers, 224
  and the Sublime, 271
Allen, Matthew, Clare's doctor at Epping, 65
American Revolution, The, 5–10, 22, 105, 196–7
Anson, (Lord) George, circumnavigator and author, 3, 86
Anti-Jacobin, The, 10–11, 164
Architecture, 11–14
Aristotle, Greek philosopher, 14, 78
Arkwright, Richard, inventor, 144
Arnold, Matthew, Victorian poet and critic, 251
Arnold, Thomas, headmaster of Rugby, 296
Association, Theory of, 14–16
  and Rogers, 224
  and Wordsworth, 304
Austen, Jane, 16–18, 148, 234
  the novels, 16–18
  and Crabbe, 92
  Mansfield Park and Lovers' Vows, 96, 148
  and the Gothic novel, 127
  Persuasion and the navy, 191

Bage, Robert, radical novelist, 146
  Man as He Is, 147
  Hermsprong, 147, 149
Ballantyne, James, printer and publisher of Scott, 229
Balzac, Honoré de, French novelist, 230
  Preface to La Comédie humaine, praises Scott, 236
Barlow, Joel, American poet, 22
  The Vision of Columbus, 22
Barnard, Lord, patron of Smart, 260
Barralet, James, eighteenth-century painter of the Sublime, 271
Barry, James, historical painter, 200
Bastille, storming of, 107
Baudelaire, Charles, French nineteenth-century poet, 176
Beattie, James, Scottish poet, 18–19
  influenced Clare, 63, 180
  and Gray, 130
  The Minstrel, 130, 180
Beattie, John, essayist on the Sublime, 270
Beaumont, Sir George, patron and connoisseur,
  and Constable, 80, 83
  and Wordsworth, 292
Beaupuy, Michel, friend of Wordsworth, 291
Beckford, William, author of Vathek, 13
  Vathek, 193
  and Chatterton, 60
  Dreams, Waking Thoughts and

Incidents, 104
and Piranesi, 104
Bedford, Grosvenor Charles,
    friend of Southey, 262
Bentham, Jeremy, nineteenth-
    century utilitarian, 136, 216
Fragment on Government, 216
Bentley, Richard, friend of
    Walpole and Gray, 13
Berkeley, George, Bishop,
    philosopher, 294
Berlioz, Hector, composer, 47
Billaud-Varenne, Jacques Nicolas,
    French revolutionary
    politician, 110
Bion, Greek poet, 252
Blackmore, (Sir) Richard,
    eighteenth-century poet, 178
Blair, Hugh, Scottish critic, 244
Blair, Robert, author of *The Grave*
    (1743), 308
Blake, William, 19–31, 170, 241,
    260, 261, 284
America, 10, 20, 22–3
The Book of Ahania, 20–3
Europe, 20
The First Book of Urizen, 20, 24
The French Revolution, 20, 22, 111
Jerusalem, 20, 21, 25–6, 181, 260
The Marriage of Heaven and Hell,
    26–8, 44, 111
Milton, 20, 28–9, 260
The Book of Los, 20
Poetical Sketches, 20
Songs of Innocence, 20
*Songs of Innocence and of
    Experience*, 3, 20, 30–1, 181
The Song of Los, 20
The Book of Thel, 20
Vala/The Four Zoas, 20, 23–5
Visions of the Daughters of Albion,
    20

and anti-slavery, 3
and the French Revolution, 20,
    22, 34, 111
and Burns, 44
and Fuseli, 170
and Imagination, 141

and Landscape, 170
and Palmer, 203
and Religious Thought, 219–20
Bloomfield, Robert, rural poet, 62,
    63
Blücher, Marshal, Prussian
    general, 281, 283
Bonaparte, Napoleon, 31–7, 58,
    111, 190, 281, 283
abdication, 2
admired Ossian, 171
and Jacques-Louis David,
    painter, 200
and Southey, 267
Bowles, William Lisle, eighteenth-
    century poet, 74
Brand, John, antiquarian, 153
Brawne, Fanny, friend of Keats,
    151, 154, 158
Bridgeman, Charles, landscape
    gardener, 169
Brissot, Jacques Pierre, Girondist,
    109
Brontë, Charlotte, 18, 148
Brooks, Charlotte, Irish
    medievalist, 179
Brothers, Richard, prophet, 220
Brougham, (Lord) Henry, lawyer
    and writer, 209
Brown, Charles, friend of Keats,
    151, 158
Brown, Lancelot, 'Capability',
    landscape gardener, 169
and the Picturesque, 214
Browning, Elizabeth Barrett, 146
Browning, Robert, 224
Buffon, Georges Louis Leclerc,
    French naturalist and writer,
    205
Bunyan, John, 203, 206
Bürger, Gottfried August, German
    poet and ballad-writer, 113
'Lenore', 114
translated by William Taylor,
    210
translated by Walter Scott, 230
Burke, Edmund, 37–42; 136, 196,
    218
A Philosophical Enquiry into the

Origin of our Ideas of the Sublime and Beautiful, 38–9, 103, 123, 177, 270, 307
Reflections on the Revolution in France, 38, 111, 147, 198, 217, 285, 287
Thoughts on the Cause of the Present Discontents, 39
A Vindication of Natural Society, 38

and the American Revolution, 9
and architecture, 13
and Coleridge, 69
and Crabbe, 92
and the Gothic Novel, 123
and medievalism, 178
and Paine, 198
influenced Uvedale Price and the Picturesque, 214
Burlington, Lord, friend of Pope, 11
Burnet, Thomas, divine and scientific writer, 168, 273
Sacred Theory of the Earth, 168
Burns, Robert, 42–4; 138, 145, 179
'Tam o' Shanter', 174
anticipates Wordsworth, 299
Burton, Robert, author of *The Anatomy of Melancholy*, 153, 158
Butler, Joseph, Bishop of Durham, 219
Butts, Thomas, patron and friend of Blake, 21
Byron, George Gordon, Lord, 45–55; 17, 139, 174, 181–2, 186, 191, 194–5, 201, 207, 210–11, 223, 228, 241, 247–8, 250
Beppo, 45–6, 53
The Bride of Abydos, 45, 52
Cain, 45, 52, 130, 144, 46, 51–2
Childe Harold's Pilgrimage, 45, 46–8, 52, 130–1, 180, 194, 228, 240, 282, 308
The Corsair, 45, 52, 194
The Deformed Transform'd, 46, 51
Don Juan, 4, 45, 48–9, 53, 131, 143, 182, 188, 240, 282, 284

English Bards and Scotch Reviewers, 92, 180
The Giaour, 45, 48, 52, 130, 193–4, 224
Heaven and Earth, 51
Hints from Horace, 180
The Island, 46. 53
Manfred, 46, 50–1, 97
Marino Faliero, 45, 50, 97
Mazeppa, 45, 53
Parisina, 45, 53
The Prisoner of Chillon, 46, 53, 127
Sardanapalus, 46, 50
The Siege of Corinth, 45, 52–3, 130, 194
The Two Foscari, 45, 50, 54
The Vision of Judgment, 45, 53–5, 266, 282
Werner, 50–1

and anti-slavery, 4
and Napoleon, 35–6
on Burns, 44

dislike of Castlereagh, 284
and Coleridge, 74, 76
and the Convention of Cintra, 35
and Crabbe, 92–3
contributed to *The Examiner*, 211
mocks *The Excursion*, 292, 294
and the French Revolution, 111
and Greece, 130–2
and Thomas Moore, 183–4, 194
and Rogers, 223
admired Rousseau, 226
in Mary Shelley's *The Last Man*, 248
and Percy Bysshe Shelley, 250–1
and Southey, 266
illustrated by Turner, 276
and Waterloo, 282, 284
opinion of *Waverley*, 233
dislike of Wellington, 284

Cabanis, Pierre Jean Georges, French philosopher and doctor, 205

Calderon de la Barca, Don Pedro,
    Spanish poet, 260
Calvert, Edward, friend of Palmer,
    203
Calvert, William, 288
Campbell, Thomas, 56; 191, 201, 248
    Gertrude of Wyoming, 56, 180
    The Pleasures of Hope, 56, 180,
        224

    and Nelson, 188
    as periodical editor, 210
Canning, George, Tory politician,
    10–11, 136, 211
Carlyle, Thomas, 68, 72, 210, 221
    History of the French Revolution,
        112
    Life of Schiller, 115
    Sartor Resartus, 115

    on Cobbett, 68
    influenced by Coleridge, 72
    and German literature, 115–16
    on Scott's novels, 236
Carnot, Lazare, French politician,
    110
Chambers, Sir William, architect
    and gardener, 12, 192
    *Dissertation on Oriental
        Gardening*, 169, 192
Chatterton, Thomas, 58–61; 280
Chaucer, Geoffrey, 59, 149
Cintra, Convention of, 35
Clairmont, Claire, step-daughter
    of Godwin, 120, 247, 250
Clare, John, 61–6; 19, 210, 260–1
    Child Harold, 62, 65
    Don Juan, 62, 65
    The Midsummer Cushion, 64–5
    The Parish, 62, 64–5, 180
    The Shepherd's Calendar, 61–2, 64
    The Rural Muse, 61–2, 64
    The Village Minstrel, 61–4, 180
    later poems, 65–6
    sonnets, 181

    and James Beattie, 63
Clarke, Charles Cowden, friend of
    Keats, 149

Clarkson, Thomas, campaigner
    against the slave trade, 1, 4
Claude, landscape painter, *see*
    Lorrain
Clough, Arthur Hugh, Victorian
    poet, 224
Cobbett, William, 66–9; 218, 282
    Rural Rides, 68
Coleridge, Samuel Taylor, 69–77;
        19, 82, 85, 92, 103, 125, 162,
        207, 250, 284
    poems:
    The Ancient Mariner, 77, 127,
        173, 181, 241, 248, 299
    Christabel, 73–4, 127, 153, 173,
        181, 231, 242, 290
    The Conversation Poems, 74–5,
        180, 239
    Dejection, an Ode, 75–6, 186, 301
    The Fall of Robespierre, 111
    'Frost at Midnight', 90, 237
    'Kubla Khan', 76–7, 98, 173,
        193, 195
    'The Pains of Sleep', 98, 101
    Poems (1796), 162

    drama:
    Remorse, 96

    prose works
    Aids to Reflection, 206, 221
    Biographia Literaria, 16, 71,
        141–2, 173, 203–4, 298
    The Friend, 71, 217
    Notebooks, 238–9
    *On the Constitution of the Church
        and State*, 217
    The Statesman's Manual, 217

    attacked in *The Anti-Jacobin*, 10
    and Association, 15–16
    and the French Revolution, 111
    and German philosophy, 116–17
    and Godwin, 119
    and Hazlitt, 134, 137
    and Imagination, 19, 141–2
    and Lamb, 162–3, 165
    share of *Lyrical Ballads*, 173,
        298–300

on Napoleon, 33
and Nature, 186–7
and Pantheism, 205–6
and Pantisocracy, 9–10, 262
and periodicals, 209–11
and Schiller, 114
and the self, 237–42
and Southey, 262–5
and Charlotte Smith, 125
and Waterloo, 282
and Dorothy Wordsworth,
    289–90
and William Wordsworth,
    291–3, 300, 303
Collins, William, 78–80; 59, 272, 279
    Oriental Eclogues, 193
Collot d'Herbois, Jean Marie,
    French revolutionary
    politician, 110
Colman, George, the younger,
    dramatist, 96
Conder, Josiah, journalist and
    hymn writer, 74
Constable, Archibald, Edinburgh
    publisher, 229
Constable, John, 80–3; 199, 276
Cookson, Revd William, uncle of
    the Wordsworths, 288
Corday, Charlotte, killer of Marat,
    109
Cottle, Joseph, Bristol publisher,
    friend of Coleridge, 61, 298
Cowley, Abraham, seventeenth-
    century poet, 181
Cowper, William, 83–91, 261, 294
    The Castaway, 86–9
    Olney Hymns, 84
    The Task, 1, 3, 89–91, 180, 186
    other poems, 1, 3, 85–8

    and anti-slavery, 1, 3
    and Coleridge, 74, 85
    admired by Constable, 82
    and Nature, 186
    Southey's *Life*, 268
Cozens, John Robert, painter, 169,
    215
Crabbe, George, 92–4; 180, 211,
    221, 294

The Borough, 93
The Library, 92
The Newspaper, 92
The Parish Register, 92
'Peter Grimes', 93, 174
Tales (1812), 92, 94
Tales of the Hall, 92
The Village, 92
Croker, John Wilson, reviewer,
    210
Crompton, Samuel, inventor, 144
Crowe, William, eighteenth-
    century poet, 74
    Lewesdon Hill as an influence on
    Coleridge, 74
Cumberland, George, friend of
    Blake, 25
Cumberland, Richard, dramatist,
    96, 123

Dante Alighieri, 25, 151, 260
Danton, Georges Jacques, French
    revolutionary politician, 109,
    222
Darby, Abraham, ironmaster, 144
David, Jacques-Louis, French
    painter, 32, 200
Davy, Humphry, scientist, 73
Defoe, Daniel, novelist and
    journalist, 168
Delacroix, Eugene, French painter,
    133
De la motte Fouqué, Friedrich,
    German writer, 113
    Undine, 113
De Loutherbourg, Philip James,
    painter, 145, 202, 215
Denham, Sir John, seventeenth-
    century poet, 74
    Cooper's Hill, 74, 273
Dennis, John, eighteenth-century
    critic, writer on landscape,
    168, 270
    *The Grounds of Criticism in
    Poetry*, 270
D'Epinay, Madame, friend of
    Rousseau, 225
De Quincey, Thomas, 94–5,
    98–104; 15, 137, 143, 167

Confessions of an English Opium-Eater, 98–104, 210

and German philosophy, 116
and Jean Paul (Richter), 115
and Landor, 166
De Warens, Madame, benefactor
of Rousseau, 225
D'Holbach, Paul Henri Dietrich,
Baron, eighteenth-century
philosopher, 225
friend of Rousseau, 225
influenced Shelley, 250
Dickens, Charles, 128, 140, 166
Bleak House, 140, 236
Pictures in Italy illustrated by
Palmer, 203
Diderot, Denis, French
encyclopedist, 106, 225
Dodsley, Robert, eighteenth-
century publisher, 179
Drury, Edward, bookseller, 61
Dryden, John, 158, 229, 280, 307
Duck, Stephen, rural poet, 62
Dyer, John, eighteenth-century
poet, 145
The Fleece, 145

Edgeworth, Maria, 175, 232, 234
Edgeworth, R. L. (father of
Maria), 232
Egremont, Lord (George
Wyndham), patron of Turner,
275
Elgin, Lord (Thomas Bruce),
purchaser of 'Elgin Marbles',
131
Eliot, George, Victorian novelist,
236
influenced by Scott, 236
Eliot, Thomas Stearns, twentieth-
century poet, 28
Four Quartets, 28
Ellis, George, periodical editor, 10,
209
Emerson, Ralph Waldo, American
transcendentalist, 204–5, 219
Evangelical religion, 84, 220–1

Evans, Evan, collector of ancient
Welsh poetry, 179

Fawkes, Walter, patron of Turner,
275
Fenwick, Isabella, friend of the
later Wordsworth, 296, 301
Fichte, Johann Gottlieb, German
philosopher, 71, 116
Fielding, Henry, novelist, 244
Fielding, Sarah, novelist, 244
Flaxman, John, painter, 20
Foscolo, Ugo, Italian poet, 223
Fouquier-Tinville, Antoine
Quentin, revolutionary public
prosecutor, 110
Fox, Charles James, statesman, 2,
34, 40–1, 191, 196
Franklin, Benjamin, American
statesman and philosopher, 8
Frend, William, radical and tutor
to Coleridge, 69, 147
Frere, J. H., author of *Whistlecraft*,
181, 211
Freud, Sigmund, psychoanalyst,
100
Fricker, Edith, wife of Southey,
262
Fricker, Sara, wife of Coleridge, 69
Fuseli, Henry, painter, 102, 104–5,
170, 201

Gainsborough, Thomas, painter,
200, 215
Galt, John, Scottish novelist, 210
The Ayrshire Legatees, 210
Garrick, David, actor, 95–6
Gaskell, Elizabeth, Victorian
novelist, 236
Gay, John, eighteenth-century
playwright, 272
George III, King of England, 6–7,
19, 41, 53–4, 263
George, Prince Regent, later
George IV, 139
Gerard, Alexander, eighteenth-
century critic, 307
Essay on Taste, 307

Géricault, Jean Louis, French painter, 173
Gibbon, Edward, historian, 262
Gifford, William, periodical editor, 10, 139, 209
Gilbert, William Schwenk, comic poet, 207
Gillman, James, doctor and biographer of Coleridge, 71, 73
Gillray, James, cartoonist, 57–8, 163
Gilpin, William, writer on the Picturesque, 13, 213–14
Girardin, René Louis, of Ermenonville, friend of Rousseau, 226
Girtin, Thomas, painter, 275
Gisborne, Maria, friend of Shelley, 251
Godwin, William, 119–20; 136, 146, 210, 216–17, 219, 221, 246, 248, 284
  Caleb Williams, 120, 148–9, dramatised, 96
  Enquiry Concerning Political Justice, 111–19, 216, 262
  St Leon, 177

  attacked by The Anti-Jacobin, 10
  and Association, 15
  and Burke, 41
  and Coleridge, 69, 221
  and Napoleon, 33–4
  and Shelley, 250–1
  and Waterloo, 282
  and Mary Wollstonecraft, 284, 286
  and Wordsworth, 291, 304
Goethe, J. W. von, 49, 55, 113–15, 171, 176, 260
  Faust, 50–1, 114, 176
  Young Werther, 114–15
Goldsmith, Oliver, 121–2; 43, 63, 207
  The Deserted Village, 121–2
  The Good-Natured Man, 244
  She Stoops to Conquer, 95, 122

The Vicar of Wakefield, 121–2
  and Burns, 43
  and Clare, 63
Gothic
  The Gothic novel, 123–8
  Gothic architecture, 11–14
  Gothic follies, 68
Gray, Thomas, 128–30; 169, 178, 213, 261, 279
  Elegy in a Country Churchyard, 129–30
  Odes, 129, 178

  and architecture, 12
  and Chatterton, 60
  and Landscape, 168
  and Macpherson, 171
  and the Marvellous, 172
  'The Bard' painted by Martin, 201
  admired by Constable and Turner, 276
Greece, 130–3; 46, 48, 152
Greene, Revd Joseph, biographer of Shakespeare, 179
Grenville, Lord (Richard Chandos), anti-slavery politician, 2, 6
Guest, Charlotte, translator of the Mabinogion, 208

Hakluyt, Richard, explorer and writer, 77
Hamilton, Emma (Lady), 190
Hamilton, William, 190
Hanmer, (Sir) Thomas, antiquarian, 78
Hargreaves, James, of Blackburn, inventor, 144
Hartley, David, 14–16
  and Coleridge, 70, 75
  and Rogers, 224
  and Wordsworth, 304
Hartley, David, junior, 1
Hastings, Warren, Governor-General of India, 39
Haydon, Benjamin Robert, historical painter, 200–1

Hayley, William, patron of Blake, author, 20, 86
Hazlitt, William, 134–7; 163, 259
  Characters of Shakespeare's Plays, 135
  Lectures on the English Poets, 135
  Liber Amoris, 135
  The Spirit of the Age, 135–7
  Table Talk, 135

  on Cobbett, 68
  on Coleridge, 73
  on Crabbe, 92
  on Kean, 95
  and German literature, 114
  and German philosophy, 116
  on Keats, 191
  and Lamb, 166
  and Napoleon, 34–5
  and Periodicals, 209–11
  response to Waterloo, 281–2
  and Wordsworth's *Excursion*, 294
Hébert, Jacques René, French journalist and revolutionary politician, 110, 222
Herder, Johann Gottfried von, German critic, 118, 171
Hobhouse, John Cam, friend of Byron, 34, 282
Hoche, Louis, Lazare, French general, 58, 110
Hoffmann, Ernst Theodore Amadeus, German novelist, 128
Hogarth, William, painter, 57, 200, 270
  Analysis of Beauty, 270
Hogg, James, Scottish writer, 137–8
  The Private Memoirs and Confessions of a Justified Sinner, 101, 137–8, 174, 242
  and Leigh Hunt, 139
  and periodicals, 210
Holcroft, Thomas, radical, novelist and playwright, 10–11, 34, 120, 147
  Anna St Ives, 147–8

  The Road to Ruin, 96
  A Tale of Mystery, 96

  his biography written by Hazlitt, 134
Hole, Richard, poet and antiquary, 178
Holland, (Henry Richard Vasall Fox) Lord, 33–4, 183
Hollar, Wenceslaus, seventeenth-century topographical engraver, 170
Home, John, Scottish dramatist, 78, 123, 171
Hood, Thomas, poet, 210
Hopkins, Gerard Manley, nineteenth-century poet, 181
Horner, Francis, Scottish politician, lawyer, and man of letters, 209
Hugo, Victor, French poet and novelist, 128
Hume, David, philosopher, 171, 226
  befriended Rousseau, 226
  and Sensibility, 243–4
Hunt, John, periodical editor, 25, 48, 54, 134, 139–40
Hunt, John, 'Orator' of Peterloo, 211–13
Hunt, Leigh, poet and periodical editor, 138–40; 25, 34, 54, 207, 250, 282
  The Examiner, 138–9, 150, 163–4, 211–12
  and Hazlitt, 134
  and Keats, 149–54
  and Lamb, 164
  and Shelley, 251
Hurd, Richard, Bishop, eighteenth-century critic, 80
  Letters on Chivalry and Romance, 172, 178
Hutchinson, Mary, wife of Wordsworth, 291–2

Imagination, 141–3; 72 (Coleridge)
Imlay, Gilbert, husband of Mary Wollstonecraft, 286

Inchbald, Elizabeth, actress, dramatist and novelist, 34, 95–6, 147–8
  Every One Has His Faults, 96
  Lovers' Vows, 96
  A Simple Story, 147–8
Industrial Revolution, The, 143–6
Irving, Edward, millennarian preacher, 220

Jacobins, 108–11, 222
  cartooned, 58
  The Jacobin novel, 147–9
James, John, landscapist, 169
  The Theory and Practice of Gardening, 169
Jean Paul, *see* Richter, J. P. F.
Jeffrey, Francis, periodical editor and reviewer, 48–9, 94, 187, 209–10, 234, 294
  Edinburgh Review, 209–10
  and Byron, 48–9
  and Crabbe, 94
  and Scott, 234
  and Wordsworth, 210, 294
Jerdan, William, editor of the *Literary Gazette*, 211
Johnson, Joseph, radical publisher, 209, 284
Johnson, Samuel, essayist, critic, novelist, 36, 74, 81, 92, 129, 246, 269–70
  Life of Savage, 246
  The Vanity of Human Wishes, 34
  Rasselas, 192

  and Goldsmith, 121
  and Macpherson, 171
  and orientalism, 192
  and Thomson, 273
  and the Wartons, 280
Joyce, James, novelist, 237

Kant, Immanuel, philosopher, 70, 116–18, 141
  influence on Coleridge, 70, 72–3
  Hazlitt on, 73
  and imagination, 141
  and Swedenborg, 220

Kean, Edmund, actor, 96, 134
Keats, John, 149–62; 46, 48, 61, 139, 145–6, 165, 171, 185, 200–1, 211
  'La belle dame sans merci', 98, 152
  Endymion, 150–1, 152–3, 180
  The Eve of St Agnes, 153–5, 180
  The Fall of Hyperion, 97–9, 151, 180, 241
  Hyperion, 155–8, 180–2
  Lamia, 151, 158–60, 173, 180
  The Odes, 151, 160–2, 165

  and Byron, 48
  and Chatterton, 61
  and dreams, 97
  and Greek myths, 131
  and Leigh Hunt, 139
  attacked by Lockhart, 139
  and nature, 185
  negative capability, 187–9
  and periodicals, 210–11
  and Peterloo, 213
  and the self, 237–8
  and Shelley, 251–2
  and Wordsworth, 151, 294
Keble, John, religious poet, 221
Kemble, John Philip, actor, 95–6, 200
Kent, William, landscape gardener, 169
Kingsbury, Henrietta, wife of Maturin, 175
Kip, Johannes, engraver, 170
Knight, Henry Gally, writer on architecture, 194
Knight, Richard Payne, connoisseur, 213–14
Knyff, Leonard, topographical painter, 170
Kotzebue, August Friedrich von, German dramatist, 96, 113
Krause, Karl Christian Friedrich, German philosopher, 206

Lafontaine, August Heinrich Jules, German novelist, 113
Lamartine, Alphonse de, French poet and politician, 47

Lamb, Charles, 162–6; 10, 55, 69, 173, 201, 206, 251, 292
Essays of Elia, 164–6
Last Essays of Elia, 164–6

attacked by *Anti-Jacobin*, 10
and Byron, 55
and Coleridge, 69, 74–5, 173
and Hazlitt, 134–5, 165
and Leigh Hunt, 139
and periodicals, 210–11
and Wordsworth, 292, 294
Lamb, Mary Anne, 139, 163, 251
Lambert, George, landscape painter, 170
Landor, Walter Savage, poet and prose writer, 167–8; 5, 193, 207
Gebir, 193
Imaginary Conversations, 207
Landscape, 168–70; 213–15
(Picturesque), 213–15
(Sublime), 269–71
Langhorne, John, poet, 294
Lawrence, D. H., novelist and critic, 18
Lawrence, Sir Thomas, painter, 81, 200
Lee, Harriet, novelist and dramatist, 51
Lee, Sophia, novelist, sister of Harriet, 124
Leibniz, Gottfried Wilhelm, German mathematician and philosopher, 70
Lemprière, John, classical scholar, 131
Lessing, Gotthold Ephraim, 70
Levasseur, Thérèse, washerwoman, mistress of Rousseau, 225
Lewis, Matthew G., 'Monk', 50, 96, 114, 124–7, 174–5, 177, 242
The Castle Spectre, 96
The Monk, 126, 177, 242
and German literature, 114
and the self, 242
Linnell, Hannah, married Samuel Palmer, 203

Linnell, John, painter, 21, 203
Lloyd, Charles, poet, 10–11
Locke, John, 14, 70, 186, 216, 273, 305
Essay Concerning Human Understanding, 14
Lockhart, John Gibson, critic and biographer, 139, 210
Lofft, Capel, poet, 34
Longinus, Cassius, 269
Peri Hupsous (On the Sublime), 269
Lorrain, Claude, landscape painter, 169, 170, 273, 277
Louis XVI, king of France, 106–9, 197

Macadam, John Loudon, civil engineer, 144
Macaulay, Thomas Babington, Victorian journalist and historian, 267
Mackenzie, Henry, Scottish author, 43, 113–14, 171
Julia de Roubigné, 245–6
The Man of Feeling, 43, 122, 244–5
The Man of the World, 245

and Burns, 43
and German theatre, 113–14
and Goldsmith, 122
and James Macpherson, 171
Mackintosh, Sir James, apologist for the French Revolution, 136
Macpherson, James, 'Ossian', 171–2; 43, 59, 178, 271
Fragments of Ancient Poetry, 171
History of Great Britain from the Restoration, 171

and Burns, 43
and Chatterton, 59
Macready, William Charles, actor-manager, 96
Mallet, David, eighteenth-century poet, 5, 271
The Excursion, 5, 178, 271
Malthus, Thomas Robert, demographic and political

writer, 136, 217–18
Mant, Richard, Bishop, 281
Marat, Jean Paul, French
    revolutionary, 109
Marie Antoinette, Queen of
    France, 40–1, 107–9, 222
Martin, John, painter, 170, 201–2
    The Bard, 201
    Belshazzar's Feast, 201
    The Great Day of His Wrath, 170
Martineau, Harriet, nineteenth-
    century writer, 219
Martineau, James, Unitarian
    preacher and writer, 219
Marvellous and Occult, The, 172–4
Mason, William, friend of Gray,
    213
Maturin, Charles Robert, Gothic
    novelist, 175–7; 127
    The Fatal Revenge, 127, 175–6
    Melmoth the Wanderer, 127, 175–7
    other novels, 175–6
Maurice, Frederick Denison,
    Victorian religious and social
    critic, 220
Medievalism, 177–9; 58–61
    (Chatterton)
Meredith, George, Victorian
    novelist, 206
Methodism, 57, 67, 219–21
Metre and Form, 180–2
Mill, James, Utilitarian
    philosopher, 209, 218
Mill, John Stuart, philosopher and
    critic, 209, 287
Miller, Sanderson, creator of
    Gothic ruins, 12
Milton, John, 28–9, 76, 78, 82, 151,
    168, 180, 192, 201, 237, 271,
    279
    Paradise Lost, 29, 76, 168, 192,
    203
Mitford, Mary Russell, novelist
    and dramatist, 211
Moncrieff, William Thomas,
    dramatist, 96
Montesquieu, Charles de
    Secondat, Baron, French
    philosophe, 106

Montgomery, James, journalist
    and hymn writer, 65
Moore, Thomas, 182–4; 18, 76–7,
    139, 180, 194–5, 223
    Lalla Rookh, 180

    and Byron, 194
    and Coleridge, 76–7
    and Leigh Hunt, 139
    and orientalism, 194–5
    and Rogers, 223
    and Waterloo, 282
More, Hannah, religious writer,
    123
Morgan, Sydney, Lady, novelist,
    175
Moschus, Greek poet, 252
Murray, John, publisher, 48, 54,
    183–4

Nature, 185–7; 89–91 (Cowper)
Necker, Jacques, Swiss financier to
    Louis XVI, 106–7
Negative Capability, 187–9
Nelson, Horatio, Lord, 189–91; 32,
    200
Neo-classicism, 11–12
Newcomen, Thomas, inventor, 144
Newman, John Henry, Cardinal,
    219, 221
Newton (Sir) Isaac,
    mathematician, 186, 273
Newton, John, evangelical
    preacher, 84, 90
North, Lord, second Earl of
    Guilford, Prime Minister, 6,
    39, 171
North, Frederick, fifth Earl of
    Guilford, philhellene, 130
Novalis (Hardenberg, Friedrich
    Leopold von), German poet,
    118

Orientalism, 192–5; 76–7
    (Coleridge), 266–7 (Southey);
    see also Moore
Ossian, see Macpherson, James
Owen, Robert, mill-owner and
    philanthropist, 217, 263

Paine, Thomas, 196–9; 8–9, 10, 41–2, 57, 216–18, 265, 284
  Common Sense, 8, 9, 196–7
  Rights of Man, 111, 119, 216–17
  and the American Revolution, 8–10
  and Burke, 41–2
  and Southey, 265
Painting, 199–202; 80–3 (Constable), 275–8 (Turner), 202–4 (Palmer)
Palmer, Samuel, 202–4; 199
Pantheism, 204–6
Peacock, Thomas Love, 206–8; 251
  Nightmare Abbey, 118
Peel, Sir Robert, Prime Minister, 283
Percy, Thomas, Bishop, editor of Reliques of Ancient English Poetry, 178–9, 181, 209, 230, 265, 299
Periodicals, 209–11; 138–40 (Leigh Hunt), 162–6 (Lamb)
Peterloo, 211–13
Petrarch, Francis (Francesco Petrarca), Italian poet, 259
Philips, Ambrose, poet, friend of Addison, 192
Phillips, Richard, radical bookseller, 210
Pickersgill, Joshua, playwright, 51
Picturesque, the, 213–15; see also 167–70 (Landscape)
  Dr Syntax's picturesque travels cartooned, 57
Pinkerton, John, Scottish antiquary and historian, 179
Piranesi, Giovanni Battista, Italian architectural designer, 76, 103–4
Pitt, William (Pitt the younger), Prime Minister, 9, 33, 40, 42, 58, 67, 119, 136, 191
Planché, James Robinson, dramatist, 96
Poe, Edgar Allan, American short-story writer and poet, 128
Polidori, John William, Dr, friend of Byron and the Shelleys, 174, 247
Political Thought, 215–18; see also 37–42 (Burke)
Pope, Alexander, 1, 44, 78, 129, 168, 178, 279–80
  and the anti-slavery movement, 1
  'Eloisa to Abelard', 178
  Moral Essays, 306
  Pastorals, 78
  and Joseph Warton, 279–80
Prévost, Antoine-François, Abbé, 245
Price, Richard, dissenting minister, 147, 216, 284–5
  Discourse upon the Love of our Country, 216
Price, (Sir) Uvedale, writer on the picturesque, 169, 213–14, 270
Priestley, Joseph, dissenter and scientist, 16, 69, 119, 284
  Coleridge writes poem on, 69
Purchas, Samuel, seventeenth-century travel writer, 76–7, 195
Pushkin, Alexander, Russian poet, 49

Rabelais, François, French author, 206
Radcliffe, Ann, Gothic novelist, 125–7, 153, 175–6
  The Italian, 125
  The Mysteries of Udolpho, 125–6
Raeburn, (Sir) Henry, Scottish portrait painter, 200, 233
Reeve, Clara, Gothic novelist, 124, 271
  The Old English Baron, 124
Religious Thought, 219–21
Repton, Humphry, landscape gardener, 169, 214
Réveillère-Lepeaux, French reformer, 10
Reynolds, John Hamilton, friend of Keats, 158, 193
  Sofie, an Eastern Tale, 193
Reynolds, (Sir) Joshua, President of the Royal Academy, 170

Richardson, Samuel, novelist, 244,
    307
    Pamela, 244
Richter, Johann Paul Friedrich,
    'Jean Paul', German novelist,
    113
Ritson, Joseph, antiquarian, 281
Robertson, William, Scottish
    historian, 244
Robespierre, Maximilien, French
    revolutionary leader, 221–3;
    10, 31, 109–10, 291
Robinson, Henry Crabb, lawyer
    and diarist, 116, 119, 166
Rockingham, Lord (Charles
    Watson Wentworth), 6, 39, 57
Rogers, Samuel, banker and poet,
    223–4; 56, 284
    Italy, 276
    The Pleasures of Memory, 15, 180
Rosa, Salvator, seventeenth-
    century landscape painter,
    169, 271, 279
Rossetti, Dante Gabriel,
    nineteenth-century poet-
    painter, 61
Rousseau, Jean-Jacques, 225–7; 3,
    106, 245, 259
    *Discours sur les Sciences et les
        Arts*, 3
    *Discours sur l'Origine de
        l'Inégalité*, 3
    Emile, 226, 284
    La Nouvelle Héloïse, 227, 246
    and the French Revolution, 106
    admired by Mary
        Wollstonecraft, 284
    in Shelley's 'The Triumph of
        Life', 259
Rowe, Nicholas, playwright, 179
Rowlandson, Thomas, cartoonist
    and artist, 57
Runciman, Alexander, Scottish
    painter, 271
Ruskin, John, art critic and
    moralist, 39, 81, 276
Russell, Lord John, Prime
    Minister, 184

Rutherford, Anne, mother of Sir
    Walter Scott, 228

Saint-Just, Antoine Louis Léon de,
    associate of Robespierre,
    109–10
Salmon, Thomas, eighteenth-
    century historian of the
    world, 193
Sandby, Paul, painter, 170, 271
Savage, Richard, eighteenth-
    century poet, 294
Schelling, Friedrich Wilhelm
    Joseph, German philosopher,
    70, 72, 116–17, 141
Schiller, Friedrich, German poet,
    113–14
    The Robbers, 114, 171
Schlegel, August Wilhelm,
    German aesthetician, 116–17
Schlegel, Friedrich, German
    aesthetician, 116
Schumann, Robert, composer, 60
Scott, Paul, topographer, 170
Scott, John, periodical editor, 135,
    164, 211
Scott, (Sir) Walter, poet and
    novelist, 228–36; 47, 93, 175–6,
    182, 191, 208, 209–10, 218, 223
    The Doom of Devorgoyl, 97, 229
    The Lay of the Last Minstrel, 153,
        173–4, 181
    Marmion, 10, 28, 231
    The Lady of the Lake, 228, 231
    other poems, 228, 266, 282,
        230–2

    Ivanhoe, 208, 235
    Kenilworth, 124, 236
    Waverley, 17, 228, 232–4
    The Waverley novels, 128, 228–9,
        234–6

    builds Abbotsford, 14
    and Jane Austen, 17
    and Byron, 49
    translates German literature, 114
    and the Gothic novel, 124

and Hazlitt, 135–6
and Hogg, 137
and Keats, 154
on Napoleon, 34, 37
and periodicals, 209
dislike of the theatre, 96
illustrated by Turner, 276
and Walpole, 172
and Waterloo, 282
Self, the, 237–42
Sensibility, 242–6
Severn, Joseph, painter, friend of
    Keats, 151
Shaftesbury, Lord, (Anthony
    Ashley Cooper, third Earl), 5,
    270
and Nature, 168
and Sensibility, 243
and the Sublime, 168
and Thomson, 273
Shakespeare, William, 94, 149,
    179, 237, 271
King Lear, 95
Macbeth and Fuseli, 201
Othello, 95, 306
Romeo and Juliet, 153
and Constable, 82
and Keats, 154, 187
Sharp, Granville, anti-slave trade
    leader, 1
Shelley, Mary, 246–9; 119–20, 201
Frankenstein, 119, 127, 174, 242,
    247–9, 286
Shelley, Percy Bysshe, 249–60; 33,
    36, 50, 103, 111–12, 120, 127,
    142–3, 167, 174, 229, 246–7
poems:
Adonais, 180, 251–3
Alastor, 180, 253–4, 264
The Cenci, 97, 180
Hellas, 131–2
'Mont Blanc', 239
Queen Mab, 145, 204, 218, 250,
    264
Prometheus Unbound, 97, 99, 101,
    112, 142, 180, 186, 239–40,
    242, 257–8
'The Triumph of Life', 226–7,
    259–60

the Odes, 254–5
other poems, 255–6

prose:
A Defence of Poetry, 251
Zastrozzi, 127

and Byron, 45–7, 50, 54, 250–1
dislike of Castlereagh, 284
and the French Revolution,
    111–12
and Godwin, 120, 217–18
and Greece, 131–2
and Leigh Hunt, 139, 250–1
and Napoleon, 33, 36
and Nature, 186
and Pantheism, 204
and Peacock, 207
and Peterloo, 211–13
influence of Plato on, 186
and Rousseau, 226–7
and Mary Shelley, 246–7, 250
and Southey, 54, 264
and Waterloo, 282
and Wordsworth, 294
Shelvocke, George, travel-writer, 77
Shenstone, William, eighteenth-
    century poet and landscape
    gardener, 43, 178, 279, 294
Sheridan, Richard Brinsley,
    playwright and politician, 96,
    184, 207
The Rivals, 96
The School for Scandal, 96
his life written by Thomas
    Moore, 184
Siddons, Sarah, actress, 95
Sieyès, Emmanuel Joseph, Abbé,
    French reformer, 106
Smart, Christopher, eighteenth-
    century poet, 260–2
Smith, Adam, economist, 218,
    243–5
Theory of Moral Sentiments, 243
Smith, Charlotte, poet and
    novelist, 124–5
Smith, Horace, novelist, 251
Smith, Sydney, periodical writer
    and wit, 209

Smollett, Tobias, novelist, 213
Humphry Clinker, 213
Sotheby, William, friend of
    Coleridge, 75
Southcott, Joanna, prophetess, 221
Southey, Robert, 262–9; 53–4, 114,
    167, 207, 223
  poems:
  Botany Bay Eclogues, 264–5, 300
  The Curse of Kehama, 194, 264,
    266
  English Eclogues, 265, 300
  The Fall of Robespierre, 111, 265
  Joan of Arc, 262, 264–5
  Madoc, 180, 264, 266
  Roderick, 266
  Thalaba, 174, 181, 194, 264, 266
  A Vision of Judgement, 263, 266
  Wat Tyler, 265
  other poems, 127, 262–3, 264–7,
    282

  prose:
  The Book of the Church, 268
  History of Brazil, 268
  History of the Peninsular War, 268
  Life of Cowper, 268
  Life of Nelson, 268
  Life of Wesley, 268
  Letters from Spain and Portugal,
    263, 268
  The Doctor, 268–9
  'The Three Bears', 269
  other prose works, 267–9

  and the slave trade, 3
  attacked by *The Anti-Jacobin*, 10
  and Blake, 26
  and Byron, 53–4, 194, 266
  and Chatterton, 61
  and Coleridge, 69–70, 73, 77,
    111, 265
  and the Gothic novel, 127
  and Napoleon, 34–5
  and Nelson, 191
  and Orientalism, 194
  and periodicals, 209–11
  and Waterloo, 282
Spenser, Edmund, 78, 149, 153, 173

The Faerie Queene, 153, 178, 180
  the Spenserian stanza, 153, 180
  and Collins, 78
  and Thomas Warton, 280
Spinoza, Benedict de,
    philosopher, 205
Staël, Anne Louise Germaine,
    Madame de, French writer,
    113, 115, 116, 175–6
  Corinne, 175–6
  De l'Allemagne, 113, 116
Steele (Sir) Richard, dramatist and
    essayist, 244
  The Conscious Lovers, 244
Stephenson, George, engineer,
    144–5
Sterling, John, friend of Carlyle
    and Maurice, 72
Sterne, Laurence, novelist, 244, 269
  A Sentimental Journey, 244
  Tristram Shandy, 269
Stothard, Thomas, painter and
    illustrator, 223
Stuart, Charles Edward, the
    'Young Pretender', 232
Sublime, the, 269–71; 169, 275–8
    (Turner)
Swedenborg, Emmanuel, religious
    writer, 27–8, 220
Swift, Jonathan, 66
  The Tale of a Tub, 66
  Scott edits, 229
Swinburne, Algernon Charles,
    Victorian poet, 167
Switzer, Stephen, garden theorist,
    169
  Iconographia Rustica, 169
Switzerland, 11, 34, 47

Tatham, Frederick, friend of
    Palmer, 203
Taylor, John, publisher, 61, 63, 64,
    155–6, 164, 211
Taylor, Tom, Victorian man of
    letters, 201
Taylor, William, translator of
    German, 209–10
Tchaikovsky, Peter Ilich, Russian
    composer, 50

Telford, Thomas, engineer, 144
Temple, William, diplomat and
    writer, 169
    Upon the Garden of Epicurus, 169
Tennyson, Alfred, 140
    *Ode on the death of the Duke of
    Wellington*, 283
Tennyson, Charles, 140
Thackeray, William Makepeace, 236
Thelwall, John, radical writer, 34,
    76, 147
Thistlewood, Arthur, conspirator,
    212
Thomson, James, 272–5; 279, 294
    Alfred, 272
    The Castle of Indolence, 180, 272
    Liberty, 272
    The Seasons, 82, 89, 146, 180,
    271, 273–5

    and the anti-slavery movement,
        1
    and Clare, 63
    and Collins, 78
    and Constable, 82
    and Cowper, 89
    and Turner, 276
Toland, John, Deist and Pantheist,
    204
Tolstoy, (Count) Lev Nikolaevich,
    230
Tooke, Horne, writer on language,
    120, 136, 284
Toussaint L'Ouverture, François,
    slave leader, 2
Towne, Francis, painter, 215
Townsend, Charles, Chancellor of
    the Exchequer, 6–7
Trelawny, Edward John, friend of
    Shelley, 132, 250–1
    *Records of Shelley, Byron and the
    Author*, 250
Turner, Joseph William Mallord,
    275–8; 170, 199, 223
    compared with Constable, 81
Tyler, Elizabeth, aunt of Southey,
    262
Tyrwhitt, Thomas, editor and
    critic, 178

Vallon, Annette, lover of
    Wordsworth, 290–1
Vane, Anne, 260
Venice, 32, 45, 47, 48, 247, 250,
    277
    Beppo (Byron), 45
    Sonnet 'On the Extinction of the
    Venetian Republic'
    (Wordsworth), 32
Vergniaud, Pierre Victurnien,
    Girondist, 109
Vienna, Congress of, 2
Virgil, P. Vergilius Maro, 203, 279,
    284
    Eclogues, 203, 279
    Georgics, 279, 294
Voltaire, François-Marie Arouet,
    69, 106
    Dictionnaire philosophique, 69

Walpole, Horace, 12–13, 60, 95,
    103–4, 123–5, 169, 172, 178,
    271
    Anecdotes of Painting in England,
    12, 103
    The Castle of Otranto, 13, 103,
    123, 125, 172, 178, 271
    On Modern Gardening, 169
    The Mysterious Mother, 95, 103

    and Gray, 128–9
    and Piranesi, 103–4
    alters Strawberry Hill, 13
    and Mary Wollstonecraft, 286
Walpole, (Sir) Robert, Prime
    Minister, 306
Warburton, William, Bishop and
    eighteenth-century critic, 178
Ward, James, painter, 170
Warton, Joseph, 279–80; 12, 78–9
    *Essay on the Genius and Writings
    of Pope*, 279
Warton, Thomas, 280–1; 12, 78,
    171, 178–9, 224, 244
    Observations on the Faerie Queene,
    178, 280
    History of English Poetry, 179, 281
    The Pleasures of Melancholy, 224,
    244, 280

Washington, George, 4, 8, 23, 196
Waterloo, 281–2; 33, 45, 47
Watson, Richard, Bishop of
    Llandaff, 42, 291
    Wordsworth's 'Letter' to, 42,
    291
Watt, James, inventor, 144
Watts, Isaac, hymn writer, 30
    Divine and Moral Songs, 30
Wedgwood, Josiah, potter, 144
Wellington, Arthur Wellesley, first
    Duke of, 283–4; 32–3, 200,
    281–2
Wesley, Charles, Methodist hymn
    writer, 220
Wesley, John, founder of
    Methodism, 219–20
West, Benjamin, painter,
    President of the Royal
    Academy, 200
West, Richard, friend of Gray,
    128–9
Westbrook, Harriet, first wife of
    Shelley, 249–50
Whitbread, Samuel, radical
    politician, 34
Whitefield, George, Calvinistic
    Methodist, 220
Wilberforce, William, anti-slavery
    leader, 1, 4–5, 67
Wilkie (Sir) David, painter, 201–2,
    278
Williams, Helen Maria, poet, 10
Wilson, John, periodical editor,
    210
Wilson, Richard, painter, 170, 277
Wollstonecraft, Mary, 284–8; 41,
    209, 216, 246
    *A Vindication of the Rights of
    Woman*, 216

    and Fuseli, 105
    marries Godwin, 119
Woodhouse, Richard, friend of
    Keats, 155
Wootton, John, landscape painter,
    170
Wordsworth, Dorothy, 288–90;
    70–1, 75, 77, 291

The 'Alfoxden Journal', 289
The 'Grasmere Journal', 289

    and Coleridge, 70–1, 75, 77,
    289–90
Wordsworth, John, brother of
    William and Dorothy, 292
Wordsworth, William, 290–305; 5,
    32, 44, 85, 92, 121–2, 124–5,
    173–4, 201, 207, 223, 283, 284
poems:
    An Evening Walk, 180
    The Borderers, 97
    Descriptive Sketches, 180
    The Excursion, 146, 151, 153, 164,
    205, 292–4
    Lyrical Ballads, 142, 181, 289,
    291, 298–300
    'Peter Bell', 292
    Poems in Two Volumes, 291–2,
    302
    The Prelude, 3, 42, 19, 111, 142,
    180, 186–7, 205, 226, 237–9,
    242, 289, 291–2, 293, 302–5
    'Lines . . . Tintern Abbey',
    185–6, 205, 238, 240, 296–8
    'Ode . . . Intimations of
    Immortality', 260, 300–2
    other poems, 72, 173, 180–1,
    185–7, 188, 192, 291–2,
    298–300

prose:
    Guide to the Lakes, 289, 292,
    295–6
    'Letter to the Bishop of
    Llandaff', 42, 291
    On the Convention of Cintra, 283
    Preface to *Lyrical Ballads*, 42, 85,
    73, 288

    and anti-slavery, 4
    and Association, 15
    praises Burns, 44
    and Burke, 42
    and Byron, 47, 49, 292
    and Chatterton, 61
    friendship with Coleridge, 70–1,
    75, 77, 291–3, 297, 300, 303;

on 'Christabel', 73
and Crabbe, 92
and the French Revolution, 111
and German Philosophy, 117–18
and Godwin, 119
and Goldsmith, 121–2
praises Gray, 129
and Hazlitt, 134, 136
and Imagination, 142–3
and Keats, 151, 154
and Lamb, 162, 166, 292, 294
and Nature, 185–7
and Nelson, 190
and Orientalism, 192
and Pantheism, 205
and Periodicals, 210–11

dislike of Robespierre, 222
and Rousseau, 226
and the Self, 237–40
and Waterloo, 282
Wortley Montagu, (Lady) Mary,
    friend of Pope, 245
Wright, Joseph, of Derby, painter,
    145, 202, 215
Wyatt, James, architect, 13
Wynn, C. W. Watkins, friend of
    Southey, 262

Young, Edward, 306–8
    Night Thoughts, 20, 24, 186, 271,
        307
    and Blake, 307